CHEMICAL COMPOUNDS IN THE ATMOSPHERE

CHEMICAL COMPOUNDS
IN THE ATMOSPHERE

T. E. Graedel

Bell Laboratories
Murray Hill,
New Jersey

ACADEMIC PRESS   New York   San Francisco   London   1978
A Subsidiary of Harcourt Brace Jovanovich, Publishers

ACADEMIC PRESS, INC.
111 Fifth Avenue, New York, New York 10003

United Kingdom Edition published by
ACADEMIC PRESS, INC. (LONDON) LTD.
24/28 Oval Road, London NW1   7DX

T. E. Graedel

Library of Congress Cataloging in Publication Data

Graedel, T. E.
     Chemical compounds in the atmosphere.

     Bibliography: p.
     Includes index.
     1. Atmospheric chemistry.   I.   Title.
QC879.6.G73   551.5'1   78-12264
ISBN 0-12-294480-1

79 010174

PRINTED IN THE UNITED STATES OF AMERICA

In memory of
Philip E. Graedel and R. Dennis Graedel

CONTENTS

PREFACE

Atmospheric compounds are numerous and chemically diverse, and information concerning them is scattered very widely throughout the scientific literature. This became apparent to me in 1974 when I was formulating a set of equations to describe hydrocarbon chemistry in urban atmospheres. It soon became obvious that the number of different hydrocarbon compounds present in the troposphere was well over 100 and that other chemical groups were well represented. Intrigued, I began to assemble such information in an orderly and comprehensive way. The result is this book, which contains references from the literature of aeronomy, agriculture, biology, botany, chemistry, environmental science, forestry, the geological sciences, horticulture, medicine, metallurgy, meteorology, and physics to more than 1600 compounds.

The information that has been assembled is presented here in several different ways, with the intent of making it conveniently available to potential users with a variety of interests. The principal organization is tabular, with compounds being divided into groups on the bases of their chemical constituents or chemical structure. For each compound, the tables provide information on its known sources and their relative importance, its presence and concentrations in ambient atmospheres, its chemical reactions, and its lifetime. Such information is of direct use to those engaged in atmospheric measurements, theoretical and laboratory studies of chemical parameters relevant to the atmosphere, and air quality assessments. The source and concentration data may prove to be of direct application to research on the effects of trace contaminants on materials and equipment.

Two cross-reference tables are included in the book. The first reorders the tabular data by source instead of by compound. For each source (e.g., automobiles, refuse combustion), each known emitted compound is listed, together with a numerical reference to the appropriate basic table. The information indexed in this way is expected to be of use in air quality assessment studies, particularly in the preparation of environmental impact assessments for new sources.

A second cross-reference table, ordered alphabetically by chemical compound, lists the chemical precursors for those compounds formed in the troposphere by chemical reactions. (They may, of course, also be directly emitted from anthropogenic or natural sources.) These data will be useful to those involved in ambient measurement and air quality assessment (as production mechanisms in lieu of or in addition to specific physical sources), and to chemists studying the reactive interplay of atmospheric compounds.

The uses mentioned above are primarily related to individual compounds or to individual sources. The totality of the information, however, has relevance to more comprehensive atmospheric studies. In particular, tropospheric oxidation processes, acid rain, atmospheric chemical lifetimes, the division of compounds between gas and aerosol phases and their transitions between those phases, and global atmospheric cycles of elements and compounds can be addressed by statistical and schematic analyses of the data. These topics are discussed in the final chapter.

In the preface to his book "Ice Physics," Peter Hobbs remarks that a book is never really *finished;* it is finally *abandoned.* This comment is particularly relevant to the present volume, which is more a collection of information than an exposition. Except for a few important additions, the literature search for the book was terminated in December, 1977. In this active field the continued flow of data to supplement that presented here will doubtless be substantial, and I will very much appreciate receiving any relevant information that is not included here.

I am indebted to a number of people for their help with this project. E. A. Chandross, L. A. Farrow, J. D. Sinclair, and C. J. Weschler have reviewed the manuscript and contributed many valuable comments. The tables and text were phototypeset at Bell Laboratories using computer programs developed by B. Kernighan, M. Lesk, and J. F. Ossanna; J. C. Blinn has been of great assistance in teaching me many of the tricks of their use. Specific suggestions by R. S. Freund, C. M. Melliar-Smith, and T. A. Weber on the inclusion and presentation of data have been incorporated into the tables, and the book is the better as a result. S. M. Kunen contributed valuable data prior to publication, for which I am most grateful. The typing was accurately and cheerfully performed by D. K. Mehring (text) and A. Palys (tables). I am grateful for the support of my mother, Helen Graedel, and my parents-in-law, Berrien and Jean Ketchum. Finally, I thank my wife Susannah and my daughters Laura and Martha for their love and assistance throughout.

1

INTRODUCTION

1.1. ATMOSPHERIC COMPOSITION
AND ATMOSPHERIC CHEMISTRY

More than 99.9% of the molecules comprising the Earth's atmosphere are nitrogen (N_2), oxygen (O_2), or one of the rare gases. For the chemist and the environmentalist, these species are of only secondary importance, however. Far more significant in their view are the chemically diverse and reactive trace species. Although many different trace molecules are found in the atmosphere and are capable of influencing or controlling certain atmospheric processes, their total concentration is very low. Carbon dioxide, which is an important factor in the Earth's radiation balance but is chemically unreactive in the troposphere, has an average concentration of about 325 parts per million (ppm). The most abundant of the reactive gases is methane, which comprises only slightly more than one part per million of the tropospheric gas. Other reactive species are still less prevalent; the combined concentration of all of the reactive trace gases om the atmosphere seldom totals 10 ppm.

Atmospheric aerosols are present throughout the troposphere. Their diverse generation processes and their interactions with the reactive gases render their chemical composition and chemical processes of great interest. Aerosols have been less well characterized than gases; though several features of their composition have been studied, much remains to be done to establish which species are of primary importance in the different regimes and conditions found in the Earth's atmosphere.

Because of low and fluctuating concentrations, large numbers of anthropogenic and natural sources, and differing importance in a variety of physical and chemical processes, information concerning the trace gases of the Earth's atmosphere is of varying comprehensiveness, organization, and availability. This book's first purpose is to bring together that information, to arrange the compounds in chemical groups, and to explore the sources and concentrations of the atmosphere's trace gases.

A second purpose, of equal importance with the first, is to study the chemical reactions and ultimate fates of these compounds. This information is vital to the study of a variety of problems. A number of compounds (such as vinyl chloride and some of the polynuclear aromatic hydrocarbons) are biologically harmful and their chemical pathways are thus of toxicologic interest. Other compounds (such as propene) are themselves benign but are involved in the chemical production of undesirable products (ozone, in the case of propene). Still others (such as dichlorodifluoromethane) are chemically inert in the troposphere but have harmful effects upon diffusion into the stratosphere. In problems such as the assessment of the sources and effects of atmospheric sulfate, data on an entire chemical class of compounds may be relevant. The chemical fates of atmospheric molecules must therefore be rather well understood if molecular impacts on environmental systems are to be accurately assessed.

With few exceptions, the chemical data and the chemical discussions in this book refer to reactions in the gas phase. This is not meant to imply that aerosol chemistry is any less important; but it does reflect the dearth of knowledge about chemical processes in and on aerosols. The very great number of compounds detected in aerosols and the evidence for their participation in chemical chains (see Chapter 10) are ample reasons to regard aerosol chemistry as worthy of study; the field will doubtless advance rapidly in the coming years.

1.2 THE TABLES

The chemical compounds in the atmosphere have been divided into groups containing compounds of similar chemical structure and behavior. For each of these groups two separate sets of tables are presented, one for *Emission and Detection* of the compounds, one for *Reactions and Products*. (Omission of the latter table indicates that no reaction or product information is available for that chemical group.) The format, guidelines, and explanations of the entries in these tables are presented below.

1.2-1. Species Entries

Chemical nomenclature is often used in the literature with less rigor than might ideally be the case, and a compiler is thereby required to identify identical compounds that are reported under different names and to present them under a unified system. Unless extensive usage dictates otherwise, as, for example, with most pesticides, the recommendations of the IUPAC Commission on the Nomenclature of Organic Chemistry (1001) have been adopted. Inorganic compounds infrequently possess nomenclature options, and no significant difficulties have been encountered in their designation.

The structural formulas are intended to accomplish the following: (a) to illustrate the structure of uncommon (and common) species; (b) to demonstrate the structural similarities among species appearing in a single table; and (c) to augment the identifying information furnished by the species name. I have preferred to utilize standard chemical line notation for the simpler compounds and schematic structural formulas for the more complex.

Reference material is not always sufficiently specific about the structure of a compound being discussed. This is particularly true for substituted organic compounds, where nomenclature such as "dimethyl indane " occurs. The experimental techniques are often incapable of locating the attachment position of ligands, and the absence of designations such as "1,2-dimethylindane" is not, therefore, necessarily an oversight by the author. The compound designations reported in the tables are generally those given in the references. If additional information is available to suggest the most common ligand attachment positions, however, I have occasionally assigned them to otherwise complete identifications in the interest of uniformity with other references to ligand-specific compound identification.

Another problem concerns the listing of minerals, which are occasionally identified specifically in aerosols (e.g., 468). Mineral formulae are generally not structural , permitting substitution of individual atoms. In *muscovite* [$H_2KAl_3(SiO_4)_3$] for example, Mn, Rb, Li, Cs, Ba, Mg, Fe, Cr, Ti, and V may appear as substituents. The rule followed herein is that minerals with specific and unique chemical formulas (e.g. *halite* = NaCl) are included in the tables; those that do not meet those criteria (e.g., *muscovite*) are excluded.

Each compound appears only once in these tables, even in the case where its difunctional properties make it eligible for inclusion in more than one grouping. Organic compounds containing nitrogen, sulfur, or halogen atoms are placed in Chapters 6,7, or 8 regardless of the presence of oxygen-containing functional groups. Oxygenated difunctional hydrocarbons are placed in Chapters 4 and 5 in the table representing their most reactive functional group, in the order

aldehyde > ketone > ether   > oxide > alcohol > ester >   acid
(1131). Inorganic compounds qualifying for more than one group are
placed in Chapter 2 in the order halogen>sulfur> nitrogen.

Two groups of compounds that are included in these tables
require special mention. The first is the emittants from tobacco smoke.
The literature on such compounds is extensive, and its incorporation
here reflects the widespread incidence of vegetation combustion, not
only in cigarette and cigar smoking but also in refuse combustion, slash
burning of agricultural lands, etc. The tobacco literature thus provides
insight into a wide variety of processes with potential atmospheric
impact.

A second group of compounds deserving comment is the natural
fragrance materials. These compounds are generally regarded as desir-
able trace species in the atmosphere. Their study will ultimately help to
define the magnitude of the role of natural vegetative volatiles on
interactive atmospheric chemistry.

1.2-2. Sources Entries

The aim of these columns is to tabulate what is known of the origins of the chemical species. Many of the species are known to be
emitted by specific processes, or from specific sources. These processes
or sources may be natural, but are more often anthropogenic. The
source lists are intended to demonstrate the known diversity of origins,
and the references are illustrative rather than exhaustive; no more than
two references for each type of source are given. The source descrip-
tions are brief: adding "manufacturing" (as in "charcoal manufacturing")
or "facility" (as in "rendering facility") often makes them more under-
standable. Further information is provided by the reference titles, and,
of course, by the references themselves.

A compound may be emitted either as an aerosol (indicated by
"a") or as a gas. (In the former case it is more properly regarded as "a
component of the liquid or solid aerosol"). A few compounds, such as
some of the higher alkanes, are apparently emitted in both phases,
because of the wide range of source temperatures. The literature is
sometimes unspecific as to the state of the compound upon emission,
often because analyses of atmospheric samples collected on filter
material are not able to determine the natural state of the detected
compounds. The division of emission into gaseous and solid states is
thus one that should not be regarded as rigorous. In many cases for
plant volatiles, the listed compounds have been identified in the "essen-
tial" (i.e., volatile) oil of the plants rather than in the air above them;
such identifications carry a letter e after the reference number (i.e.,
552e).

The numerical order in which the sources are listed represents an attempt to distinguish, on a globally averaged basis, the relative importance of the emission fluxes. Such an ordering can seldom be definitive, since emission flux data for most compounds and processes are inadequate, but the *controlling* processes are often known. Sources for which information is insufficient to permit them to be ranked in this way are listed alphabetically.

It is beyond the scope of this work to critically evaluate each atmospheric measurement cited. The listing of two references to the detection of a specific compound is intended to serve as independent confirmation of that detection. If only one reference is given, however, the detection should be regarded as tentative. In some cases, authors themselves describe identification as tentative, in which circumstance a lowercase t follows the reference number (e.g., 121t).

Concentrations of the compounds as they are emitted from their several sources are not presented in these tables. Such concentrations are extremely dependent on the characteristics of the specific source being measured, on the proximity of the detector to the source, and on the techniques and methodology utilized for measurement. Inclusion of emission stream concentrations in the tables would therefore require substantial additional information to be presented as well; it seems preferable to have interested readers examine the references themselves. To aid in this examination, source references that contain concentration data in addition to species identification are denoted in the tables with an asterisk (e.g., 257*).

1.2-3. Detection Entries

The distinction between measurement of a given species from a source and its detection as a gas or aerosol(a) is that in the latter case the detection is made in the absence of any known or suspected sources in the vicinity of the measurement. Authors are sometimes vague on this point; I have placed entries in the Sources columns in those cases where a choice based on inadequate evidence was required, since it is more chemically conservative not to assume a compound to be long-lived enough to be present in the ambient atmosphere. The division between gas phase and aerosol phase is often useful but, as noted above, such a distinction is sometimes difficult and those presented here must not be regarded as rigorous. Tentative ambient detections are noted by a lowercase t following the reference number (e.g., 121t).

Concentrations of compounds in the ambient troposphere are listed in the tables where such information is available. A range of concentrations is often appropriate, since ambient concentrations in urban, rural, and ocean atmospheres differ dramatically even in the absence of proximate sources. Concentrations in the gas phase are expressed in parts per million (about 2.5×10^{13}molec/cm^3 at sea level) or parts per billion (about 2.5×10^{10}molec/cm^3). Concentrations in the aerosol phase are given in micrograms or nanograms per cubic meter of air. Details of the concentration data are available in the references marked with an asterisk (e.g., 257*), as well as in reviews by Bach (1002) and Heicklen (1003).

For a small number of compounds and ions, sufficient information is available to permit a distributional analysis of the concentration data to be performed (1004). Each of these species contains an ampersand (&) in its concentration column; the results are presented in the text and figures accompanying the table.

1.2-4. Chemical Reactions Entries

The chemical fates of the compounds in the atmosphere are analyzed in these columns. The reactivity of all compounds with solar photons, hydroxyl radicals (HO·), oxygen atoms [O(^{3}P)], hydroperoxyl radicals (HO$_2$·), ozone molecules (O$_3$), and with other species known from the literature to be pertinent is reviewed. Where possible, a selected value of the rate constant is listed (error limits are those of the experimenters). Unless otherwise noted, the rate constants refer to reactions in the gas phase, and, since this book is principally concerned with the chemistry of the troposphere, evaluated at 25°C . Although measured rate parameters predominate, a few values derived from careful theoretical studies have been included in cases where no measurements exist. The rate constants are given in cm-molecule-sec units; photosensitive reactions are in units of sec^{-1} and are evaluated for a noontime summer sun at 40°N latitude. Unimolecular decomposition reactions are indicated by a U in the reactant column. Although every effort has been made to select the best available values for the rate constants, the principal use of the entries here is to assess relative rates of reaction and to estimate tropospheric lifetimes. Rate constant measurements are perhaps evolving more rapidly than any other area touched upon in this book, and readers needing the most accurate determinations are best referred to the recent kinetic literature.

Rate constants for the various chemical loss mechanisms have been used to determine the most rapid of the chemical sinks, using the typical average tropospheric diurnal concentrations for the reactants as indicated in Table 1.1. The species lifetime in the troposphere (in seconds) is then computed as the inverse of the product of the reactant concentration and the appropriate rate constant. Stable reaction products are specified if they are known or if they can accurately be predicted to occur in oxygen-rich atmospheres at a pressure of 760 torr. The products may be directly produced by the reaction indicated, or, if sufficient information is available, they may be known to be subsequent products of the reaction sequence initiated by that reaction.

TABLE 1.1. TYPICAL AVERAGE CONCENTRATIONS FOR REACTIVE ATMOSPHERIC SPECIES

Specie	Concentration[a]	Reference
$h\nu$	Fct. of λ	1005,1009
HO·	4.1×10^5	1199
O(^{3}P)	2.5×10^4	1007
HO$_2$·	6.5×10^8	1007
O$_3$	1.0×10^{12}	1008
H$_2$O	2.5×10^{17}	294
O(^{1}D)	5.0×10^{-1}	1007
O$_2$	5.0×10^{18}	294
NO	1.5×10^{11}	294
O$_2$($^1\Delta$)	2.0×10^7	1229
Cl·	2.0×10^4	1125

[a] Molecules cm^{-3}

1.2-5. Remarks Entries

Occasional items of interest that are not included elsewhere are given here. The most common is remark B, which indicates that the compound has a known or suspected chemical precursor; the specific information is contained in the precursor cross-reference table, Appendix B.

1.3 THE TEXT INFORMATION

A brief discussion is provided with each of the tables. The intent of this textual material is to place the compounds of the table into proper perspective with the atmosphere as a whole, and to explore their most common sources and removal mechanisms. The first section, *Identified Compounds* , presents information on the structural similarities of the compounds in the table and on the most common sources of the compounds. The second section, *Ambient Concentrations* , discusses the abundances of the compounds in the lower atmosphere of the Earth. A concluding section, *Chemistry* , explores the most important or most probable atmospheric chemical reactions of the compounds.

Atmospheric chemistry has progressed rapidly in the past few years and many of the important chemical processes have now been specified. Such information forms a good base for many of the chemical reaction sequences outlined in succeeding chapters. Conversely, in a substantial number of cases relevant laboratory work remains to be performed. The structural and chemical similarity of many systems to other systems already studied often enables one to proceed by analogy, however. Such an approach is used freely throughout this volume to examine the probable fates of many of the atmospheric compounds.

The basic approach herein is taxonomic rather than pedagogical. As a result, some readers may find the need for additional information on topics covered here only in passing. Among the many useful chemistry references available are the treatise on photochemistry by Calvert and Pitts (1231) and those on chemical kinetics by Benson(1269,1270). Heicklen (1003) describes the interaction of light with molecules in the atmosphere and presents more complete discussions of many topics in atmospheric chemistry than are appropriate in this work. An overview of air quality and its analysis and control is provided by Stern's five-volume treatise(1271). In a field evolving as rapidly as is atmospheric chemistry, frequent reference to the current literature is also of great value.

2

INORGANIC COMPOUNDS

2.0. INTRODUCTION

The atmosphere is chiefly composed of the inorganic compounds that are the subject of this chapter. Although the number of such compounds is few compared with the organic species, the chemical reactions of the inorganic compounds are central to virtually all atmospheric transformation processes. Many of the most important of these atmospheric reactions have received detailed study in the laboratory; it seems fair to say that the inorganic component of tropospheric chemistry is much better understood than is its organic analog.

Tropospheric inorganic compounds are divided herein into five groups: rare gases and compounds of oxygen and hydrogen, nitrogen-containing compounds, sulfur-containing compounds, halogen-containing compounds, and all else (largely oxides and carbonates).

2.1. RARE GASES AND COMPOUNDS OF OXYGEN AND HYDROGEN

2.1.1. Identified Compounds

The principal constituents of the atmosphere are the stable gases O_2 and N_2. The small amounts of the five rare gases He, Ne, Ar, Kr, and Xe are largely remnants of the Earth's prenebula. However, both helium and argon, the more abundant of the five, have been detected in volcanic eruptions; thus their concentrations are to some extent dynamically established. The evolution of the atmosphere over long time scales is outside the scope of this book, which assumes the existence of a natural atmosphere with a number of major and minor constituents and attempts to classify the processes that perturb those constituents over time spans of a few seconds to a few years.

2.1.2. Concentrations

The concentrations of the rare gases and of molecular oxygen are rather precisely known and are negligibly affected by the sources listed in Table 2.1. The concentrations of the other species on the table are dynamically determined. Molecular hydrogen is emitted from a wide variety of microbial and combustion processes; its concentration varies with proximity to these sources. Water vapor is ubiquitous but widely variable because of its dependence on insolation and advection. Several of the rare gases have been identified in volcano effluent, as have O_2 and H_2O . Xenon is a minor emittant from nuclear power plants. Radon outgases from the soil and contributes to radioactive processes in the atmosphere. For ozone, enough information is available to permit a statistical summary of median hourly average values for all hours and all seasons. The result, shown in Fig. 2.1-1, has a mean value of 18 ppb. The measured H_2O_2 concentrations appear to be good indicators of photochemical activity.

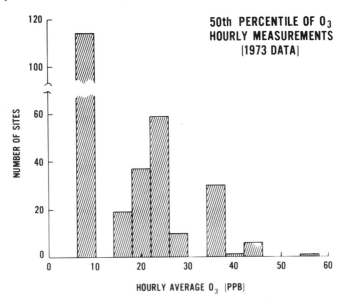

Fig. 2.1-1.   Distribution of 50th percentiles of O_3 hourly measurements at air quality monitoring sites within the continental United States, 1973 data (from reference 1004).

Concentration measurements for the hydroxyl radical are preliminary as this is written and are subject to revision as more data become available. It is clear, however, that the hydroxyl radical is sufficiently abundant that its catholic taste in reaction partners and rapid reaction rate constants render it the principal scavenger of most atmospheric trace species (1011).

2.1.3. Chemistry

The rare gases are not known to be involved chemically in any of the atmospheric processes. This is not the case with the other compounds of Table 2.1, however. Molecular hydrogen enters into atmospheric chemistry by reaction with the hydroxyl radical:

$$H_2 + HO\cdot \rightarrow H_2O + H\cdot,$$

and is generated by formaldehyde photolysis:

$$HCHO \overset{h\nu}{\rightarrow} H_2 + CO.$$

The rates of these reactions are influenced by man's activity, but the cumulative effects on the global concentrations of H_2 remain to be determined (1188).

Molecular oxygen plays a very important role in the chemistry of the atmosphere. Most trace species undergo oxidation during their atmospheric residence time, and O_2 is ultimately the species responsible for these processes. The oxidation is generally not direct, however, but proceeds through peroxy radicals. The formation of these radicals, e.g.,

$$H\cdot + O_2 \overset{M}{\rightarrow} HO_2\cdot$$

$$CH_3\cdot + O_2 \overset{M}{\rightarrow} CH_3O_2\cdot,$$

constitutes the chemical loss mechanism for O_2 in the atmosphere. Despite its importance for atmospheric chemistry, the O_2 loss is so small relative to the concentration of O_2 that its effect on atmospheric oxygen concentrations is negligible (1181).

Ozone has perhaps the most interesting chemistry of any of the atmospheric compounds. Except for minor amounts of generation by lightning and power lines, there are no physical sources of O_3. It is produced by photochemical processes in both the troposphere and stratosphere and reacts with a wide variety of atmospheric trace constituents. During darkness, the principal chemical removal path for a variety of hydrocarbon species is reaction with ozone (1007, 1010). The regeneration of ozone by photolysis of NO_2 and by a variety of other processes is rapid (1007), and the study of ozone formation and removal cannot be separated from studies of virtually all of the major chemical processes of the atmosphere.

The presence of the hydroxyl radical in the atmosphere is thought to limit the concentrations of ammonia, hydrogen sulfide, methane, and a wide variety of other trace species. In urban areas, HO· is formed

TABLE 2.1 RARE GASES AND COMPOUNDS OF OXYGEN AND HYDROGEN
Emission and Detection

Species Number	Name	Chemical Formula	Emission		Detection	Ambient conc.
			Source	Ref.	Ref.	
2.1-1	Helium	He	natural gas volcano	456* 73	294*,558*	5.24ppm
2.1-2	Neon	Ne			294*,558*	18.18ppm
2.1-3	Argon	Ar	natural gas volcano	456* 106*,234	294*,558*	9340ppm
2.1-4	Krypton	Kr			294*,558*	1.14ppm
2.1-5	Xenon	Xe	nuclear power	391*	294*,558*	87ppb
2.1-6	Radon	Rn	soil volcano	467,483 430*	494*(a)	
2.1-7	Hydrogen	H_2	1.microbes 2.auto forest fire geothermal steam natural gas oceans rocket vegetation volcano	210 401* 354 372,402* 423*,456* 291 536 461 106*,246	291*,294*	540-810ppb
2.1-8	Oxygen	O_2	volcano	73	294*,558*	2.09×10^5 ppm
2.1-9	Ozone	O_3	lightning power trans.	318 318	283*,410*	&,B
2.1-10	Water	H_2O	volcano	106*	294	

TABLE 2.1 RARE GASES AND COMPOUNDS OF OXYGEN AND HYDROGEN
Emission and Detection

Species Number	Name	Chemical Formula	Emission		Detection Ref.	Ambient conc.
			Source	Ref.		
2.1-11	Hydrogen peroxide	H_2O_2			90,319*	10-180ppb
2.1-12	Hydroxyl radical	HO·			259*,287*	0.1-5.5×10^7 molec cm^{-3}

TABLE 2.1 RARE GASES AND COMPOUNDS OF OXYGEN AND HYDROGEN
Reactions and Products

Species Number	Reactant	Chemical reactions			Products	Ref.	Remarks
		k	Ref.	Lifetime			
2.1-7	HO·	$(7.0\pm0.7)\times10^{-15}$	1211	3.2×10^{8}			B
	HO$_2$·	1.5×10^{-18}	1152				
	Cl·	$(1.8\pm0.1)\times10^{-14}$	1247				
2.1-9	NO	$(1.6\pm0.1)\times10^{-14}$	1170	4.2×10^{2}	NO$_2$	1170	B
	hν	4.8×10^{-4}	1125				
	HO$_2$·	$(1.0\pm0.1)\times10^{-15}$	1170				
	HO·	$(5.2\pm1.4)\times10^{-14}$	1170				
	O	$(8.4\pm1.1)\times10^{-15}$	1212				
	O(^{1}D)	$(1.2\pm0.1)\times10^{-10}$	1170				
2.1-10	O(^{1}D)	$(2.3\pm0.2)\times10^{-10}$	1170				
2.1-11	hν	9.7×10^{-6}	1176	1.0×10^{5}	HO·	319	
	HO·	$(8.5\pm1.8)\times10^{-13}$	1120				
	Cl·	$(3.5\pm0.5)\times10^{-13}$	1042				
2.1-12	HO$_2$·	$(5.1\pm1.6)\times10^{-11}$	1119	3.0×10^{1}	H$_2$O,O$_2$	1170	B
	*						

*Reacts with many atmospheric species, generally abstracting a hydrogen atom.

primarily by (1007)

$$HO_2 \cdot + NO \rightarrow HO \cdot + NO_2.$$

In remote areas where NO concentrations are very low, the principal formation process is

$$O_3 \overset{h\nu}{\rightarrow} O_2 + O(^1D)$$

$$O(^1D) + H_2O \rightarrow HO \cdot + HO \cdot.$$

Hydrogen peroxide acts as a bridge compound between the "odd hydrogen" radicals $HO \cdot$ and $HO_2 \cdot$. It is formed by

$$HO_2 \cdot + HO_2 \cdot \rightarrow H_2O_2 + O_2$$

and removed by

$$H_2O_2 \overset{h\nu}{\rightarrow} HO \cdot + HO \cdot.$$

Its principal chemical function is thus as a holding and transfer tank in the odd hydrogen chemistry. The hydroperoxyl radical ($HO_2 \cdot$) must certainly be present in the troposphere, but no ambient detection of this hard-to-measure species has been reported.

2.2. INORGANIC NITROGEN COMPOUNDS

2.2.1. Identified Compounds

Nitrogen in the atmosphere is dominated by N_2, which does not participate actively in tropospheric chemistry. The remaining compounds, termed the "odd nitrogen" compounds, are of concern here. Nitrogen in the reduced state is represented by ammonia, which is produced by a host of natural bacterial decomposition processes and by a number of industrial operations. NO and NO_2 (and, to a lesser extent, other oxides of nitrogen) are created during combustion and thus have a wide range of anthropogenic sources. Nitrous oxide, in addition to arising during combustion, is produced during the nitrogen fixation process (1267). Nitrate salts are common soil compounds and their presence in aerosols doubtless results at least in part from mechanical suspension of soil particles by the wind.

2.2.2. Ambient Concentrations

The concentrations of several of the inorganic nitrogen compounds in air have been well studied; as a result, statistical summaries of the data can be presented. Fig. 2.2-1 shows the distribution of annual median hourly average data for nitric oxide concentrations. The mean value of these data is 24 ppb, and applies to urban areas. Distributional data for nitrogen dioxide are presented in Fig. 2.2-2. The mean value, again applicable to urban areas, is 37 ppb. The global concentrations of the oxides of nitrogen are of the order of $0.1-1.0$ ppb (1013).

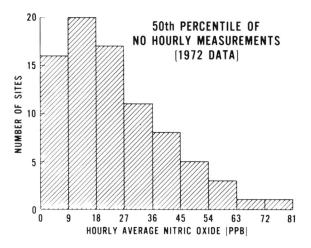

Fig. 2.2-1.   Distribution of 50th percentiles of NO hourly measurements at air quality monitoring sites within the continental United States; 1972 data except for 1967 data from six sites (from reference 1004).

Another substantial set of data is for the ammonium ion (NH_4^+) in particulate matter (Fig. 2.2-3). These data almost certainly reflect the rapid incorporation of ammonia gas into the ambient aerosol. The mean value, largely applicable to urban areas, is $0.25 \ \mu g/m^3$.

Atmospheric concentration information on other inorganic nitrogen compounds is indicated in Table 2.2.

2.2.3. Chemistry

The chemistry of atmospheric nitrogen is amazingly varied, in part because nitrogen appears in valence states ranging from $+5$ (HNO_3) to -3 (NH_3). and its importance to atmospheric chemistry is very great. One manifestation of this involvement is the rapid scavenging of ozone

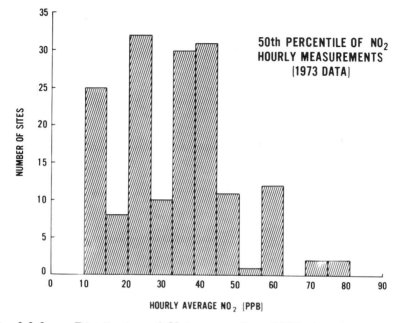

Fig. 2.2-2.   Distribution of 50th percentiles of NO_2 hourly measurements at air quality monitoring sites within the continental United States, 1973 data (from reference 1004).

by nitric oxide:

$$NO + O_3 \rightarrow NO_2 + O_2.$$

This process removes one of the principal oxidizing reactants from the atmosphere. In the presence of sunlight, however, nitrogen dioxide photodissociates to product atomic oxygen, another of the central reactants:

$$NO_2 \xrightarrow{h\nu} NO + O.$$

With the exception of ozone photolysis, this reaction is the sole tropospheric source of atomic oxygen.

In addition to its role as a precursor to nitrogen dioxide through ozone scavenging, NO is the principal reducing species in the lower atmosphere. This latter function occurs through reaction with a variety of peroxy radicals:

$$RO_2{\cdot} + NO \rightarrow RO{\cdot} + NO_2,$$

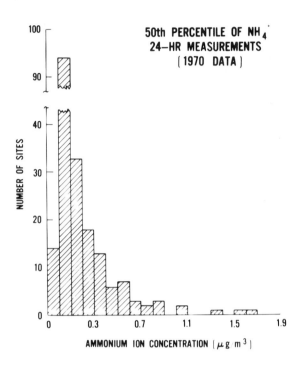

Fig. 2.2-3.   Distribution of 50th percentiles of ammonium ion particulate matter daily measurements at air quality monitoring sites within the continental United States, 1970 data
(from reference 1004).

where R refers to any organic fragment. The most important of this
group of reactions is that involving the odd hydrogen radicals:

$$HO_2\cdot + NO \rightarrow HO\cdot + NO_2.$$

This reaction is the crucial step in establishing the $HO_2\cdot/HO\cdot$ concentration ratio in urban atmospheres. Since $HO\cdot$ is the primary reactant for
most gases emitted into the atmosphere, air quality is closely associated
with emissions of oxides of nitrogen (1104).

Nitrous oxide is chemically inert in the lower atmosphere, but
apparently participates in stratospheric chemistry by serving as a natural
source of NO. This process and its possible consequences have been
discussed extensively in the literature (e.g., 1267), to which the
interested reader is referred for further details.

Nitrous acid is not known to be emitted into the atmosphere. It is
formed by the reaction

TABLE 2.2 INORGANIC NITROGEN COMPOUNDS
Emission and Detection

Species Number	Name	Chemical Formula	Emission		Detection Ref.	Ambient conc.
			Source	Ref.		
2.2-1	Nitrogen	N_2	geothermal steam	372,402*	294*,558*	7.85×10^5 ppm
			propellant	559,563		
			volcano	73,106*		
2.2-2	Ammonia	NH_3	1. microbes	302	23*,342*	<0.1-50 ppb
			2. animal waste	141,160		
			3. sewage tmt.	174*		
			ammonia mfr.	58*		
			auto	565,581		
			coke mfr.	58*		
			fertilizer mfr.	58*,256*		
			fish meal mfr.	110,255*		
			foundry	279*		
			geothermal steam	372,402*		
			glue vapor	11		
			lacquer mfr.	427,428		
			Na_2CO_3 mfr.	58*		
			petroleum mfr.	58*,221		
			plastics comb.	46,354		
			refrigeration	559		
			refuse comb.	395		
			rocket	536		
			tobacco smoke	396		
			volcano	73		
2.2-3	Potassium cyanide	KCN	steel mfr.	225(a)		
2.2-4	Ammonium hydroxide	NH_4OH	petrol. vapor	58*		

TABLE 2.2 INORGANIC NITROGEN COMPOUNDS
Emission and Detection

Species Number	Name	Chemical Formula	Emission		Detection Ref.	Ambient conc.
			Source	Ref.		
2.2-5	Nitric oxide	NO	1.auto 2.combustion diesel microbes	282,444* 282 136* 210,302	282* 399(a)	&
2.2-6	Nitrogen dioxide	NO_2	1.auto 2.combustion diesel tobacco smoke volcano	282,444* 282 136* 4* 234	282* 399(a)	&
2.2-7	Nitrous oxide	N_2O	1.microbes 2.ocean 3.combustion plant volatile propellant	210,300* 297* 353* 547 562,563	296*,299* 399(a)	250-336 ppb
2.2-8	Dinitrogen pentoxide	N_2O_5	fertilizer mfr.	278*		
2.2-9	Nitrous acid	HNO_2			320*	0.4-11 ppb

TABLE 2.2 INORGANIC NITROGEN COMPOUNDS
Emission and Detection

Species Number	Name	Chemical Formula	Emission		Detection	Ambient conc.
			Source	Ref.	Ref.	
2.2-10	Nitric acid	HNO_3	HNO_3 mfr. explosives mfr. rocket	35,58* 58* 536	90,226* 151(a),213(a)	1-11ppb
2.2-11	Sodium nitrate	$NaNO_3$			213(a),292(a)	
2.2-12	Ammonium nitrate	NH_4NO_3	rocket	536	81(a),214*(a)	10-770 ng/m^3
2.2-13	Potassium nitrate	KNO_3			292(a)	

TABLE 2.2 INORGANIC NITROGEN COMPOUNDS
Reactions and Products

Species Number	Reactant	Chemical reactions			Products	Ref.	Remarks
		Ref.	k	Lifetime			
2.2-2	HCl	1219	1.9×10^{-17}	2.1×10^{6}	NH_4Cl	1219	
	HO·	1213	$(1.6 \pm 0.2) \times 10^{-13}$		NO	1192	
	HNO_3				NH_4NO_3	1086	
	HO_2·				$NH_3;HO_2$	1057	B
2.2-5	O_3	1170	$(1.6 \pm 0.1) \times 10^{-14}$	6.3×10	NO_2	1170	
	HO_2·	1223	$(8.1 \pm 1.5) \times 10^{-12}$		NO_2	1170	
	NO_3·	1021	$(8.7 \pm 1.2) \times 10^{-12}$		NO_2	1021	
	HO·	1211	$(6.1 \pm 1.0) \times 10^{-12}$		HNO_2	1021	
	O	1170	$(2.8 \pm 0.2) \times 10^{-12}$		NO_2	1170	
2.2-6	$h\nu$	1125	9.9×10^{-3}	1.0×10^{2}	NO,O_3	1021	B
	HO_2·	1225	$(2.5 \pm 0.6) \times 10^{-13}$		HO_2,NO_2	1064	
	O_3	1170	$(3.2 \pm 0.3) \times 10^{-17}$				
	HO·	1224	1.6×10^{-11}		HNO_3	1170	
	NO_3·	1226	$(8.0 \pm 1.8) \times 10^{-15}$		N_2O_5	1021	
	O	1170	$(9.1 \pm 0.3) \times 10^{-12}$				
2.2-7	$O(^1D)$	1170	$(1.1 \pm 0.1) \times 10^{-10}$	1.8×10^{10}	NO	1170	
	HO·	1227	$\leqslant 4 \times 10^{-16}$				
2.2-8	U	1226	2.0×10^{-1}	5.0			
	O	1254	$\leqslant 3 \times 10^{-16}$				
	H_2O				HNO_3	1021	

TABLE 2.2 INORGANIC NITROGEN COMPOUNDS
Reactions and Products

Species Number	Reactant	Chemical reactions			Products	Ref.	Remarks
		k	Ref.	Lifetime			
2.2-9	$h\nu$	2.8×10^{-3}	1125	3.6×10^{2}	$NO,HO\cdot$	1021	B
	$HO\cdot$	$(6.6\pm0.3)\times10^{-12}$	1133				
	O_3	$\leqslant1.0\times10^{-19}$	1253				
	$(CH_3)_2NH$	5.3×10^{-17}	1054		$(CH_3)_2NNO$	1054	
2.2-10	$h\nu$	4.4×10^{-7}	1125	2.3×10^{6}	$NO_2,HO\cdot$	1021	B
	$HO\cdot$	$(8.0\pm0.4)\times10^{-14}$	1170				
	NH_3				NH_4NO_3	1086	
2.2-12							B

$$NO + HO\cdot \overset{M}{\rightarrow} HNO_2$$

and thereafter plays an intermediate role in tropospheric photochemistry by releasing the hydroxyl radical during photodissociation:

$$HNO_2 \overset{h\nu}{\rightarrow} NO + HO\cdot.$$

Nitrous acid thus shares two important characteristics with ozone: it is formed entirely by a gas phase chemical reaction, and it photodissociates to produce a highly reactive intermediate species.

Nitric acid and the nitrate salts are found primarily in the aerosol phase. It appears that efficient gas-to-particle conversion is involved in aerosol nitrate generation (1238), but the details of the conversion remain obscure.

2.3. INORGANIC SULFUR COMPOUNDS

2.3.1. Identified Compounds

Although inorganic sulfur compounds are apparently ubiquitous in the earth's troposphere, their interactions with other atmospheric species appear to be of only moderate importance. This is due in part to the fact that none of the compounds photodissociate in the lower troposphere and thus no reactive fragment radicals are produced. The principal reduced form of sulfur in the gas phase is probably hydrogen sulfide, which is a natural product of microbial decomposition and is also produced by a variety of industrial processes. The total amounts are relatively small, however, and reactions of H_2S with other atmospheric species proceed slowly (1014).

Sulfide dioxide is the most abundant of the atmospheric sulfur compounds. It is produced by anthropogenic combustion processes, and is also an intermediate in the oxygenation chains of the reduced sulfur species (1014). The chemical removal processes for SO_2 lead to higher oxidation states (1016, 1017) and probably to sulfuric acid or related acidic compounds.

The ammonium sulfate salts are ubiquitous atmospheric aerosol compounds (1018), and are probably formed by the reaction of ammonia with sulfuric acid. Some of the variety of other inorganic sulfates that have been detected in aerosols have industrial sources; some are naturally occurring soil compounds.

2.3.2. Ambient Concentrations

The information currently available on ambient concentrations of the reduced inorganic sulfur compounds is sparse, and it is difficult to assess its broad applicability or the relative importance of the compounds to the global sulfur balance. A wealth of data is available for SO_2, however. For urban areas, a statistical distribution of the hourly average medians has a mean value of 70 ppb (see Fig. 2.3-1). Concentrations of SO_2 in nonurban areas and over the oceans are much lower, perhaps of the order of 0.1 ppb (1019,1020).

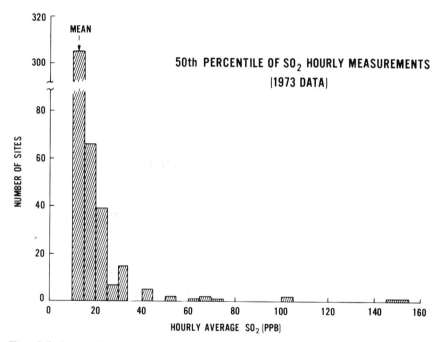

Fig. 2.3-1.　Distribution of 50th percentiles of SO_2 hourly measurements at air quality monitoring sites within the continental United States, 1973 data (from reference 1004).

2.3.3. Chemistry

Most of the compounds of Table 2.3 are sulfide and sulfate salts that have been detected in aerosols. The principal exceptions are H_2S and SO_2. For hydrogen sulfide, the initiating reaction is hydrogen abstraction by HO· (1014, 1248) followed by oxidation to SO_2 (1249), perhaps through an SO intermediate:

TABLE 2.3 INORGANIC SULFUR COMPOUNDS
Emission and Detection

Species Number	Name	Chemical Formula	Emission Source	Emission Ref.	Detection Ref.	Ambient conc.
2.3-1	Hydrogen sulfide	H_2S	1.microbes	210,302	23*,83*	.08-24ppb
			2.petroleum mfr.	127*		
			animal waste	157,522		
			auto	418*		
			fish meal mfr.	58*,255*		
			geothermal steam	372,402*		
			natural gas	423*,456*		
			plastics comb.	354		
			rubber mfr.	119*		
			sewage tmt.	174,204		
			starch mfr.	523*		
			synthetic fibre mfr.	58*,191		
			tobacco smoke	158,396		
			volcano	106*,246		
			water treatment	185		
			wood pulping	19,58*		
2.3-2	Sodium sulfide	Na_2S	wood pulping	53(a),256*(a)		
2.3-3	Zinc sulfide	ZnS	zinc mfr.	256*(a)		
2.3-4	Lead sulfide	PbS	industrial	101(a)		
			lead smelting	256*(a)		
2.3-5	Arsenopyrite	$FeAsS$	iron pellet mfr.	328(a)		
2.3-6	Flowers of sulfur	S_8	foundry	279(a)	213(a),589(a)	
			sulfur mfr.	58*		
			titanium mfr.	412*(a)		
			volcano	538		

TABLE 2.3 INORGANIC SULFUR COMPOUNDS
Emission and Detection

Species Number	Name	Chemical Formula	Emission		Detection Ref.	Ambient conc.
			Source	Ref.		
2.3-7	Sulfur dioxide	SO_2	1.combustion	316	289(a),399(a)	&
			auto	541		
			coke mfr.	316		
			diesel	58*		
			H_2SO_4 mfr.	316		
			ore smelting	316		
			petroleum mfr.	316		
			plastics comb.	46		
			refuse comb.	316		
			starch mfr.	523*		
			turbine	71		
			volcano	28*,235		
2.3-8	Sulfur trioxide	SO_3	auto	541		
			battery mfr.	379		
			brick mfr.	268		
			cement mfr.	539		
			fuel oil comb.	317*,356		
			H_2SO_4 mfr.	93		
2.3-9	Sulfurous acid	H_2SO_3	combustion	203		
			metal mfr.	150		
2.3-10	Ammonium sulfite	$(NH_4)_2SO_3$			213(a)	
2.3-11	Sulfuric acid	H_2SO_4	auto	58*,520*	22(a),257(a)	
			combustion	580*		
			explosives mfr.	58*		
			furnace soot	51*(a)		
			H_2SO_4 mfr.	35,58*		
			steel mfr.	142*(a)		
			volcano	28*		

TABLE 2.3 INORGANIC SULFUR COMPOUNDS
Emission and Detection

Species Number	Name	Chemical Formula	Emission		Detection	Ambient conc.
			Source	Ref.	Ref.	
2.3-12	Ammonium hydrogen-sulfate	NH_4HSO_4			257(a),292(a)	
2.3-13	Ammonium sulfate	$(NH_4)_2SO_4$			22(a),292(a)	
2.3-14	Triammonium hydrogen-bis(sulfate)	$(NH_4)_3H(SO_4)_2$			36(a)	
2.3-15	Sodium hydrogen-sulfate	$NaHSO_4$			213(a),589(a)	
2.3-16	Sodium sulfate	Na_2SO_4	wood pulping	45(a),53(a)	292(a)	
2.3-17	Sodium sulfate decahydrate	$Na_2SO_4 \cdot 10H_2O$	wood pulping	10(a)		
2.3-18	Calcium sulfate	$CaSO_4$	lime mfr. wood pulping	256*(a) 45(a)	280(a)	
2.3-19	Potassium sulfate	K_2SO_4	cement mfr.	468(a)		
2.3-20	Nickel sulfate	$NiSO_4$			280(a)	
2.3-21	Zinc diammonium-(bis)sulfate	$Zn(NH_4)_2(SO_4)_2$			107(a)	
2.3-22	Zinc sulfate	$ZnSO_4$			107(a)	
2.3-23	Lead sulfate	$PbSO_4$	industrial refuse comb.	101(a) 196(a)	280(a)	

TABLE 2.3 INORGANIC SULFUR COMPOUNDS
Reactions and Products

Species Number	Chemical reactions						Remarks
	Reactant	k	Ref.	Lifetime	Products	Ref.	
2.3-1	HO·	$(5.3\pm0.5)\times10^{-12}$	1213	4.1×10^{5}	SO_2	1014	
	O	$(3.0\pm0.2)\times10^{-14}$	1214				
	O_3	$<2.0\times10^{-20}$	1059				
2.3-7	$HO_2\cdot$	$(9.0\pm1.8)\times10^{-16}$	1021	3.6×10^{6}	SO_3	1014	B
	HO·	6.0×10^{-13}	1215		H_2SO_4	1016	
	O	$(5.0\pm0.2)\times10^{-14}$	1216				
	O_3	$<1.0\times10^{-22}$	1061				
2.3-8	H_2O	1.0×10^{-12}	1217	4.0×10^{-6}	H_2SO_4	1016	B
2.3-11							B

$$H_2S \xrightarrow{HO\cdot} HS\cdot \xrightarrow{O_2} SO \xrightarrow{O_2} SO_2 .$$

Sulfur dioxide reacts with both the HO· and HO_2· radicals:

$$SO_2 + HO\cdot \rightarrow HSO_3\cdot$$

$$SO_2 + HO_2\cdot \rightarrow SO_3 + HO\cdot .$$

The sulfur trioxide goes rapidly to sulfuric acid in the atmosphere:

$$SO_3 + H_2O \rightarrow H_2SO_4.$$

The precise fate of the HSO_3· radical is uncertain, but a recent review of sulfur thermochemistry (1250) predicts that a variety of oxygenated sulfates and sulfoxynitrates will be produced. These compounds, together with sulfuric acid and the ammonium sulfates, constitute the photochemically produced "sulfate aerosol" upon their incorporation into atmospheric particulate matter.

Since sulfur dioxide absorbs solar radiation but does not photodissociate at tropospheric wavelengths, considerable experimental effort has been expended on excited state SO_2 chemistry (1076-1080). Many possible reactions must be considered and their efforts explored in detail. At this writing, it appears that SO_2 photochemistry is not of major atmospheric importance but may be of significance in certain regimes (such as plume chemistry) or in the creation of certain atmospheric products (such as the sulfinic acids).

2.4. INORGANIC HALOGENATED COMPOUNDS

2.4.1. Identified Compounds

These compounds can be divided into three groups: gaseous halogen molecules and their acids, petroleum additive derivatives, and inorganic halogen salts. Industrial sources of the acids and of Cl_2 are numerous. HCl is also produced in great quantity by oceanic processes. Volcanos are minor sources of the halogen acids. Alkyl lead compounds are used as antiknock additives in many gasoline formulations; the halogenated lead salts in Table 2.4 doubtless arise from that practice. Most of the other halogen salts arise principally from industrial processing involving natural ore. The halogen salts are virtually always present as aerosol constituents and seem unlikely to have any role in gas phase tropochemistry.

TABLE 2.4 INORGANIC HALOGENATED COMPOUNDS
Emission and Detection

Species Number	Name	Chemical Formula	Emission		Detection Ref.	Ambient conc.
			Source	Ref.		
2.4-1	Chlorine	Cl_2	1.chlorine mfr.	58*,76		
			aluminum mfr.	58*,457		
			refuse comb.	395		
			titanium mfr.	115		
			wood pulping	76		
			zinc mfr.	256*		
2.4-2	Iodine	I_2	iodine mfr.	398	213(a),589(a)	
2.4-3	Hydrogen fluoride	HF	1.fertilizer mfr.	58*,93	44,271	
			aluminum mfr.	115,256*		
			brick mfr.	58*		
			ceramics mfr.	417*		
			HF mfr.	58*		
			lacquer mfr.	427		
			phosph. acid mfr.	58*		
			rocket	536		
			steel mfr.	58*		
			volcano	30*,234		
2.4-4	Hydrogen chloride	HCl	auto	386*	129*	1ppb
			combustion	76		
			forest fire	354		
			HCl mfr.	58*		
			lacquer mfr.	427		
			polymer comb.	304,354		
			refuse comb.	31,76		
			rocket	536		
			sea salt	470		
			titanium mfr.	115		
			volcano	30*,234		

TABLE 2.4 INORGANIC HALOGENATED COMPOUNDS
Emission and Detection

Species Number	Name	Chemical Formula	Emission		Detection Ref.	Ambient conc.
			Source	Ref.		
2.4-5	Hydrogen bromide	HBr	auto volcano	386* 152		
2.4-6	Hydrogen iodide	HI	volcano	152		
2.4-7	Fluorosilicic acid	H_2SiF_6	chemical mfr. fertilizer mfr.	93 256*(a)		
2.4-8	Boron trifluoride	BF_3			114	
2.4-9	Ammonium chloride	NH_4Cl	fertilizer mfr. fertilizer mfr. zinc mfr.	58*,93 256*(a) 256*(a)	213*(a),214*(a)	30-1020ng/m^3
2.4-10	Ammonium perchlorate	NH_4ClO_4	rocket	536		
2.4-11	Sodium fluoride	NaF	aluminum mfr.	256*(a)		
2.4-12	Sodium chloride	$NaCl$	oceans wood pulping	585(a) 256*(a)	280(a),292(a)	
2.4-13	Ammonium chloride lead bromochloride	$NH_4Cl \cdot 2PbBrCl$	auto	96(a),311(a)		
2.4-14	Diammonium chloride lead bromochloride	$2NH_4Cl \cdot PbBrCl$	auto	96(a)		
2.4-15	Magnesium chloride	$MgCl_2$	titanium mfr.	115		
2.4-16	Aluminum fluoride	AlF_3	aluminum mfr. lead mfr.	256*(a),492(a) 115(a)		
2.4-17	Aluminum chloride	$AlCl_3$	aluminum mfr. rocket	256*(a) 536	460t(a)	
2.4-18	Trisodium aluminum-hexafluoride	Na_3AlF_6	aluminum mfr.	256*(a),492(a)		

TABLE 2.4 INORGANIC HALOGENATED COMPOUNDS
Emission and Detection

Species Number	Name	Chemical Formula	Emission Source	Emission Ref.	Detection Ref.	Ambient conc.
2.4-19	Sulfur hexafluoride	SF_6	dispersion tracer	590,593	37*,135*	
2.4-20	Silicon tetrafluoride	SiF_4	aluminum mfr. fertilizer mfr. phosph. acid mfr.	256* 58*,93 58*	12*(a)	$8 \mu g/m^3$
2.4-21	Calcium fluoride	CaF_2	aluminum mfr.	256*(a),492(a)		
2.4-22	Titanium tetrachloride	$TiCl_4$	titanium mfr.	115		
2.4-23	Lead bromochloride	$PbBrCl$	auto	96(a),311(a)		
2.4-24	Lead chloride	$PbCl_2$	industrial	101(a)		
2.4-25	Mercuric chloride	$HgCl_2$			388	
2.4-26	Ferrous chloride	$FeCl_2$			409(a)	
2.4-27	Zinc chloride	$ZnCl_2$	zinc mfr.	425(a)		
2.4-28	Sulfur dioxide difluoride	SO_2F_2	industrial mfr.	400		

TABLE 2.4 INORGANIC HALOGENATED COMPOUNDS
Reactions and Products

Species Number	Chemical reactions						Remarks
	Reactant	k	Ref.	Lifetime	Products	Ref.	
2.4-1	$h\nu$	2.3×10^{-3}	1235	4.4×10^2	HCl	1096	
	O	$(4.8\pm0.5)\times10^{-14}$	1221				
2.4-4	NH_3	1.9×10^{-17}	1219	2.1×10^6	NH_4Cl	1219	B
	HO·	$(7.6\pm0.4)\times10^{-13}$	1170				
	O	$(2.0\pm1.0)\times10^{-14}$	1069				
	O_3	$(2.7\pm1.0)\times10^{-25}$	1069				
2.4-5	HO·	$(5.1\pm1.0)\times10^{-12}$	1067	4.2×10^5			
	O	$(3.3\pm0.3)\times10^{-14}$	1068				
2.4-6	O	$(1.7\pm0.2)\times10^{-12}$	1068	2.4×10^7			
2.4-9							B

2.4.2. Concentrations

HCl is found in the atmosphere at concentrations of about one part per billion, reflecting the balance struck between its generation from oceanic processes and its loss to the hydroxyl radical. Ammonium chloride is a frequently seen component of the atmospheric aerosol; the wide range of observed concentrations presumably reflects the varying amounts of ambient ammonia available to neutralize the HCl. Sparse data are available on concentrations of a few of the halogen salts. The levels of SF_6 reflect its occasional use as a tracer of atmospheric motions; it is good for such a purpose since it is anthropogenic in origin and has no significant chemical decay processes.

2.4.3. Chemistry

The diatomic halogen molecules (X_2, where $X=F$, Cl, Br, or I) all undergo photodissociation to atoms at tropospheric wavelengths,

$$X_2 \xrightarrow{h\nu} X\cdot + X\cdot,$$

and the resulting atoms will abstract hydrogen atoms from virtually all atmospheric organic compounds to form the halogen acid:

$$X\cdot + RH \rightarrow HX + R\cdot.$$

Only HI is subject to photodissociation in the troposphere, and it seems probable that most of the highly soluble HX molecules will be lost to aerosol or ground surfaces, either in the acid forms or as ammonium salts.

The chemistry of the lead halides in the atmosphere is much less well understood. Evidence has been presented (1255, 1256) to indicate that halogen atoms are lost from aerosols, perhaps by diffusion to the surface and subsequent vaporization. The halogen atoms are likely to exist in the HX form and thus will be subject to the processes described above.

2.5. INORGANIC ELEMENTS, HYDRIDES, OXIDES, AND CARBONATES

2.5.1. Identified Compounds

The oxides of 24 different elements constitute a major portion of Table 2.5. Inorganic oxides are, of course, the major constituents of the Earth's crust. Analyses of crustal rock suggest than ten oxides constitute virtually all of the crustal mass (Table 2.5-2); each has been identified as a component of aerosols or an emittant. Although the compounds of Table 2.5 are emitted from a variety of industrial processes, most of the processes involve soil or rock in some way, the emissions being "pass-through" or slightly processed by-products.

TABLE 2.5.2. AVERAGE CHEMICAL COMPOSITION OF THE CONTINENTAL CRUST[a]

Compound	Wt. percent	Aerosol constituent
SiO_2	60.18	X
Al_2O_3	15.61	X
CaO	5.17	X
Na_2O	3.91	X
FeO	3.88	X
MgO	3.56	X
K_2O	3.19	X
Fe_2O_3	3.14	X
TiO_2	1.06	X
P_2O_5	0.30	X

[a] Information in columns 1 and 2 is from reference 473.

The concerns of this book are principally with chemical compounds, but it is interesting to note in passing that 84 elements have been detected in ambient aerosol samples, although the compounds of which they are a part remain to be specified. Three of the elements are products of nuclear weapons testing; the remainder are apparently natural. References for detection of the elements, together with their abundance in crustal rocks, is presented in Table 2.5-3. A detailed discussion of the elemental analyses of ambient aerosols and the implications of such work for aerosol chemistry is given by Rahn (476).

2.5.2. Concentrations

CO and CO_2 are among the best measured of the atmospheric trace gases and the references of Table 2.5 provide extensive information on their global concentrations. Carbon monoxide has also been monitored diligently in urban atmospheres and a statistical summary of its median hourly averages (displayed as Fig. 2.5-1) has a mean value of 2.5 ppm.

TABLE 2.5-3. AVERAGE ELEMENTAL COMPOSITION
OF CRUSTAL ROCKS

At. no.	Symbol	Concentration[a]	Aerosol constituent	Ref.
1	H	1,400	X	280,476
2	He	0.003		
3	Li	30	X	280,476
4	Be	2	X	280,476
5	B	3	X	280,476
6	C	320	X	280,476
7	N	46	X	280,476
8	O	446,000	X	280,476
9	F	700	X	280,476
10	Ne	<0.001		
11	Na	28,300	X	280,476
12	Mg	20,900	X	280,476
13	Al	81,300	X	280,476
14	Si	277,200	X	280,476
15	P	1180	X	280,476
16	S	520	X	280,476
17	Cl	200	X	280,476
18	Ar	0.04		
19	K	25,900	X	280,476
20	Ca	36,300	X	280,476
21	Sc	5	X	280,476
22	Ti	4,400	X	280,476
23	V	110	X	280,476
24	Cr	200	X	280,476
25	Mn	1,000	X	280,476
26	Fe	50,000	X	280,476
27	Co	23	X	280,476
28	Ni	80	X	280,476
29	Cu	45	X	280,476
30	Zn	65	X	280,476
31	Ga	15	X	280,476
32	Ge	2	X	280,476
33	As	2	X	280,476
34	Se	0.09	X	280,476
35	Br	3	X	280,476
36	Kr	<0.001		
37	Rb	120	X	280,476
38	Sr	450	X	280,476
39	Y	40	X	280,476
40	Zr	160	X	280,476
41	Nb	24	X	280,476

TABLE 2.5-3.   AVERAGE ELEMENTAL COMPOSITION
OF CRUSTAL ROCKS

At. no.	Symbol	Concentration[a]	Aerosol constituent	Ref.
42	Mo	1	X	280,476
43	Tc	<0.001	b	482
44	Ru	0.001	X	476
45	Rh	0.001	X	476
46	Pd	0.01	X	476
47	Ag	0.1	X	280,476
48	Cd	0.2	X	280,476
49	In	0.1	X	476,477
50	Sn	3	X	280,476
51	Sb	0.2	X	280,476
52	Te	0.002	X	280,476
53	I	0.3	X	280,476
54	Xe	<0.001		
55	Cs	1	X	280,476
56	Ba	400	X	280,476
57	La	18	X	280,476
58	Ce	46	X	280,476
59	Pr	6	X	280,476
60	Nd	24	X	280,476
61	Pm	<0.001	b	484
62	Sm	7	X	280,476
63	Eu	1	X	280,476
64	Gd	6	X	280,476
65	Tb	0.9	X	280,476
66	Dy	5	X	280,476
67	Ho	1	X	280,476
68	Er	3	X	280,476
69	Tm	0.2	X	280,476
70	Yb	3	X	280,476
71	Lu	0.8	X	280,476
72	Hf	5	X	280,476
73	Ta	2	X	280,476
74	W	1	X	280,476
75	Re	0.001	X	280,476
76	Os	0.001	X	476
77	Ir	0.001	X	476,478
78	Pt	0.005	X,c	476,544
79	Au	0.005	X	476,478
80	Hg	0.5	X	280,476
81	Tl	1	X	280,476
82	Pb	15	X	280,476

TABLE 2.5-3. AVERAGE ELEMENTAL COMPOSITION OF CRUSTAL ROCKS

At. no.	Symbol	Concentration[a]	Aerosol constituent	Ref.
83	Bi	0.2	X	280,476
84	Po	<0.001	X	479
85	At	<0.001		
86	Rn	<0.001	X	345,483
87	Fr	<0.001		
88	Ra	<0.001	X	483
89	Ac	<0.001		
90	Th	10	X	280,483
91	Pa	<0.001		
92	U	2	X	280,476
93	Np	<0.001		
94	Pu	<0.001	b	481,485

_

[a]Concentrations are in parts per million by weight (473)

[b]Present only as a result of nuclear weapons testing

[c]Emitted from automotive catalytic converters.

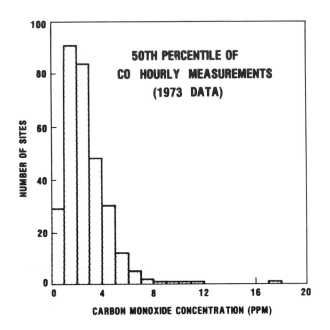

Fig. 2.5-1.   Distribution of 50th percentiles of CO hourly measure-
             ments at air quality monitoring sites within the continen-
             tal United States, 1973 data (from reference 1004).

2.5.3. Chemistry

No significant atmospheric chemical processes are known for most
of these compounds, although it has been suggested that some of them
may be catalytically active in aerosol chemical transformations, particularly of SO_2 to sulfate (1113). Carbon monoxide is an important constituent in atmospheric chemistry, however, reacting with the hydroxyl
radical to form CO_2 and an odd hydrogen radical:

$$CO + HO\cdot \rightarrow CO_2 + H\cdot$$

$$H\cdot + O_2 \xrightarrow{M} HO_2\cdot$$

The atmospheric abundance of carbon monoxide appears sufficient to
make it one of the principal sinks of HO· through this process.

TABLE 2.5 INORGANIC ELEMENTS, HYDRIDES, OXIDES, and CARBONATES
Emission and Detection

Species Number	Name	Chemical Formula	Emission		Detection	Ambient conc.
			Source	Ref.	Ref.	
2.5-1	Carbon	C	1.combustion	256*(a)		
			aluminum mfr.	256*(a)		
			foundry	279(a)		
			titanium mfr.	412*(a)		
2.5-2	Mercury	Hg	chlorine mfr.	27,58*	286,529*	$0.2\text{-}2\times10^5 \text{ng/m}^3$
			coal comb.	490*		
			geothermal steam	489		
			mercury mines	286		
			paint	66,222		
			refuse comb.	403*		
			sewage tmt.	529*		
			volcano	61		
2.5-3	Phosphine	PH_3			147	
2.5-4	Arsine	AsH_3	aluminum mfr.	116		
2.5-5	Beryllium oxide	BeO	rocket	536		
2.5-6	Boric oxide	B_2O_3	rocket	536		

TABLE 2.5 INORGANIC ELEMENTS, HYDRIDES, OXIDES, and CARBONATES
Emission and Detection

Species Number	Name	Chemical Formula	Emission		Detection	Ambient conc.
			Source	Ref.	Ref.	
2.5-7	Carbon monoxide	CO	1.auto	281,444*	290*,293*	&
			combustion	281,395	399(a)	
			diesel	136*		
			microbes	210,302		
			natural gas	423*		
			plastics comb.	46		
			rocket	536		
			tobacco smoke	4*,168		
			trees	199		
			turbine	71,594		
			volcano	106*		
			wood pulping	19		
2.5-8	Carbon dioxide	CO_2	1.combustion	281	295*,305*	290-500ppm
			2.auto	281,592*	399(a)	
			diesel	136*		
			foaming agent	562		
			geothermal steam	372,402*		
			natural gas	456*		
			plastics comb.	46		
			propellant	559,563		
			rocket	536		
			tobacco smoke	396,445		
			volcano	73,106*		
2.5-9	Sodium oxide	Na_2O	aluminum mfr.	256*(a)		
			cement mfr.	256*(a)		
			combustion	256*(a)		
			refuse comb.	256*(a)		
			rock dust	473(a)		
			zinc mfr.	256*(a)		

TABLE 2.5 INORGANIC ELEMENTS, HYDRIDES, OXIDES, and CARBONATES
Emission and Detection

Species Number	Name	Chemical Formula	Emission		Detection Ref.	Ambient conc.
			Source	Ref.		
2.5-10	Magnesium oxide	MgO	cement mfr.	256*(a),539(a)		
			combustion	256*(a)		
			fertilizer mfr.	256*(a)		
			foundry	279(a)		
			lime mfr.	256*(a)		
			magnesium mfr.	58*(a)		
			refuse comb.	256*(a)		
			steel mfr.	161(a),256*(a)		
			titanium mfr.	412*(a)		
			zinc mfr.	256*(a)		
2.5-11	Aluminum oxide	Al_2O_3	aluminum mfr.	256*(a)	280(a),460t(a)	
			cement mfr.	256*(a)		
			combustion	256*(a)		
			fertilizer mfr.	256*(a)		
			foundry	279(a)		
			lead mfr.	115(a)		
			lime mfr.	256*(a)		
			refuse comb.	256*(a)		
			rock dust	473(a)		
			rocket	536(a)		
			steel mfr.	256*(a)		
			titanium mfr.	412*(a)		
			wood pulping	256*(a)		
			zinc mfr.	256*(a)		
2.5-12	Aluminum oxide trihydrate	$Al_2O_3 \cdot 3H_2O$			460t(a)	

TABLE 2.5 INORGANIC ELEMENTS, HYDRIDES, OXIDES, and CARBONATES
Emission and Detection

Species Number	Name	Chemical Formula	Emission		Detection Ref.	Ambient conc.
			Source	Ref.		
2.5-13	Silicon dioxide	SiO_2	aluminum mfr.	256*(a)	280(a), 468(a)	
			cement mfr.	256*(a), 539(a)		
			combustion	256*(a)		
			fertilizer mfr.	256*(a)		
			foundry	279(a)		
			refuse comb.	256*(a)		
			rock dust	473(a)		
			steel mfr.	256*(a)		
			titanium mfr.	412*(a)		
			wood pulping	256*(a)		
			zinc mfr.	256*(a)		
2.5-14	Phosphorus pentoxide	P_2O_5	combustion	256*(a)		
			fertilizer mfr.	187*		
			fertilizer mfr.	256*(a)		
			foundry	279(a)		
			phosph. acid mfr.	58*, 93		
			rock dust	473(a)		
			steel mfr.	256*(a)		
2.5-15	Potassium oxide	K_2O	cement mfr.	256*(a)	280(a)	
			combustion	256*(a)		
			refuse comb.	256*(a)		
			rock dust	473(a)		

TABLE 2.5 INORGANIC ELEMENTS, HYDRIDES, OXIDES, and CARBONATES Emission and Detection

Species Number	Name	Chemical Formula	Emission		Detection Ref.	Ambient conc.
			Source	Ref.		
2.5-16	Calcium oxide	CaO	cement mfr.	256*(a),539(a)		
			combustion	256*(a)		
			fertilizer mfr.	256*(a)		
			foundry	279(a)		
			lime mfr.	256*(a)		
			refuse comb.	256*(a)		
			rock dust	473(a)		
			steel mfr.	161(a),256*(a)		
			titanium mfr.	412*(a)		
			wood pulping	45(a),256*(a)		
			zinc mfr.	256*(a)		
2.5-17	Wollastinite	$CaSiO_3$			459(a)	
2.5-18	Titanium dioxide	TiO_2	combustion	256*(a)		
			foundry	279(a)		
			refuse comb.	256*(a)		
			rock dust	473(a)		
			titanium mfr.	115(a),412*(a)		
2.5-19	Vanadium pentoxide	V_2O_5	foundry	279(a)		
			titanium mfr.	412*(a)		
2.5-20	Chromiic oxide	Cr_2O_3	foundry	279(a)		
			steel mfr.	256*(a)		
			titanium mfr.	412*(a)		
2.5-21	Manganous oxide	MnO	foundry	279(a)		
			rock dust	473(a)		
			steel mfr.	65(a),256*(a)		
2.5-22	Manganese dioxide	MnO_2	titanium mfr.	412*(a)		

TABLE 2.5 INORGANIC ELEMENTS, HYDRIDES, OXIDES, and CARBONATES
Emission and Detection

Species Number	Name	Chemical Formula	Emission		Detection	Ambient conc.
			Source	Ref.	Ref.	
2.5-23	Ferrous oxide	FeO	combustion	256*(a)	280(a)	
			foundry	279(a)		
			rock dust	473(a)		
			steel mfr.	161(a),256*(a)		
2.5-24	Ferric oxide	Fe_2O_3	aluminum mfr.	256*(a)	280(a),336(a)	
			cement mfr.	256*(a),539(a)		
			combustion	256*(a)		
			fertilizer mfr.	256*(a)		
			foundry	279(a)		
			lime mfr.	256*(a)		
			refuse comb.	256*(a)		
			rock dust	473(a)		
			steel mfr.	161(a),256*(a)		
			titanium mfr.	412*(a)		
			wood pulping	256*(a)		
2.5-25	Iron tetroxide	Fe_3O_4			280(a),336(a)	
2.5-26	Cobalt oxide	Co_3O_4	foundry	279(a)		
2.5-27	Nickel oxide	NiO	foundry	279(a)		
			steel mfr.	256*(a)		
			zinc mfr.	256*(a)		
2.5-28	Cupric oxide	CuO	steel mfr.	256*(a)		
2.5-29	Zinc oxide	ZnO	copper mfr.	256*(a)		
			lead mfr.	256*(a)		
			steel mfr.	256*(a)		
			zinc mfr.	256*(a),350(a)		

46

TABLE 2.5 INORGANIC ELEMENTS, HYDRIDES, OXIDES, and CARBONATES
Emission and Detection

Species Number	Name	Chemical Formula	Emission		Detection	Ambient conc.
			Source	Ref.	Ref.	
2.5-30	Arsenious oxide	As_2O_3	copper mfr.	302	113*	0.4-4.1ng/m³
			copper mfr.	256*(a)	213(a),589(a)	
			gold mfr.	328(a)		
			lead mfr.	115		
			lead mfr.	256*(a)		
2.5-31	Selenium dioxide	SeO_2	lead mfr.	256*(a)	213(a),589(a)	
			refuse comb.	220*,530		
2.5-32	Molybdenum trioxide	MoO_3	foundry	279(a)		
2.5-33	Cadmium oxide	CdO	lead mfr.	256*(a)		
2.5-34	Stannous oxide	SnO	lead mfr.	256*(a)		
2.5-35	Stannic oxide	SnO_2	zinc mfr.	256*(a)		
2.5-36	Antimonious oxide	Sb_2O_3	copper mfr.	256*(a)		
			lead mfr.	115		
2.5-37	Tellurium dioxide	TeO_2	lead mfr.	256*(a)		
2.5-38	Lead oxide	PbO	copper mfr.	256*(a)	280(a)	
			lead mfr.	256*(a)		
			steel mfr.	101(a),256*(a)		
2.5-39	Lead dioxide	PbO_2	industrial	101(a)		
2.5-40	Lead tetroxide	Pb_3O_4	lead mfr.	256*(a)		
2.5-41	Sodium carbonate	Na_2CO_3	lead mfr.	115(a)		
			lime mfr.	256*(a)		
			wood pulping	45(a),256*(a)		
2.5-42	Magnesium carbonate	$MgCO_3$	lime mfr.	256*(a)		
			wood pulping	256*(a)		

TABLE 2.5 INORGANIC ELEMENTS, HYDRIDES, OXIDES, and CARBONATES
Emission and Detection

Species Number	Name	Chemical Formula	Emission		Detection	Ambient
			Source	Ref.	Ref.	conc.
2.5-43	Calcium carbonate	$CaCO_3$	lime mfr.	256*(a)	280(a),468(a)	
			wood pulping	10(a),256*(a)		
2.5-44	Chromic acid	H_2CrO_4	industrial	58*,108		
2.5-45	Ammonium carbonate	$(NH_4)_2CO_3$	petrochem. mfr.	337(a)		
2.5-46	Boric acid	H_3BO_3	fiberglass mfr.	348	366	
			geothermal steam	372		
			oceans	586		
			volcano	349		
2.5-47	Phosphoric acid	H_3PO_4	lacquer mfr.	428		

TABLE 2.5 INORGANIC ELEMENTS, HYDRIDES, OXIDES, AND CARBONATES
Reactions and Products

Species Number	Reactant	Chemical reactions						Remarks
		k	Ref.	Lifetime	Products	Ref.		
2.5-7	HO·	$(3.3\pm0.1)\times10^{-13}$	1218	6.6×10^6	CO_2	1170		B
	$HO_2\cdot$	1.1×10^{-16}	1152					
2.5-8								B

3

HYDROCARBONS

3.0. INTRODUCTION

The hydrocarbons are the most numerous of the major atmospheric chemical groups; well over 500 compounds are listed in the following pages. A very large fraction of these compounds is derived from the combustion of fossil fuels. Many of the molecules (such as the alkanes listed in Table 3.1) are initially present in the fuel and are found in exhausts because of incomplete fuel combustion. Others (such as the polynuclear aromatic hydrocarbons listed in Table 3.7) are produced by chemical reactions during combustion.

The atmospheric chemistry of the hydrocarbons has been studied extensively. As a result the principal reactions of many of the compounds have been identified, as have a number of the reaction products. Many questions remain unanswered, however, particularly those pertaining to the chemistry of hydrocarbons with $C \geqslant 5$. As will be seen, large groups of hydrocarbons are present in the air in both the gas and aerosol phases. The details of the hydrocarbon reaction chains in both these phases and of the transition processes between the phases will be active research topics in the coming years.

3.1. ALKANES

3.1.1. Identified Compounds

More than 100 different compounds in this group have been identified as emittants into or constituents of ambient air. The list includes all of the straight chain alkanes $(C_x H_2 x + 2)$ from $C_1 C_{37}$, and many of the associated 2-methyl isomers. For the $x \leqslant 6$ compounds, most of the possible multimethyl substituted derivatives have also been identified. Compounds with $x < 6$ are generally found in gaseous form, $x > 20$ as components of particulate matter, $6 \leqslant x \leqslant 20$ in either or both forms. This characteristic represents, of course, the decrease in vapor pressure of the compounds with increasing carbon chain length. The most commonly identified sources of the alkanes are motor vehicles and tobacco smoke. Many of the lighter compounds are also constituents of vapor emissions from gasoline and natural gas.

3.1.2. Ambient Concentrations

The atmospheric concentrations of methane have been extensively measured for some years. The global background value in the troposphere is about 1.4 ppm (1106). Concentrations as high as 6 ppm have been measured in urban areas.

Total nonmethane hydrocarbons have been measured at a number of urban sites. These measurements are nonspecific, but the alkanes are probably the major component (6). The distribution of median values is shown on Fig. 3.1-1; the mean of these measurements is 1370 μg/m^3.

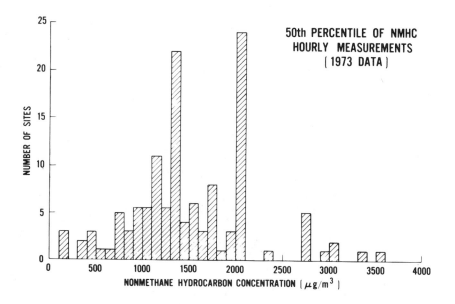

Fig. 3.1-1.   Distribution of 50th percentiles of nonmethane hydrocarbon hourly measurements at air quality monitoring sites within the continental United States, 1973 data (from reference 1004).

Individual concentrations of nonmethane alkanes have been determined only on irregular occasions, usually at urban sites. Of the compounds so studied, two characteristics stand out. The first is the very wide concentration ranges that can be found, usually high in urban areas and low in nonurban areas. The second is the trend of decreasing abundance with increasing carbon number, which illustrates the tendency of sources to emit the more volatile fragments of the petroleum precursor.

3.1.3. Chemistry

The alkanes are relatively unreactive compounds in the atmosphere. The principal chemical loss mechanism is abstraction of the most weakly bound hydrogen atom by the hydroxyl radical

$$C_xH_{2x+2} + HO\cdot \rightarrow C_xH_{2x+1}\cdot + H_2O.$$

The resulting alkyl radical forms an alkylperoxy radical by three-body addition of an oxygen molecule

$$C_xH_{2x+1}\cdot + O_2 \overset{M}{\rightarrow} C_xH_{2x+1}O_2\cdot .$$

Alkylperoxy radicals may become components of $NO-NO_2-O_3$ chemistry by acting as oxidants for nitric oxide,

$$C_xH_{2x+1}O_2\cdot + NO \rightarrow C_xH_{2x+1}O\cdot + NO_2,$$

or they may participate in atmospheric sulfur chemistry by oxidizing SO_2 (1108):

$$C_xH_{2x+1}O_2\cdot + SO_2 \rightarrow C_xH_{2x+1}O\cdot + SO_3 .$$

The fate of the alkoxy radical in the atmosphere will be the formation of the corresponding aldehyde, with concomitant liberation of an $HO_2\cdot$ radical

$$C_xH_{2x+1}O\cdot + O_2 \rightarrow C_{x-1}H_{2x-1}CHO + HO_2\cdot.$$

For alkylperoxy or alkoxy radicals with $x \geqslant 4$, it appears that internal hydrogen shift isomerization may compete with the above processes (1027). Thus, if the NO concentrations are low enough (perhaps of the order of 0.1 ppb, i.e., near the global background), the alkylperoxy isomerization reaction

$$C_xH_{2x-1}O_2\cdot \rightarrow C_xH_{2x-2}O_2H\cdot$$

would be competitive. Similarly, for the alkoxy radical,

$$C_xH_{2x-1}O\cdot \rightarrow C_xH_{2x-2}OH\cdot$$

would be important. The details of these processes, and of their influence on atmospheric chemistry, remain to be determined.

TABLE 3.1 ALKANES
Emission and Detection

Species Number	Name	Chemical Formula	Emission		Detection	Ambient conc.
			Source	Ref.	Ref.	
3.1-1	Methane	CH_4	1.microbes	301,302	6*,232*	1.3-4.0ppm
			2.auto	159*,526		
			diesel	435*		
			forest fire	183,232		
			foundry	279*		
			geothermal steam	372,402*		
			insects	471		
			natural gas	232,423*		
			petroleum mfr.	232,543*		
			polymer comb.	118,354		
			refuse comb.	439		
			tobacco smoke	396,448		
			turbine	359*,414		
			volcano	106*,234		
			water treatment	185		
			wood pulping	19		

TABLE 3.1 ALKANES
Emission and Detection

Species Number	Name	Chemical Formula	Emission Source	Emission Ref.	Detection Ref.	Ambient conc.
3.1-2	Ethane	CH_3CH_3	1.auto	159*,465*	123*,178	.05-95ppb
			diesel	434*	399(a)	
			forest fire	232,439		
			gasoline vapor	465*		
			geothermal steam	402*		
			insects	471		
			microbes	302		
			natural gas	232,423*		
			petroleum mfr.	232,543*		
			polymer comb.	118,304		
			tobacco smoke	396,453		
			turbine	359*,414		
			volcano	234		
			water treatment	185		
			wood pulping	19		
3.1-3	Propane		1.auto	57,170*	170*,472*	12-94ppb
			2.natural gas	232,423*	399(a)	
			diesel	136*,433*		
			forest fire	232,439		
			gasoline vapor	232,465*		
			insects	471		
			microbes	302		
			petroleum mfr.	232,543*		
			polymer comb.	118,304		
			propellant	559,563		
			tobacco smoke	396,453		
			turbine	414		
			volcano	234		
			water treatment	185		
			wood pulping	19		

TABLE 3.1 ALKANES
Emission and Detection

Species Number	Name	Chemical Formula	Emission		Detection Ref.	Ambient conc.
			Source	Ref.		
3.1-4	Butane		1.auto	159*,465*	6*,170*	.01-182ppb
			2.natural gas	423*,465*	399(a)	
			building mtls.	82		
			diesel	136*,435*		
			forest fire	232,524		
			gasoline vapor	232,465*		
			microbes	302		
			petroleum mfr.	232,221		
			polymer comb.	118,304		
			propellant	559		
			tobacco smoke	396		
			turbine	414		
			water treatment	185		
3.1-5	Isobutane		1.auto	47,170*	6*,170*	.06-35ppb
			diesel	434*,435*		
			forest fire	232,524		
			gasoline vapor	232,465*		
			microbes	302		
			natural gas	423*,465*		
			petroleum mfr.	232,543*		
			propellant	559,563		
			tobacco smoke	396		
			turbine	414		
			volcano	234		
3.1-6	2,2-Dimethylbutane		auto	47,153*	164,263	
			forest fire	232,524		
			gasoline vapor	232		
			petroleum mfr.	232		

TABLE 3.1 ALKANES
Emission and Detection

Species Number	Name	Chemical Formula	Emission		Detection Ref.	Ambient conc.
			Source	Ref.		
3.1-7	2,3-Dimethylbutane		auto	153*,159*	232*,524*	3.8ppb
			forest fire	232		
			gasoline vapor	232,486*		
3.1-8	2,2,3-Trimethyl-butane		auto	47,526		
			turbine	414		
3.1-9	2,2,3,3-Tetramethyl-butane		turbine	414		
3.1-10	Pentane		auto	159*,465*	6*,170*	.023-64ppb
			building mtls.	82	364(a)	
			diesel	136*,435*		
			forest fire	232,524		
			gasoline vapor	232,465*		
			natural gas	423*,465*		
			petroleum mfr.	58*,232		
			polymer comb.	304		
			syn. rubber mfr.	58*		
			turbine	414		
			volcano	234		
			water treatment	185		
3.1-11	Isopentane		auto	159*,465*	6*,170*	0.1-101ppb
			building mtls.	82	399(a)	
			diesel	136*,434*		
			forest fire	232,524		
			gasoline vapor	232,465*		
			natural gas	423*,465*		
			petroleum mfr.	58*,232		
			plant volatile	552e		
			turbine	414		

TABLE 3.1 ALKANES
Emission and Detection

Species Number	Name	Chemical Formula	Emission Source	Emission Ref.	Detection Ref.	Ambient conc.
3.1-12	Neopentane		turbine	414		
3.1-13	2-Methylpentane		auto	153*,159*	178,232*	5-12ppb
			forest fire	232	399(a)	
			gasoline vapor	232,465*		
			natural gas'	465*		
			petroleum mfr.	232		
			plant volatile	552e		
			turbine	414		
3.1-14	3-Methylpentane		auto	153*,159*	164,232*	3-11ppb
			forest fire	232		
			gasoline vapor	232,465*		
			natural gas	465*		
			petroleum mfr.	232		
			plant volatile	552e		
3.1-15	2,2-Dimethylpentane		auto	284	232*,582	1-2ppb
3.1-16	2,3-Dimethylpentane		auto	284,465*	164,232*	2ppb
			gasoline vapor	465*		
			natural gas	465*		
			turbine	414		
3.1-17	2,4-Dimethylpentane		auto	159*,170*	170*,263	0.3-13ppb
			gasoline vapor	486		
3.1-18	3,3-Dimethylpentane		auto	284	582	

TABLE 3.1 ALKANES
Emission and Detection

Species Number	Name	Chemical Formula	Emission		Detection	Ambient conc.
			Source	Ref.	Ref.	
3.1-19	2,2,3-Trimethyl-pentane		auto	47,526	374	
			turbine	414		
3.1-20	2,2,4-Trimethyl-pentane		auto	159*,170*	170*,263	17ppb
			gasoline vapor	465*		
			natural gas	465*		
			turbine	414		
3.1-21	2,3,3-Trimethyl-pentane		auto	526		
			gasoline vapor	486		
			turbine	414		
3.1-22	2,3,4-Trimethyl-pentane		auto	159*,526	374,525	
			gasoline vapor	486		
3.1-23	2,2,3,3-Tetramethyl-pentane		turbine	414	374,582	
3.1-24	2,2,3,4-Tetramethyl pentane		turbine	414		
3.1-25	3-Ethylpentane				582	
3.1-26	3-Ethyl-2-methyl-pentane				582	
3.1-27	3-Ethyl-3-methyl-pentane				582	
3.1-28	3-Ethyl-2,3-dimethyl-pentane		turbine	414		

TABLE 3.1 ALKANES
Emission and Detection

Species Number	Name	Chemical Formula	Emission Source	Emission Ref.	Detection Ref.	Ambient conc.
3.1-29	Hexane		auto	159*,465*	164,170*	4-27ppb
			building mtls.	82	399(a),458(a)	
			chemical mfr.	566*		
			diesel	435*		
			forest fire	232,524		
			gasoline vapor	232,465*		
			natural gas	465*		
			petroleum mfr.	58*,232		
			plant volatile	552e		
			polymer comb.	304		
			turbine	414		
3.1-30	2-Methylhexane		auto	159*,170*	86*,170*	10-28ppb
			turbine	414	399(a)	
3.1-31	3-Methylhexane		auto	47,153*	57,123*	10ppb
			gasoline vapor	465*,486		
			natural gas	465*		
			turbine	414		
3.1-32	2,2-Dimethylhexane		auto	284,465*	582	
			gasoline vapor	465*		
			natural gas	465*		
			turbine	414		
3.1-33	2,3-Dimethylhexane		auto	284		
3.1-34	2,4-Dimethylhexane		auto	159*	374,582	
			turbine	414		
3.1-35	2,5-Dimethylhexane				582,591	
3.1-36	3,3-Dimethylhexane		turbine	414	582	

TABLE 3.1 ALKANES
Emission and Detection

Species Number	Name	Chemical Formula	Emission		Detection Ref.	Ambient conc.
			Source	Ref.		
3.1-37	3,4-Dimethylhexane				374	
3.1-38	2,2,5-Trimethyl-hexane		auto	47,486*	374	
			gasoline vapor	486		
			turbine	414		
3.1-39	2,3,4-Trimethyl-hexane		turbine	414		
3.1-40	2,3,5-Trimethyl-hexane		auto	486*		
			gasoline vapor	486		
3.1-41	2-Methyl-4-ethyl-hexane		turbine	414		
3.1-42	3-Methyl-4-ethyl hexane		turbine	414		
3.1-43	2,2-Dimethyl-4-ethylhexane		turbine	414		
3.1-44	Heptane		auto	159*,465*	86*,591*	0.2-34ppb
			building mtls.	82	399(a),458(a)	
			gasoline vapor	58*,465*		
			natural gas	465*		
			petroleum mfr.	543*		
			solvent	78		
			turbine	414		

TABLE 3.1 ALKANES
Emission and Detection

Species Number	Name	Chemical Formula	Emission		Detection	Ambient
			Source	Ref.	Ref.	conc.
3.1-45	2-Methylheptane		auto gasoline vapor natural gas turbine	284,465* 58*,465* 465* 414t	86*,582 399(a)	3.4ppb
3.1-46	3-Methylheptane		auto gasoline vapor tobacco smoke	284,486* 486 421	263,582	
3.1-47	4-Methylheptane		auto	284	582	
3.1-48	2,2-Dimethylheptane		turbine	414		
3.1-49	2,4-Dimethylheptane		auto gasoline vapor syn. rubber mfr. turbine	486* 486 58* 414	591	
3.1-50	2,5-Dimethylheptane		turbine	414	582	
3.1-51	3,3-Dimethylheptane		turbine	414		
3.1-52	3,4-Dimethylheptane		turbine	414		
3.1-53	3-Ethylheptane		tobacco smoke turbine	421 414		
3.1-54	3-Ethyl-4-methyl-heptane		turbine	414		

TABLE 3.1 ALKANES
Emission and Detection

Species Number	Name	Chemical Formula	Emission		Detection Ref.	Ambient conc.
			Source	Ref.		
3.1-55	3-Ethyl-6-methyl-heptane		turbine	414		
3.1-56	4-Ethylheptane		turbine	414		
3.1-57	Octane	$CH_3(CH_2)_6CH_3$	auto	47,465*	86*,200	.04-3.4ppb
			brewing	144	364(a),458(a)	
			building mtls.	82		
			gasoline vapor	465*,486		
			landfill	365		
			natural gas	465*		
			solvent	134		
			tobacco smoke	421		
			turbine	414		
3.1-58	2-Methyloctane	$\searrow$(CH$_2$)$_5$CH$_3$	auto	465*	86,525	
			gasoline vapor	465*	399(a)	
			natural gas	465*		
3.1-59	3-Methyloctane		tobacco smoke	421	582	
			turbine	414		
3.1-60	4-Methyloctane		turbine	414	582	
3.1-61	2,3-Dimethyloctane		turbine	414		
3.1-62	2,6-Dimethyloctane		turbine	414		
3.1-63	2,3,4-Trimethyl-octane		turbine	414		

TABLE 3.1 ALKANES
Emission and Detection

Species Number	Name	Chemical Formula	Emission		Detection Ref.	Ambient conc.
			Source	Ref.		
3.1-64	3,3,5-Trimethyl-octane		turbine	414		
3.1-65	Nonane	$CH_3(CH_2)_7CH_3$	auto	465*,486*	525*,591*	0.1-9.0ppb
			brewing	144	364(a),458(a)	
			gasoline vapor	486		
			petroleum mfr.	543		
			tobacco smoke	421		
			turbine	414		
3.1-66	2-Methylnonane	$(CH_2)_6CH_3$	turbine	414	86,200	
					399(a)	
3.1-67	2,5-Dimethylnonane		turbine	414		
3.1-68	4,5-Dimethylnonane		turbine	414		
3.1-69	Decane	$CH_3(CH_2)_8CH_3$	auto	47,526	200*,525*	1.0-11.2ppb
			brewing	144	364(a),458(a)	
			diesel	176		
			tobacco smoke	421		
			turbine	359*,414		
3.1-70	2-Methyldecane	$(CH_2)_7CH_3$	tobacco smoke	421	86,200	
					399(a)	
3.1-71	3-Methyldecane	$(CH_2)_6CH_3$	diesel	309(a)		
			tobacco smoke	421		
			turbine	414		
3.1-72	4-Methyldecane	$(CH_2)_5CH_3$	turbine	414		
3.1-73	5-Methyldecane	$(CH_2)_4CH_3$	tobacco smoke	421		

63

TABLE 3.1 ALKANES
Emission and Detection

Species Number	Name	Chemical Formula	Emission		Detection	Ambient conc.
			Source	Ref.	Ref.	
3.1-74	Undecane	$CH_3(CH_2)_9CH_3$	brewing tobacco smoke turbine	144 421 414	86*,200* 364(a),458(a)	0.95-8.8ppb
3.1-75	2-Methylundecane	$(CH_2)_8CH_3$			86,200 399(a)	
3.1-76	3-Methylundecane	$(CH_2)_7CH_3$	diesel	309(a)		
3.1-77	Dodecane	$CH_3(CH_2)_{10}CH_3$	brewing tobacco smoke	144 421	86*,200* 364(a),458(a)	1.3-5.1ppb
3.1-78	2-Methyldodecane	$(CH_2)_9CH_3$			86,200 399(a)	
3.1-79	Tridecane	$CH_3(CH_2)_{11}CH_3$	brewing tobacco smoke turbine	144 421 414	86,525 364(a),458(a)	
3.1-80	2-Methyltridecane	$(CH_2)_{10}CH_3$			86,200	
3.1-81	Tetradecane	$CH_3(CH_2)_{12}CH_3$	brewing diesel	144 309(a)	86,525 310(a),363(a)	
3.1-82	2-Methyltetradecane	$(CH_2)_{11}CH_3$			399(a)	
3.1-83	3-Methyltetradecane	$(CH_2)_9CH_3$	diesel	309(a)		
3.1-84	Pentadecane	$CH_3(CH_2)_{13}CH_3$	brewing diesel	144 309(a)	86 ,525 310(a),363(a)	

TABLE 3.1 ALKANES
Emission and Detection

Species Number	Name	Chemical Formula	Emission Source	Emission Ref.	Detection Ref.	Ambient conc.
3.1-85	2-Methylpentadecane	$\diagup(CH_2)_{12}CH_3$	diesel	309(a)	399(a)	
3.1-86	Hexadecane	$CH_3(CH_2)_{14}CH_3$	auto / diesel	311(a) / 309(a)	86,200* / 310*(a),363(a) / 399(a)	0.16-1.0ppb
3.1-87	2-Methylhexadecane	$\diagup(CH_2)_{13}CH_3$	diesel	309(a)	399(a)	
3.1-88	Heptadecane	$CH_3(CH_2)_{15}CH_3$	auto / diesel	310(a) / 309(a)	86,200 / 310*(a),363(a) / 399(a)	0.1ng/m^3
3.1-89	2-Methylheptadecane	$\diagup(CH_2)_{14}CH_3$	diesel	309(a)		
3.1-90	Octadecane	$CH_3(CH_2)_{16}CH_3$	auto / diesel	310(a) / 309(a)	86,200 / 310*(a),363(a) / 399(a)	0.8ng/m^3
3.1-91	2-Methyloctadecane	$\diagup(CH_2)_{15}CH_3$	diesel		399(a)	
3.1-92	Nonadecane	$CH_3(CH_2)_{17}CH_3$	auto / diesel	310(a) / 309(a)	86 / 310*(a),363(a)	4ng/m^3
3.1-93	Eicosane	$CH_3CH_2)_{18}CH_3$	auto / diesel	310(a) / 309(a)	86 / 310*(a),363(a)	8ng/m^3
3.1-94	Heneicosane	$CH_3(CH_2)_{19}CH_3$	auto / diesel	310(a),311(a) / 310(a)	310*(a),363(a)	11ng/m^3
3.1-95	Docosane	$CH_3(CH_2)_{20}CH_3$	auto / diesel	310(a),311(a) / 310(a)	310*(a),363(a)	3-18ng/m^3
3.1-96	Tricosane	$CH_3(CH_2)_{21}CH_3$	auto / diesel	310(a),311(a) / 310(a)	310*(a),363(a)	5-28ng/m^3

TABLE 3.1 ALKANES
Emission and Detection

Species Number	Name	Chemical Formula	Emission Source	Emission Ref.	Detection Ref.	Ambient conc.
3.1-97	Tetracosane	$CH_3CH_2)_{22}CH_3$	auto diesel	310(a),311(a)310*(a),363(a) 310(a)	310(a),311(a)310*(a),363(a)	10-34ng/m^3
3.1-98	Pentacosane	$CH_3(CH_2)_{23}CH_3$	auto diesel	310(a),311(a)310*(a),363(a) 310(a)	310(a)	11-43ng/m^3
3.1-99	Hexacosane	$CH_3(CH_2)_{24}CH_3$	auto diesel tobacco smoke	310(a)311(a) 310*(a),363(a) 310(a) 396(a)	310(a)311(a) 310*(a),363(a)	16-39ng/m^3
3.1-100	2-Methylhexacosane	⋏$(CH_2)_{23}CH_3$	tobacco smoke	396(a)		
3.1-101	Heptacosane	$CH_3(CH_2)_{25}CH_3$	auto diesel tobacco smoke	310(a),311(a)310*(a),363(a) 310(a) 396(a)	310(a),311(a)310*(a),363(a)	23-41ng/m^3
3.1-102	Octacosane	$CH_3(CH_2)_{26}CH_3$	auto diesel tobacco smoke	310(a),311(a)310*(a),363(a) 310(a) 396(a)	310(a),311(a)310*(a),363(a)	24-38ng/m^3
3.1-103	2-Methyloctacosane	⋏$(CH_2)_{25}CH_3$	tobacco smoke	396(a)		
3.1-104	Nonacosane	$CH_3(CH_2)_{27}CH_3$	auto diesel tobacco smoke	310(a),311(a)310*(a),363(a) 310(a) 396(a)	310(a),311(a)310*(a),363(a)	22-35ng/m^3
3.1-105	2-Methylnonacosane	⋏$(CH_2)_{26}CH_3$	tobacco smoke	396(a)		
3.1-106	Tricontane	$CH_3(CH_2)_{28}CH_3$	auto diesel tobacco smoke	311(a) 310(a) 396(a)	310*(a),363(a)	12-34ng/m^3
3.1-107	2-Methyltriacontane	⋏$(CH_2)_{27}CH_3$	tobacco smoke	396(a)		

TABLE 3.1 ALKANES
Emission and Detection

Species Number	Name	Chemical Formula	Emission		Detection	Ambient conc.
			Source	Ref.	Ref.	
3.1-108	Hentriacontane	$CH_3(CH_2)_{29}CH_3$	auto diesel tobacco smoke	311(a) 310(a) 396(a)	310*(a),363(a)	9-26ng/m^3
3.1-109	2-Methylhentriacontane	$(CH_2)_{28}CH_3$	tobacco smoke	396(a)		
3.1-110	Dotriacontane	$CH_3(CH_2)_{30}CH_3$	auto diesel tobacco smoke	311(a) 310(a) 396(a)	310*(a),363(a)	8-19ng/m^3
3.1-111	2-Methyl-dotriacontane	$(CH_2)_{29}CH_3$	tobacco smoke	396(a)		
3.1-112	Tritriacontane	$CH_3(CH_2)_{31}CH_3$	auto diesel tobacco smoke	311(a) 310(a) 396(a)	310*(a),363(a)	4-10ng/m^3
3.1-113	Tetratriacontane	$CH_3(CH_2)_{32}CH_3$			310*(a),363(a)	1-8ng/m^3
3.1-114	Pentatriacontane	$CH_3(CH_2)_{33}CH_3$			310(a),363(a)	
3.1-115	Hexatriacontane	$CH_3(CH_2)_{34}CH_3$	auto	311(a)	310 (a)	
3.1-116	Heptatriacontane	$CH_3(CH_2)_{35}CH_3$			310 (a)	

TABLE 3.1 ALKANES
Reactions and Products

Species Number	Reactant	Chemical reactions					Remarks
		Ref.	k	Lifetime	Products	Ref.	
3.1-1	HO·	1133	$(7.5\pm0.5)\times10^{-15}$	2.9×10^{8}	CO,HCHO	1023	
	Cl·	1155	$(1.1\pm0.3)\times10^{-13}$		H_2,HCHO	1135	
	O(^{1}D)	1149	$(1.3\pm0.3)\times10^{-10}$		HCOOH,CO	439	
	O_3	1140	1.4×10^{-24}				
	O	1228	$(1.7\pm0.7)\times10^{-17}$				
3.1-2	Cl·	1203	$(6.0\pm1.0)\times10^{-11}$				
	HO·	1150	$(2.9\pm0.6)\times10^{-13}$	8.3×10^{5}			
	HO$_2$·	1152	6.2×10^{-17}				
	O(^{1}D)	1153	$(2.1\pm0.5)\times10^{-10}$				
	O	1151	9.8×10^{-16}		CH_3OH,HCOOH	439	
	O_3	1141	9.7×10^{-24}				
3.1-3	HO·	1118	$(2.2\pm0.6)\times10^{-12}$				
	HO$_2$·	1152	1.6×10^{-16}	9.9×10^{5}			
	O	1030	$(6.5\pm1.2)\times10^{-14}$				
	O(^{1}D)	1153	$(2.7\pm0.7)\times10^{-10}$		CH_3OH,HCOOH	439	
	O_3	1154	7.9×10^{-24}				
3.1-4	HO·	1118	$(2.9\pm0.7)\times10^{-12}$				
	HO$_2$·	1152	4.0×10^{-16}	7.5×10^{5}			
	O	1156	$(3.1\pm0.3)\times10^{-14}$				
	O(^{1}D)	1153	$(3.6\pm0.9)\times10^{-10}$		CH_3OH,HCOOH	439	
	O_3	1142	9.8×10^{-24}				

TABLE 3.1 ALKANES
Reactions and Products

Species Number	Reactant	Chemical reactions					Remarks
		k	Ref.	Lifetime	Products	Ref.	
3.1-5	HO·	$(2.4\pm0.1)\times10^{-12}$	1121	9.0×10^5			
	HO$_2$·	1.0×10^{-15}	1152				
	O(^{1}D)	$(3.4\pm0.8)\times10^{-10}$	1153			439	
	O$_3$	2.0×10^{-23}	1142				
3.1-7	HO$_2$·	4.2×10^{-13}	1026	3.7×10^3	CH$_3$OH,	1026	
	HO·	$(5.2\pm0.8)\times10^{-12}$	1025				
	O	$(2.0\pm0.6)\times10^{-13}$	1228				
3.1-8	HO·	$(3.8\pm0.8)\times10^{-12}$	1025	5.8×10^5			
3.1-9	HO·	$(1.1\pm0.1)\times10^{-12}$	1121	2.0×10^6			
	O	$(1.3\pm0.5)\times10^{-14}$	1228				
3.1-10	O	$(5.8\pm1.8)\times10^{-14}$	1228	6.9×10^8		1027	
	O(^{1}D)	$(4.1\pm1.0)\times10^{-10}$	1153				
	HO·						
3.1-11	HO·	$(3.3\pm0.7)\times10^{-12}$	1085	7.4×10^5			
	O	$(1.3\pm0.4)\times10^{-13}$	1228				
3.1-12	HO·	$(8.5\pm0.1)\times10^{-13}$	1121	6.6×10^5			
	O(^{1}D)	$(4.1\pm1.0)\times10^{-10}$	1153				
	O	$(5.5\pm1.7)\times10^{-15}$	1228				
3.1-13	HO·	$(5.3\pm1.0)\times10^{-12}$	1085	4.6×10^5			

69

TABLE 3.1 ALKANES
Reactions and Products

Species Number	Chemical reactions				Products	Ref.	Remarks
	Reactant	k	Ref.	Lifetime			
3.1-14	HO·	$(7.2\pm1.5)\times10^{-12}$	1085	3.4×10^{5}			
3.1-15	O	$(1.1\pm0.3)\times10^{-13}$	1228	3.7×10^{8}			
3.1-17	O	$(1.7\pm0.5)\times10^{-13}$	1228	2.4×10^{8}			
3.1-20	HO·	$(3.8\pm0.1)\times10^{-12}$	1121	5.8×10^{5}			
	O·	$(9.2\pm3.7)\times10^{-14}$	1228				
	O(^{1}D)	$(5.2\pm1.3)\times10^{-10}$	1153				
3.1-22	O	$(5.0\pm2.0)\times10^{-14}$	1228	8.0×10^{8}			
3.1-29	HO·	$(6.3\pm1.3)\times10^{-12}$	1085	3.5×10^{5}			
3.1-44	O	$(1.3\pm0.4)\times10^{-13}$	1228	3.1×10^{8}			
3.1-46	O	4.2×10^{-13}	1024	9.5×10^{7}			
3.1-57	HO·	$(8.8\pm0.1)\times10^{-12}$	1121	2.5×10^{5}			
	O	$(1.7\pm0.5)\times10^{-13}$	1228				

The production of aldehydes from alkanes renders the alkanes significant chemical precursors in the troposphere. Because the initial reaction with HO· is not exceptionally fast, much of their chemical participation will occur relatively far from the point of emission. It is possible that some fraction of the alkanes will penetrate into the stratosphere and participate in lower stratospheric chemistry (1105).

3.2. ALKENES AND ALKYNES

3.2.1. Identified Compounds

About 100 alkenes and alkynes have been detected in ambient or source-related air. The alkenes (C_xH_{2x}) from $x=1$ to 19 have been identified. Most of these are thought to be the 1-isomers, although isomeric determinations have not always been made. More than one alkene isomer is known for C_4 (the lowest in which isomeric structure is possible) to C_8. Diene structures have been detected for C_3 to C_8. The triple-bond alkyne structure (C_xH_{2x-2}) appears in compounds from C_2 to C_9 (except for heptyne, thus far undetected). Di-alkynes are known for C_3 and C_6.

Alkenes and alkynes are volatile compounds and are generated by a wide variety of natural and anthropogenic processes. The lower alkenes are common vegetative emittants, ethene and isoprene being perhaps the most common from these sources. Combustion processes are sources of a wide variety of the unsaturated hydrocarbons.

3.2.2. Ambient Concentrations

Concentration measurements for the individual alkenes have generally been made in urban areas, although some nonurban data are available for the lighter compounds. As with the alkanes, a trend of diminishing concentration with increasing carbon number is generally seen. The alkenes are quite reactive and their concentrations are thus sensitive to both source proximity and photochemical loss processes.

The concentrations of the alkynes, which are much less reactive than are the alkenes, are more uniform. Acetylene, often used as a tracer of urban air masses (472), is typically half as abundant as ethene in urban air. Propyne (methyl acetylene) is also common, typically at concentrations of a few parts per billion.

3.2.3. Chemistry

The alkenes are extremely reactive under atmospheric conditions, and are central constituents in urban smog chemistry. The chemical processes and products are complex, and in no case have been completely specified by laboratory work. However, reaction chains in reasonable agreement with existing information have been postulated for ethene (1114), propene (1022), and 2-butene (1022).

The principal daytime reaction of the alkenes will be with the hydroxyl radical (1007, 1130). The radical may add over the double bond or may abstract a hydrogen atom. For propene, the best studied compound, the results are

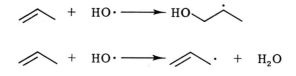

The dominant reaction for the resulting alkyl radicals is combination with molecular oxygen to form alkoxy radicals

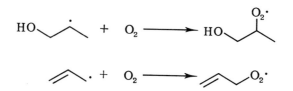

The alkoxy radicals then enter into the $NO_x - O_3$ cycle by oxidizing NO

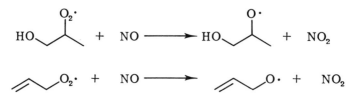

The fate of the resulting alkoxy radicals is quite dependent on their structure. The typical end products of the chains are CH_3CHO, $HCHO$, CO_2, $HCOOH$, and $HO_x\cdot$.

In the absence of the hydroxyl radicals produced by solar radiation, the principal alkene reaction is with ozone (1007). The ozone

TABLE 3.2 ALKENES AND ALKYNES
Emission and Detection

Species Number	Name	Chemical Formula	Emission		Detection	Ambient conc.
			Source	Ref.	Ref.	
3.2-1	Acetylene	HC≡CH	auto	159*,465*	90*,232*	0.2-227ppb
			calc. carbide mfr.	58*		
			diesel	136*,433*		
			forest fire	232		
			foundry	279*		
			petroleum mfr.	232,524		
			refuse comb.	439		
			tobacco smoke	396,463†		
			turbine	359*,414		
			vegetation	461		
3.2-2	Ethene	$H_2C=CH_2$	auto	159*,465*	3*,123*	0.7-700ppb
			diesel	136*,434*	399(a)	
			forest fire	183,232		
			foundry	279*		
			fruit ripening	146		
			microbes	172,210		
			petroleum mfr.	232		
			polymer comb.	118,304		
			refuse comb.	419*,439		
			solvent	134		
			tobacco smoke	396,453		
			turbine	359*,414		
			vegetation	461		
			volcano	234		
			wood pulping	19		

TABLE 3.2 ALKENES AND ALKYNES
Emission and Detection

Species Number	Name	Chemical Formula	Emission		Detection	Ambient
			Source	Ref.	Ref.	conc.
3.2-3	Propene		auto	159*,465*	6*,170	1-52ppb
			diesel	136*,433*	399(a)	
			forest fire	183,232		
			microbes	302		
			natural gas	465*		
			petroleum mfr.	232		
			polymer comb.	118,304		
			refuse comb.	439		
			tobacco smoke	396,453		
			turbine	359*,414		
			volcano	234		
			wood pulping	19		
3.2-4	Propadiene		auto	49*,486*	57,170*	2-4ppb
			diesel	433*		
			turbine	359*,414		
3.2-5	Propyne		auto	159*,465*	170*,524*	1-6ppb
			forest fire	232,524		
			petroleum mfr.	232		
			tobacco smoke	396		
			turbine	359*,414		
3.2-6	Diacetylene				399(a)	

TABLE 3.2 ALKENES AND ALKYNES
Emission and Detection

Species Number	Name	Chemical Formula	Emission		Detection Ref.	Ambient conc.
			Source	Ref.		
3.2-7	1-Butene		auto	284	232*,260	1-6ppb
			diesel	136*,223*	399(a)	
			forest fire	232,524		
			gasoline vapor	232		
			microbes	302		
			petroleum mfr.	232		
			polymer comb.	118,304		
			turbine	414		
			volcano	234		
3.2-8	2-Methyl-1-butene		auto	47,170*	123*,207	1-19ppb
			forest fire	232,524		
			gasoline vapor	232		
			turbine	359*,414		
3.2-9	3-Methyl-1-butene		auto	153*,284	582	
			diesel	433*,434*		
			turbine	359*		
			volcano	234		
3.2-10	2,3-Dimethyl-1-butene		auto	284	582	
					399(a)	
3.2-11	3,3-Dimethyl-1-butene				582	
3.2-12	2,3,3-Trimethyl-1-butene		auto	284	582	
			turbine	414		
3.2-13	2-Ethyl-1-butene		auto	284	582	
			turbine	359*		
3.2-14	1-Butyne		petroleum mfr.	371		
			syn. rubber mfr.	58*		

TABLE 3.2 ALKENES AND ALKYNES
Emission and Detection

Species Number	Name	Chemical Formula	Emission		Detection Ref.	Ambient conc.
			Source	Ref.		
3.2-15	1-Butyne-3-ene		petroleum mfr.	371		
3.2-16	Isobutene		auto	284,527	162,232* 399(a)	1-6ppb
			diesel	136*		
			forest fire	232,524		
			gasoline vapor	232		
			petroleum mfr.	232		
			syn. rubber mfr.	58*		
			tobacco smoke	396		
			turbine	414		
			volcano	234		
3.2-17	1,2-Butadiene		petroleum mfr.	371		
3.2-18	1,3-Butadiene		auto	159	170*,232* 399(a)	1-9ppb
			diesel	136,433		
			forest fire	232,524		
			petroleum mfr.	111,371		
			plastics mfr.	104		
			syn. rubber mfr.	58		
			tobacco smoke	396		
3.2-19	Isoprene		1.trees	198,207	198*,472* 399(a)	0.2-2.9ppb
			2.auto	159,465		
			gasoline vapor	465		
			rubber abrasion	331		
			tobacco smoke	396,446		
			turbine	359,414		
3.2-20	2-Ethyl-1,3-butadiene				364(a)	

76

TABLE 3.2 ALKENES AND ALKYNES
Emission and Detection

Species Number	Name	Chemical Formula	Emission		Detection	Ambient conc.
			Source	Ref.	Ref.	
3.2-21	cis-2-Butene		auto	49*,170*	123*,170*	1-11ppb
			diesel	433*		
			forest fire	232,524		
			gasoline vapor	232		
3.2-22	trans-2-Butene		auto	49,486	170*,207	1-3ppb
			diesel	136,433		
			forest fire	232,524		
			gasoline vapor	232,486		
			solvent	134		
			turbine	414		
3.2-23	2-Methyl-2-butene		auto	159,170	123*,232*	2-18ppb
			diesel	433		
			forest fire	232		
			gasoline vapor	232,486		
			turbine	359,414		
3.2-24	2,3-Dimethyl-2-butene		auto	153,284	57,582	
3.2-25	2-Butyne		turbine	359*		
3.2-26	1-Pentene		auto	47,486*	57,170*	1-12ppb
			diesel	136*,435*	364(a),458(a)	
			gasoline vapor	486		
			polymer comb.	304		
			turbine	359*,414		

TABLE 3.2 ALKENES AND ALKYNES
Emission and Detection

Species Number	Name	Chemical Formula	Emission		Detection Ref.	Ambient conc.
			Source	Ref.		
3.2-27	2-Methyl-1-pentene		auto turbine	47 414	57	
3.2-28	4-Methyl-1-pentene		auto diesel turbine	170* 433* 414	57,170* 364(a)	1-3ppb
3.2-29	2,3-Dimethyl-1-pentene		microbes	302	582	
3.2-30	2,4-Dimethyl-1-pentene		turbine	414		
3.2-31	3,4-Dimethyl-1-pentene		auto	284		
3.2-32	4,4-Dimethyl-1-pentene		turbine	414		
3.2-33	2,4,4-Trimethyl-1-pentene		auto turbine	284 414	374	
3.2-34	3-Ethyl-1-pentene		auto	526	374	
3.2-35	1-Pentyne		turbine	414		
3.2-36	1,3-Pentadiene		syn. rubber mfr. tobacco smoke	58* 396	582 399(a)	
3.2-37	2-Methyl-1,3-pentadiene				399(a)	

TABLE 3.2 ALKENES AND ALKYNES
Emission and Detection

Species Number	Name	Chemical Formula	Emission Source	Emission Ref.	Detection Ref.	Ambient conc.
3.2-38	cis-2-Pentene		auto diesel gasoline vapor turbine	47,170* 433* 486 414	57,170*	2-6ppb
3.2-39	trans-2-Pentene		auto forest fire gasoline vapor	47,153* 232,524 486	123*,232*	2-4ppb
3.2-40	2-Methyl-2-pentene		auto gasoline vapor turbine	153*,486* 486 414	582	
3.2-41	3-Methyl-2-pentene		auto turbine	153* 414	582	
3.2-42	4-Methyl-2-pentene		auto turbine	284 359*	582	
3.2-43	2,4-Dimethyl-2-pentene		turbine	414	170,582	3-10ppb
3.2-44	3,4-Dimethyl-2-pentene		auto	284		
3.2-45	4,4-Dimethyl-2-pentene		turbine	414		
3.2-46	2,4,4-Trimethyl-2-pentene		auto	284,526	374	
3.2-47	1-Hexene		auto turbine	49* 414	123*,263 458(a)	3ppb
3.2-48	2-Methyl-1-hexene		turbine	414	582	

TABLE 3.2 ALKENES AND ALKYNES
Emission and Detection

Species Number	Name	Chemical Formula	Emission		Detection Ref.	Ambient conc.
			Source	Ref.		
3.2-49	4-Methyl-1-hexene		auto turbine	47 414	399(a)	
3.2-50	5-Methyl-1-hexene		auto	47,526	374,582	
3.2-51	2,5-Dimethyl-1-hexene		tobacco smoke	421		
3.2-52	3,5,5-Trimethyl-1-hexene		turbine	359*		
3.2-53	2-Ethyl-1-hexene		auto	284		
3.2-54	Hexadiene				399(a),458t(a)	
3.2-55	2-Methylhexadiene				399(a)	
3.2-56	1-Hexyne				399(a)	
3.2-57	cis-2-Hexene		auto	49*,170*	57,170*	4-8ppb
3.2-58	trans-2-Hexene		auto	49*,526	57,374	
3.2-59	3-Methyl-2-hexene				582	
3.2-60	5-Methyl-2-hexene		auto	284		

TABLE 3.2 ALKENES AND ALKYNES
Emission and Detection

Species Number	Name	Chemical Formula	Emission Source	Emission Ref.	Detection Ref.	Ambient conc.
3.2-61	2,3-Dimethyl-2-hexene		auto	284		
			refuse comb.	26(a)		
3.2-62	2,4-Hexadiyne		turbine	414		
3.2-63	cis-3-Hexene		auto	526	57,582	
3.2-64	2-Methyl-cis-3-hexene				582	
3.2-65	trans-3-Hexene		auto	49*,526	374,582	
3.2-66	2-Methyl-trans-3-hexene		turbine	414		
3.2-67	Hexa-1,3,5-triyne		microbes	302		
3.2-68	1-Heptene		auto	49*,153*	374,591	
			turbine	359*,414	399(a),458(a)	
3.2-69	2-Methyl-1-heptene				399(a)	
3.2-70	Heptadiene				458(a)	
3.2-71	cis-2-Heptene		auto	526	582	
			turbine	359*		
3.2-72	trans-2-Heptene		auto	153*		
			turbine	359*		

TABLE 3.2 ALKENES AND ALKYNES
Emission and Detection

Species Number	Name	Chemical Formula	Emission		Detection Ref.	Ambient conc.
			Source	Ref.		
3.2-73	3-Heptene		auto turbine	284 414		
3.2-74	2,6-Dimethyl-3-heptene		auto turbine	284 414		
3.2-75	1-Octene	$(CH_2)_5CH_3$	auto brewing turbine	284 144 414	591 364(a),458(a)	
3.2-76	2-Methyl-1-octene	$(CH_2)_5CH_3$			582 399(a)	
3.2-77	2,6-Dimethyl-1-octene		turbine	414	399(a),458(a)	
3.2-78	Octadiene	$(CH_2)_4$				
3.2-79	Octyne	$(CH_2)_3CH_3$			399(a)	
3.2-80	cis-2-Octene	$(CH_2)_3CH_3$	auto	526	374	
3.2-81	trans-2-Octene	$(CH_2)_4CH_3$	auto tobacco smoke turbine	284,526 421 359*	374	
3.2-82	1-Nonene	$(CH_2)_6CH_3$	brewing tobacco smoke turbine	144 421 359*,414	591 364(a),458(a)	
3.2-83	2-Methyl-1-nonene	$(CH_2)_6CH_3$			582 399(a)	

TABLE 3.2 ALKENES AND ALKYNES
Emission and Detection

Species Number	Name	Chemical Formula	Emission		Detection Ref.	Ambient conc.
			Source	Ref.		
3.2-84	1-Nonyne	(structure)	tobacco smoke	421		
3.2-85	4-Nonene	(structure)			582	
3.2-86	1-Decene	$(CH_2)_7CH_3$	brewing tobacco smoke turbine	144 421 359*	591 364(a),458(a)	
3.2-87	2-Methyl-1-decene	$(CH_2)_7CH_3$			399(a)	
3.2-88	1-Undecene	$(CH_2)_8CH_3$	brewing turbine	144 359*	364(a),458(a)	
3.2-89	2-Methyl-1-undecene	$(CH_2)_8CH_3$			399(a)	
3.2-90	1-Dodecene	$(CH_2)_9CH_3$	auto brewing tobacco smoke	526 144 421	200 364(a),458(a)	
3.2-91	2-Methyl-1-dodecene	$(CH_2)_9CH_3$			399(a)	
3.2-92	1-Tridecene	$(CH_2)_{10}CH_3$	brewing	144	200,591 364(a),458(a)	
3.2-93	1-Tetradecene	$(CH_2)_{11}CH_3$	brewing diesel	144 309(a)	591 364(a),458(a)	
3.2-94	2-Methyl-1-tetradecene	$(CH_2)_{11}CH_3$			399(a)	
3.2-95	Pentadecene	$(CH_2)_{12}CH_3$	brewing	144	364(a),458(a)	

83

TABLE 3.2 ALKENES AND ALKYNES
Emission and Detection

Species Number	Name	Chemical Formula	Emission Source	Emission Ref.	Detection Ref.	Ambient conc.
3.2-96	1-Hexadecene	$(CH_2)_{13}CH_3$			364(a),458(a)	
3.2-97	2-Methyl-1-hexadecene	$(CH_2)_{13}CH_3$			399(a)	
3.2-98	1-Heptadecene	$(CH_2)_{11}CH_3$	diesel	309(a)	364(a),458(a)	
3.2-99	2-Methyl-1-heptadecene	$(CH_2)_{11}CH_3$			399(a)	
3.2-100	1-Octadecene	$(CH_2)_{15}CH_3$			364(a),458(a)	
3.2-101	2-Methyl-1-octadecene	$(CH_2)_{15}CH_3$			399(a)	
3.2-102	1-Nonadecene	$(CH_2)_{16}CH_3$			364(a)	

TABLE 3.2 ALKENES AND ALKYNES
Reactions and Products

Species Number	Reactant	k	Ref.	Chemical reactions Lifetime	Products	Ref.	Remarks
3.2-1	HO·	$(1.7\pm0.2)\times10^{-13}$	1159	1.3×10^{7}	HO·	1082	B
	O_3	$(7.0\pm0.3)\times10^{-20}$	1165		CO	1029	
	O						
3.2-2	HO·	$(8.7\pm1.7)\times10^{-12}$	1085	2.5×10^{5}	HCOOH	1081	B
	O_3	$(1.9\pm0.1)\times10^{-18}$	1034		HCHO,HCOOH	439	
	$NO_3·$	$(9.3\pm1.0)\times10^{-16}$	1031				
	O	$(7.8\pm0.8)\times10^{-13}$	1160		$H_2,H_2C=CO$	1236	
	$HO_2·$	1.7×10^{-17}	1152				
	$CH_5O·$	$(6.2\pm1.3)\times10^{-17}$	1040		[epoxide structure]	1040	
	$O_2(^1\Delta)$	$\leqslant1.7\times10^{-17}$	1032				
3.2-3	HO·	$(2.9\pm0.6)\times10^{-11}$	1085	7.5×10^{4}			B
	O_3	$(1.3\pm0.1)\times10^{-17}$	1034		HCHO,CH_3CHO	439	
	$NO_3·$	$(5.3\pm0.3)\times10^{-15}$	1031				
	O	$(4.4\pm0.4)\times10^{-12}$	1160				
	$O_2(^1\Delta)$	$\leqslant1.7\times10^{-17}$	1032				
3.2-4	HO·	$(9.3\pm0.9)\times10^{-12}$	1158	2.4×10^{5}	HCHO,HO·	1038,1082	
	O_3	$(4.0\pm0.4)\times10^{-19}$	1038		$H_2C=CH_2,CO$	1083	
	O	$(1.1\pm0.1)\times10^{-12}$	1162				
3.2-5	O	$(8.8\pm2.3)\times10^{-13}$	1039	5.0×10^{7}	$H_2C=CH_2,CO$	1029	B
	HO·				HCHO	1029	
	O_3				HCHO,HO·	1082	

85

TABLE 3.2 ALKENES AND ALKYNES
Reactions and Products

Species Number	Reactant	Chemical reactions			Products	Ref.	Remarks
		Ref.	k	Lifetime			
3.2-7	HO·	1161	$(2.9\pm0.1)\times10^{-11}$	7.5×10^{4}	HCHO, ⟨CHO⟩	439,1259	
	O_3	1034	$(1.2\pm0.1)\times10^{-17}$				
	$NO_3\cdot$	1031	$(7.8\pm0.8)\times10^{-15}$				
	O	1160	$(4.6\pm0.5)\times10^{-12}$				
	$O_2(^1\Delta)$	1032	$\leqslant1.7\times10^{-17}$				
3.2-8	HO·	1163	$(1.4\pm0.1)\times10^{-10}$	1.5×10^{4}	CH_3CHO, ⟨ketone⟩	439	
	O_3	1145	1.0×10^{-17}				
3.2-9	HO·	1158	$(3.1\pm0.3)\times10^{-11}$	7.1×10^{4}			
	O_3	1145	1.0×10^{-17}				
3.2-14	O	1242	8.3×10^{-13}	4.8×10^{7}	CO, ⟨alkene⟩	1242	B
3.2-16	$NO_3\cdot$	1031	$(1.1\pm0.1)\times10^{-13}$	3.0×10^{4}	HCHO, ⟨ketone⟩	1022	
	HO·	1033	$(5.1\pm1.5)\times10^{-11}$				
	O_3	1034	$(1.4\pm0.1)\times10^{-17}$		HCHO, ⟨ketone⟩	439	
	O	1160	$(1.5\pm0.2)\times10^{-11}$				
	$HO_2\cdot$	1152	1.7×10^{-16}				
3.2-17	O	1157	$(7.1\pm0.7)\times10^{-12}$	5.6×10^{6}	⟨enone/CHO structure⟩	1157	
3.2-18	HO·	1085	$(7.7\pm1.6)\times10^{-11}$	2.8×10^{4}	CH_3CHO, ⟨CHO structure⟩	439	
	O_3	1034	$(8.4\pm0.2)\times10^{-18}$				
	$O_2(^1\Delta)$	439	1.3×10^{-17}				

TABLE 3.2 ALKENES AND ALKYNES
Reactions and Products

Species Number	Reactant	Chemical reactions		Lifetime	Products	Ref.	Remarks
		k	Ref.				
3.2-19	HO·	$(7.8\pm2.3)\times10^{-11}$	1033	2.8×10^4	CHO (structure)	1234	
	$O_2(^1\Delta)$	1.6×10^{-17}	439				
	O						
3.2-21	O_3	$(1.6\pm0.1)\times10^{-16}$	1034	6.3×10^3	CH_3CHO,CH_3COOH	439	
	$NO_3·$	$(1.8\pm0.2)\times10^{-13}$	1031				
	HO·	$(6.2\pm0.6)\times10^{-11}$	1033				
	O	$(1.8\pm0.2)\times10^{-11}$	1160				
	$O_2(^1\Delta)$	$(3.8\pm0.1)\times10^{-18}$	1035				
3.2-22	O_3	$(2.6\pm0.1)\times10^{-16}$	1034	3.9×10^3	CO,CH_3CHO	1084	
	$NO_3·$	$(1.4\pm0.1)\times10^{-13}$	1031				
	HO·	$(7.0\pm0.7)\times10^{-11}$	1164				
	O	$(2.4\pm0.2)\times10^{-11}$	1160				
	$O_2(^1\Delta)$	$(1.5\pm0.1)\times10^{-18}$	1035				
3.2-23	O_3	$(4.9\pm0.2)\times10^{-16}$	1034	2.0×10^3	CH_3CHO (structure)	439	
	$NO_3·$	$(5.5\pm0.5)\times10^{-12}$	1031				
	HO·	$(7.7\pm0.8)\times10^{-11}$	1036				
	O	$(5.2\pm0.5)\times10^{-11}$	1157				
	$O_2(^1\Delta)$	$(5.0\pm0.1)\times10^{-17}$	1035				
3.2-24	$NO_3·$	$(3.7\pm0.5)\times10^{-11}$	1031	9.0×10^1	HCOOH, (structure)	439	
	O_3	$(1.5\pm0.1)\times10^{-15}$	1034				
	HO·	$(5.7\pm0.1)\times10^{-11}$	1161				
	O	$(7.6\pm0.2)\times10^{-11}$	1166				
	$O_2(^1\Delta)$	$(1.0\pm0.1)\times10^{-15}$	1035				

TABLE 3.2 ALKENES AND ALKYNES
Reactions and Products

Species Number	Reactant	Chemical reactions			Products	Ref.	Remarks
		k	Ref.	Lifetime			
3.2-26	O_3 $O_2(^1\Delta)$	$(1.1\pm0.1)\times10^{-17}$ 3.2×10^{-18}	1034 439	9.1×10^4	HCHO, ⌒⌒CHO	439	
3.2-27	O_3	1.7×10^{-17}	1145	5.9×10^4	CO_2,CH_3CHO	439	
3.2-28	O_3	1.2×10^{-17}	1145	8.3×10^4	CO,CH_3CHO	439	
3.2-36	$O_2(^1\Delta)$	$\leqslant1.7\times10^{-17}$	1032	$\geqslant2.9\times10^9$			
3.2-38	HO· O_3 O	$(1.4\pm0.1)\times10^{-10}$ 3.3×10^{-17} 1.8×10^{-11}	1163 1145 1024	1.5×10^4	CH_3CHO, ⌒⌒CHO	439	
3.2-39	O_3	3.7×10^{-17}	1145	2.7×10^4	CH_3CHO ⌒⌒CHO	439	
3.2-40	O_3	4.4×10^{-17}	1145	2.3×10^4	CO,CH_3CHO	439	
3.2-41	O_3	$(5.1\pm0.2)\times10^{-16}$	1034	2.0×10^3	CO,CH_3CHO	439	
3.2-42	O_3	3.1×10^{-17}	1145	3.2×10^4	CO,CH_3CHO	439	
3.2-47	O_3	$(1.1\pm0.1)\times10^{-17}$	1034	9.1×10^4	HCHO, ⌒⌒CHO	439	

TABLE 3.2 ALKENES AND ALKYNES
Reactions and Products

Species Number	Reactant	Chemical reactions					Remarks
		k	Ref.	Lifetime	Products	Ref.	
3.2-57	O_3	3.7×10^{-17}	1145	2.7×10^4	CO, CH_3CHO	439	
	$O_2(^1\Delta)$	1.1×10^{-17}	439				
3.2-68	$HO \cdot$	$(3.7 \pm 0.8) \times 10^{-11}$	1025	5.9×10^4			
	O_3	8.2×10^{-18}	1146				
3.2-75	O_3	8.2×10^{-18}	1146	1.2×10^5			
3.2-86	O_3	1.1×10^{-17}	1146	9.1×10^4			

adds over the double bond and the resulting ozonide cleaves to form an aldehyde and a biradical. A typical chain for propene is

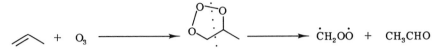

The fate of the "Criegee intermediate" $\dot{C}H_2O\dot{O}$ in the atmosphere is not certain. Although reaction with molecular oxygen to produce formaldehyde has been regarded (1022) as probable,

$$\dot{C}H_2O\dot{O} + O_2 \rightarrow HCHO + O_3 ,$$

recent results suggest the possibility of formation of a "hot" intermediate followed by decomposition to stable molecules (1262,1263):

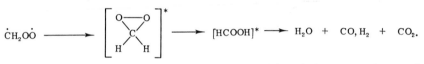

The creation of an ozonide from the intermediate is known to be possible if an abundant carbonyl compound is present (1103),

$$\dot{C}H_2O\dot{O} \quad + \quad HCHO \longrightarrow$$

but this process is expected to be unimportant in the troposphere.

The rapid generation of aldehydes by atmospheric alkene reactions is crucial to the progression of the gas-phase chemical chains of atmospheric importance (see Section 4.1). The alkenes are also potentially subject to isomerization, nitrosification, and oxidation. All these processes result in compounds with lower vapor pressures than their precursors, and thus may be significant factors in the formation of the organic fraction of the atmospheric aerosol (1115, 1116).

3.3. TERPENES

3.3.1. Identified Compounds

Terpenes are a special class of alkenes, distinctive because they are emitted in large quantities by vegetation (198). The identified sources are either natural or represent the processing or combustion of natural vegetation. More than a dozen such compounds are known to

TABLE 3.3 TERPENES
Emission and Detection

Species Number	Name	Chemical Formula	Emission		Detection Ref.	Ambient conc.
			Source	Ref.		
3.3-1	α-Pinene		1.trees	74,198	14*,198	0.93-1.20ppb
			2.solvent	439	399(a)	
			microbes	302		
			tobacco smoke	421		
			veneer drying	197		
			wood pulping	19,267		
3.3-2	β-Pinene		1.trees	74,198	14*,171	0.14-0.40ppb
			microbes	302		
			tobacco smoke	421		
			wood pulping	19,267		
3.3-3	Myrcene		microbes	302	14*,198	0.74-1.0ppb
			trees	198,461		
			wood pulping	267,545		
3.3-4	Ocimene		plant volatile	496,502		
3.3-5	Limonene		1.trees	74,198	14*,525*	0.06-5.7ppb
			microbes	302		
			tobacco smoke	396,421		
			wood pulping	19,267		
3.3-6	α-Terpinene		trees	198,499		
			wood pulping	19,267		
3.3-7	γ-Terpinene		wood pulping	19,267		

TABLE 3.3 TERPENES
Emission and Detection

Species Number	Name	Chemical Formula	Emission Source	Emission Ref.	Detection Ref.	Ambient conc.
3.3-8	Terpinolene		trees wood pulping	461 19,267		
3.3-9	Δ^3-Carene		microbes trees wood pulping	302 461,506e 19,267	472	
3.3-10	Camphene		plant volatile wood pulping	515e,517e 19,545		
3.3-11	Sabinene		plant volatile	517e		
3.3-12	α-Phellandrene		wood pulping	267,545		
3.3-13	β-Phellandrene		microbes plant volatile wood pulping	302 512e,531 19,267		
3.3-14	Bisabolene		plant volatile	514e,531		
3.3-15	Cadinene		plant volatile	506e,514e		
3.3-16	Thujopsene		plant volatile	510e		

TABLE 3.3. TERPENES
Emission and Detection

Species Number	Name	Chemical Formula	Emission		Detection Ref.	Ambient conc.
			Source	Ref.		
3.3-17	Chamazulene		plant volatile	510e		
3.3-18	α-Caryophyllene		plant volatile	499		
3.3-19	β-Caryophyllene		plant volatile	506e,531		
3.3-20	Squalene		tobacco smoke	396		
3.3-21	Isosqualene		tobacco smoke	396		

TABLE 3.3 TERPENES
Reactions and Products

Species Number	Reactant	Chemical reactions					Remarks
		k	Ref.	Lifetime	Products	Ref.	
3.3-1	O_3	$(1.4\pm0.4)\times10^{-16}$	1112	7.1×10^3		1087	
	HO·	$(5.8\pm0.9)\times10^{-11}$	1033				
	O						
3.3-2	O_3	$(4.0\pm1.0)\times10^{-17}$	1112	2.5×10^4			
	HO·	$(6.7\pm1.0)\times10^{-11}$	1033				
3.3-3	O_3	$(1.2\pm0.4)\times10^{-15}$	1112	8.3×10^2			
	HO·	$(2.3\pm0.7)\times10^{-10}$	1033				
3.3-5	O_3	$(6.4\pm1.9)\times10^{-16}$	1112	1.6×10^3			
	HO·	$(1.5\pm0.3)\times10^{-10}$	1033				
3.3-6	O_3	$(8.8\pm2.6)\times10^{-14}$	1112	1.1×10^1			
3.3-7	O_3	$(3.0\pm0.8)\times10^{-16}$	1112	3.3×10^3			
3.3-8	O_3	$(1.0\pm0.3)\times10^{-14}$	1112	1.0×10^2		1088	
3.3-9	O_3	$(1.0\pm0.3)\times10^{-16}$	1112	1.0×10^4			
	HO·	$(8.6\pm2.6)\times10^{-11}$	1033				
3.3-12	O_3	$(1.2\pm0.4)\times10^{-14}$	1112	8.3×10^1			

TABLE 3.3 TERPENES
Reactions and Products

Species Number	Chemical reactions						
	Reactant	k	Ref.	Lifetime	Products	Ref.	Remarks
3.3-13	O_3	$(1.8 \pm 0.5) \times 10^{-16}$	1112	5.6×10^3			
	$HO \cdot$	$(1.2 \pm 0.4) \times 10^{-10}$	1033				

be atmospheric constituents. The hemiterpene isoprene, listed in Table 3.2, is often emitted by the same sources that emit the monoterpenes.

3.3.2. Ambient Concentrations

Very few concentration measurements of the terpenes have been performed. The compounds are known to be very reactive and would thus be expected to decrease with distance away from their sources as removal reactions occur; such trends are seen in the data. Ambient concentrations near, but not within, forests give typical values of $0.1-1$ ppb.

3.3.3. Chemistry

The chemistry of the terpenes in the gas phase is less well determined than is the case for the smaller unsaturated hydrocarbons. The initial rates of reaction with HO· and O_3 have been studied for a number of terpenes, however. As with the small alkenes, hydroxyl reaction dominates the chemical loss processes during the day while ozone reaction is most important at night (1010).

The terminal products of terpene chemistry in the atmosphere are thought to be oxygenated fragments of large molecular weight capable of forming the "blue haze" aerosol often seen over forested areas in spring and early summer (1117). Support for this position is provided by a laboratory smog chamber study involving α-pinene (1087). A measured product of this process was pinonic acid, a compound subsequently found also on forest aerosols.

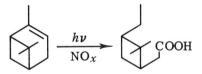

The production of oxygenated derivatives from α-pinene has been shown theoretically (1010) to be a straightforward process of atmospheric chemistry; the agreement between experiment and theory thus appears to be satisfactory.

3.4. CYCLIC HYDROCARBONS

3.4.1. Identified Compounds

More than 40 cyclic compounds are listed in Table 3.4. Except for minor atmospheric injection from natural gas, all are anthropogenically generated. Cyclopentane, cyclohexane, cyclopentene, and some of their alkyl derivatives are commonly identified constituents of motor vehicle exhaust and gasoline vaporization and are often identified in urban air samples. Other cyclic compounds arise from a variety of industrial processes. Saturated cyclic compounds, and/or their alkyl or phenyl derivatives, are known for C_3 to C_6. Unsaturated cyclic compounds with $C=5, 6, 8$, and 10 have been identified.

3.4.2. Ambient Concentrations

Ambient concentration measurements are available only for five cyclic compounds often detected in automobile exhaust: the C_5 and C_6 saturated compounds and their 1-methyl derivatives, and the C_5 unsaturated compound. The observed concentrations range from 2 to 50 ppb.

3.4.3. Chemistry

The cyclic hydrocarbons are moderately reactive atmospheric compounds. Several rate constants for reaction with $HO\cdot, O_3$, and $O_2(^1\Delta)$ have been determined and, as is often the case in the atmosphere, reaction with $HO\cdot$ is expected to dominate gas-phase removal. Hydrogen abstraction will occur in the case of the saturated compounds (1118) and hydroxyl addition will be most important in the case of unsaturated cyclic hydrocarbons (1025).

The cyclic alkanes appear unlikely to play major roles in atmospheric chemistry. A possible sequence is inaugurated by hydrogen abstraction:

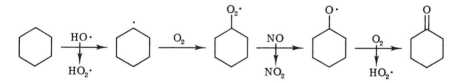

Photolysis of cyclic ketones is not rapid in the troposphere and reformation of the original ketone following ring cleavage next to the carbonyl group or disproportionation into small molecules are the typical photolytic processes (1231).

Unlike the cyclic alkanes, cyclic olefins react rapidly with ozone. Grosjean (1063) has presented evidence that the products of such reactions are difunctional organic compounds, formed by chains such as

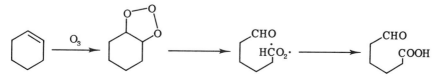

TABLE 3.4 CYCLIC HYDROCARBONS
Emission and Detection

Species Number	Name	Chemical Formula	Emission Source	Emission Ref.	Detection Ref.	Ambient conc.
3.4-1	Cyclopropane				582	
3.4-2	trans-1,2-Dimethyl-cyclopropane		turbine	414		
3.4-3	1-Methyl-1-ethyl-cyclopropane		turbine	414		
3.4-4	Isopropyl-cyclopropane		turbine	414		
3.4-5	Cyclobutane		turbine	414t		
3.4-6	Ethylcyclobutane		turbine	414		
3.4-7	Cyclopentane		auto gasoline vapor natural gas petroleum mfr. turbine	47,153* 232 456* 58*,232 414	170*,232*	2-14ppb
3.4-8	Methylcyclopentane		auto gasoline vapor natural gas petroleum mfr. plant volatile turbine	159*,465* 465* 465* 122 552e 359*	170*,263	
3.4-9	1,1-Dimethyl-cyclopentane		turbine	414		

98

TABLE 3.4 CYCLIC HYDROCARBONS
Emission and Detection

Species Number	Name	Chemical Formula	Emission		Detection	Ambient
			Source	Ref.	Ref.	conc.
3.4-10	1,3-Dimethyl-cyclopentane		turbine	414	582	
3.4-11	1,2,3-Trimethyl-cyclopentane		tobacco smoke turbine	421 414	582	
3.4-12	1,2,4-Trimethyl-cyclopentane				399(a)	
3.4-13	Methylethyl-cyclopentane				582	
3.4-14	Cyclopentene		auto forest fire	153*,170* 524	170*,524 458(a)	2-6ppb
3.4-15	1-Methyl-cyclopentene		gasoline vapor	486	364(a),458(a)	
3.4-16	Cyclopentadiene		polymer mfr.	251		
3.4-17	6,6-Dimethylfulvene		turbine	414		
3.4-18	Dicyclopentadiene		polymer mfr.	251		
3.4-19	Cyclohexane		1.auto 2.solvent petroleum mfr. plant volatile turbine volcano	47,526 78,134 58*,543* 552e 414 234	123*,170*	3-6ppb

TABLE 3.4 CYCLIC HYDROCARBONS
Emission and Detection

Species Number	Name	Chemical Formula	Emission		Detection Ref.	Ambient conc.
			Source	Ref.		
3.4-20	Methylcyclohexane		auto gasoline vapor natural gas turbine volcano	47,465* 465*,486 465* 414 234	123*,566*	3-7ppb
3.4-21	1,1-Dimethyl-cyclohexane		tobacco smoke	421	582	
3.4-22	cis-1,2-Dimethyl-cyclohexane		auto	526	374,566*	3ppb
3.4-23	trans-1,2-Dimethyl-cyclohexane		auto	526	374,582	
3.4-24	1,3-Dimethyl-cyclohexane		tobacco smoke	421		
3.4-25	1,1,3-Trimethyl-cyclohexane		tobacco smoke	421	399(a),458t(a)	
3.4-26	1,3,5-Trimethyl-cyclohexane					
3.4-27	Ethylcyclohexane		auto	284,526		
3.4-28	Diethylcyclohexane				591	
3.4-29	n-Butylcyclohexane		turbine	414		
3.4-30	tert-Butylcyclo-hexane				582	

TABLE 3.4 CYCLIC HYDROCARBONS
Emission and Detection

Species Number	Name	Chemical Formula	Emission		Detection	Ambient conc.
			Source	Ref.	Ref.	
3.4-31	Dodecylcyclohexane	$(CH_2)_{11}CH_3$	auto	311(a)	364(a)	
3.4-32	Phenylcyclohexane					
3.4-33	Ethylphenyl-cyclohexane				399(a),458(a)	
3.4-34	Cycloheptane		petroleum mfr.	543*		
3.4-35	Cyclooctane		petroleum mfr.	543*		
3.4-36	Cyclohexene		auto	153*	364(a),458(a)	
			rubber abrasion	331		
3.4-37	1-Methyl-cyclohexene		auto	284		
			rubber abrasion	331		
3.4-38	4-Methylcyclohexene		auto	527		
3.4-39	4-Vinylcyclohexene		vulcanization	308		

TABLE 3.4 CYCLIC HYDROCARBONS
Emission and Detection

Species Number	Name	Chemical Formula	Emission		Detection	Ambient conc.
			Source	Ref.	Ref.	
3.4-40	Phenylcyclohexene				364(a)	
3.4-41	1,4-Cyclohexadiene		refuse comb.	26(a)	364(a)	
3.4-42	1,5-Cyclooctadiene		vulcanization	308		
3.4-43	1,5,9-Cyclo-dodecatriene		vulcanization	308		
3.4-44	α-Gurjerene		plant volatile	497e		

TABLE 3.4 CYCLIC HYDROCARBONS
Reactions and Products

Species Number	Reactant	Chemical reactions			Products	Ref.	Remarks
		k	Ref.	Lifetime			
3.4-5	HO·	$(1.2\pm0.3)\times10^{-12}$	1118	1.9×10^{6}	(cyclobutanone structure) H_2,	1137	
	O	$(1.2\pm0.6)\times10^{-14}$	1228				
	O(^{1}D)	$(3.1\pm0.8)\times10^{-10}$	1153				
3.4-7	O	$(1.3\pm0.4)\times10^{-13}$	1228	3.2×10^{8}			
	O(^{1}D)	$(3.8\pm0.9)\times10^{-10}$	1153				
3.4-14	O$_3$	$(8.1\pm0.8)\times10^{-16}$	1034	1.2×10^{3}			
	O$_2$($^1\Delta$)	$(1.5\pm0.1)\times10^{-18}$	1035				
3.4-15	O$_2$($^1\Delta$)	$(2.0\pm0.1)\times10^{-17}$	1035	2.5×10^{9}			
3.4-16	O	5.4×10^{-11}	1041	7.4×10^{5}	(alkyne structure)	1041	
	O$_2$($^1\Delta$)	1.5×10^{-14}	439				
3.4-19	HO·	$(8.1\pm0.1)\times10^{-12}$	1121	2.7×10^{5}		1228	
	O	$(1.4\pm0.4)\times10^{-13}$	1228				
	O(^{1}D)	$(4.4\pm1.1)\times10^{-10}$	1153				
3.4-22	O$_2$($^1\Delta$)	6.6×10^{-16}	1032	7.6×10^{7}			
3.4-36	O$_3$	$(1.7\pm0.2)\times10^{-16}$	1034	5.9×10^{3}	CHO—(chain)—CHO	1087	
	HO·	$(7.8\pm1.5)\times10^{-11}$	1025				
	O	$(1.8\pm0.2)\times10^{-11}$	1157				
	O$_2$($^1\Delta$)	$<2\times10^{-19}$	1035				

103

TABLE 3.4 CYCLIC HYDROCARBONS
Reactions and Products

Species Number	Reactant	k	Chemical reactions		Products	Ref.	Remarks
			Ref.	Lifetime			
3.4-37	HO·	$(9.6 \pm 0.2) \times 10^{-11}$	1025	2.1×10^4			
	$O_2(^1\Delta)$	$(1.2 \pm 0.1) \times 10^{-18}$	1035				
3.4-41	$O_2(^1\Delta)$	$\leqslant 1.7 \times 10^{-17}$	1032	$\geqslant 2.9 \times 10^9$			

Different oxidation and abstraction routes for the ozone adduct, some of which may occur in the aerosol phase, can lead to a variety of products in which either or both of the functional groups can be $-CHO$, $-CH_2OH$, $-COOH$, or, if oxides of nitrogen are present, $-ONO_2$. The vapor pressures of these difunctional compounds are very low and they are expected to serve as nuclei for secondary aerosol formation in the troposphere or to be removed by impaction on atmospheric aerosols.

3.5. AROMATIC COMPOUNDS: BENZENE AND DERIVATIVES

3.5.1. Identified Compounds

Benzene and many of its substituted derivatives are common constituents of urban atmospheres. The one-methyl substituent (toluene) appears to be more ubiquitous than benzene. The two-methyl derivatives (the xylenes and ethyl benzene) are also widely found.

The aromatic hydrocarbons are emitted primarily from anthropogenic sources. Internal combustion engine exhaust contains high concentrations of aromatic compounds, more so since the advent of increased aromatic content of gasolines (in order to achieve satisfactory octane ratings without the use of lead additives). The light aromatics also see wide use as solvents, with volatilization into the atmosphere being not uncommon. A number of industrial processes utilize the substituted benzenes; some of these emit into the air portions of what they use.

The only natural sources of these compounds that are known with certainty are the combustion of vegetation and emission from volcanos; neither is thought to be significant. (An exception is p -cymene, known to be emitted from trees in the same manner as the terpenes.) Chatfield and co-workers (472), however, have detected toluene, xylene, and ethylbenzene in rural areas under noncombustion conditions, which suggests that vegetation may emit a variety of aromatic compounds.

3.5.2. Ambient Concentrations

Because motor vehicle exhaust hydrocarbons contain roughly equal amounts of alkanes, alkenes, and aromatics (159), it is not surprising that the concentrations of the major species in these groups should be similar. Toluene, the most abundant of the aromatics, has been seen in urban areas at concentrations as high as 129 ppb. (For comparison, note that the maximum concentrations recorded for ethane and propene are 95 and 52 ppb.) Benzene is the next most abundant after toluene, followed by the xylenes and ethyl benzene. In all, 19 of these compounds have at one time or another been detected in concentrations exceeding 1 ppb.

3.5.3. Chemistry

The principal atmospheric reactions of benzene and the substituted benzenes are with the hydroxyl radical. The reaction chains in the atmosphere are not well determined, although at room temperature addition to the aromatic ring appears to be favored (1122). The most likely sequential process is reaction with molecular oxygen to form an alcohol (1134). For benzene, toluene, and m -xylene, the reactions are

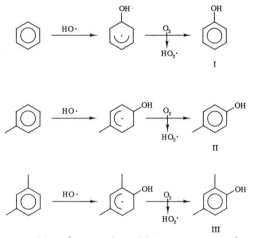

The alcohols are capable of repeating this sequence to form di- and tri-alcohols. The low vapor pressures of the compounds make it more likely, however, that they will be deposited on aerosols and surfaces rather than becoming involved in extensive chemical chains. For alkyl benzenes, a secondary sequence is likely to be initiated by hydrogen abstraction by hydroxyl from the alkyl group (1239). Thus, for toluene,

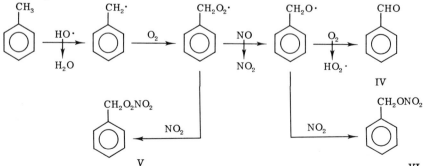

TABLE 3.5 AROMATIC HYDROCARBONS: BENZENE AND DERIVATIVES
Emission and Detection

| Species Number | Name | Chemical Formula | Emission | | Detection | Ambient |
			Source	Ref.	Ref.	conc.
3.5-1	Benzene		1.auto	159*,376	139*,170* 364(a),458(a)	.025-57ppb
			2.solvent	60,78		
			chemical mfr.	566*		
			diesel	434*		
			fish oil mfr.	110*		
			forest fire	354,439		
			gasoline vapor	58*,232		
			lacquer mfr.	428		
			landfill	365		
			petroleum mfr.	127*,543*		
			plant volatile	552e		
			polymer comb.	46,304		
			polymer mfr.	103		
			refuse comb.	439		
			refuse comb.	26(a)		
			sewage tmt.	204		
			tobacco smoke	396,446		
			turbine	359*,414		
			volcano	234		
			vulcanization	331		

TABLE 3.5 AROMATIC HYDROCARBONS: BENZENE AND DERIVATIVES
Emission and Detection

Species Number	Name	Chemical Formula	Emission		Detection	Ambient
			Source	Ref.	Ref.	conc.
3.5-2	Toluene		1.auto	159*,465*	139*,170*	.005-129ppb
			2.solvent	78,134	364(a),458(a)	
			chemical mfr.	566*		
			diesel	176		
			fish oil mfr.	110*		
			forest fire	354,439		
			gasoline vapor	58*,232		
			landfill	365		
			petroleum mfr.	127*,543*		
			plant volatile	552e		
			polymer comb.	304,354		
			polymer mfr.	103		
			tobacco smoke	396,421		
			tobacco smoke	339(a)		
			turbine	359*,414		
			volcano	234		
			vulcanization	308,331		
			wood pulping	19,267		
3.5-3	o-Xylene		auto	159*,465*	139*,566*	0.5-33ppb
			fish oil mfr.	110*		
			forest fire	439		
			gasoline vapor	232,465*		
			phthal anhyd. mfr.	64*		
			plant volatile	552e		
			polymer comb.	304		
			tobacco smoke	421		

TABLE 3.5 AROMATIC HYDROCARBONS: BENZENE AND DERIVATIVES
Emission and Detection

Species Number	Name	Chemical Formula	Emission		Detection	Ambient
			Source	Ref.	Ref.	conc.
3.5-4	m-Xylene		1.auto	159*,465*	123*,139*	1-61ppb
			2.solvent	134,439		
			diesel	176		
			fish oil mfr.	110		
			gasoline vapor	232,465*		
			polymer comb.	304		
			turbine	414		
			vulcanization	308,331		
3.5-5	p-Xylene		auto	159*,465*	139*,232*	1-25ppb
			diesel	176	458(a)	
			fish oil mfr.	110*		
			gasoline vapor	232,465*		
			lacquer mfr.	428		
			polymer comb.	304		
			turbine	414		
			vulcanization	308		
3.5-6	1,2,3-Trimethyl-benzene		auto	465*,526	232*,566*	1-2ppb
			gasoline vapor	232,465*		
			plant volatile	552e		
			turbine	414		
3.5-7	1,2,4-Trimethyl-benzene		auto	159*,465*	139*,232*	3-15.3ppb
			gasoline vapor	232,465*		
			plant volatile	552e		
			tobacco smoke	396(a)		

109

TABLE 3.5 AROMATIC HYDROCARBONS: BENZENE AND DERIVATIVES
Emission and Detection

Species Number	Name	Chemical Formula	Emission		Detection	Ambient conc.
			Source	Ref.	Ref.	
3.5-8	1,3,5-Trimethyl-benzene		auto	284,486*	123*,139*	1.3-11ppb
			diesel	309(a)	99(a)	
			gasoline vapor	232,486		
			plant volatile	552e		
			solvent	134		
			tobacco smoke	396(a)		
			turbine	359*,414		
3.5-9	1,2,3,4-Tetramethyl-benzene				86,582	
3.5-10	1,2,3,5-Tetramethyl-benzene		diesel	309(a)	200*,566*	1.3-5.3ppb
3.5-11	1,2,4,5-Tetramethyl-benzene		auto	526	86,200*	1.5-3.9ppb
3.5-12	Hexamethylbenzene				200	
3.5-13	Ethylbenzene		auto	159*,526	139*,202	.1-22ppb
			chemical mfr.	566*	399(a),458(a)	
			diesel	176		
			gasoline vapor	232		
			polymer mfr.	103		
			solvent	134,439		
			tobacco smoke	421		
			turbine	359*		
			vulcanization	308		

TABLE 3.5 AROMATIC HYDROCARBONS: BENZENE AND DERIVATIVES
Emission and Detection

Species Number	Name	Chemical Formula	Emission		Detection	Ambient conc.
			Source	Ref.	Ref.	
3.5-14	o-Ethyltoluene		auto gasoline vapor plant volatile	526 232 552e	86*,200*	0.7-2.6ppb
3.5-15	m-Ethyltoluene		auto gasoline vapor plant volatile turbine	465* 232,465* 552e 414	86*,232*	1.1-13ppb
3.5-16	p-Ethyltoluene		auto gasoline vapor plant volatile turbine	159* 232 552e 414	86*,232* 458(a)	1.1-13ppb
3.5-17	Ethyldimethyl-benzene	$(CH_3)_2$	auto gasoline vapor	465* 465*	86*,566*	0.74-2ppb
3.5-18	1,2-Diethylbenzene		turbine	414	200,374	
3.5-19	1,3-Diethylbenzene		auto	526	582	
3.5-20	Methyldiethyl-benzene	CH_3 $(C_2H_5)_2$			525t	

TABLE 3.5 AROMATIC HYDROCARBONS: BENZENE AND DERIVATIVES
Emission and Detection

Species Number	Name	Chemical Formula	Emission Source	Emission Ref.	Detection Ref.	Ambient conc.
3.5-21	Styrene		building resins	9*	123*,566*	1.5-5ppb
			chemical mfr.	566*	364(a),458(a)	
			glue vapor	11		
			polymer mfr.	251,407*		
			plant volatile	531		
			plastics mfr.	104,117*		
			river water odor	202		
			solvent	134		
			tobacco smoke	421		
			vulcanization	308		
3.5-22	α-Methylstyrene		diesel	176	591	
			solvent	134	364(a)	
			tobacco smoke	421		
3.5-23	m-Methylstyrene				399(a),458(a)	
3.5-24	p-Methylstyrene				399(a),458(a)	
3.5-25	Ethylstyrene				399(a)	
3.5-26	Propylstyrene				399(a)	

TABLE 3.5 AROMATIC HYDROCARBONS: BENZENE AND DERIVATIVES
Emission and Detection

Species Number	Name	Chemical Formula	Emission		Detection	Ambient conc.
			Source	Ref.	Ref.	
3.5-27	Vinyltoluene		polymer mfr.	407*	525	
3.5-28	Vinylxylene		polymer mfr.	407*		
3.5-29	Phenylacetylene		tobacco smoke	396(a)		
3.5-30	n-Propylbenzene		auto gasoline vapor plant volatile turbine	47,486* 232,486 552e 414	123*,139* 399(a),458(a)	1-6ppb
3.5-31	1-Methyl-2-propyl-benzene		auto	526	582	
3.5-32	4-Propyltoluene				86,525 399(a)	
3.5-33	Cumene		auto gasoline vapor plant volatile solvent tobacco smoke turbine	284,465* 232 552e 134 421 359*,414	123*,139* 399(a),458(a)	1-12ppb

TABLE 3.5 AROMATIC HYDROCARBONS: BENZENE AND DERIVATIVES
Emission and Detection

Species Number	Name	Chemical Formula	Emission		Detection	Ambient conc.
			Source	Ref.	Ref.	
3.5-34	m-Cymene		auto	526	582	
3.5-35	p-Cymene		plant volatile solvent trees wood pulping	507e,552e 134 461 19	14*,566*	0.12-2ppb
3.5-36	Ethylisopropyl-benzene				525	
3.5-37	Chavicene		plant volatile	531		
3.5-38	n-Butylbenzene		auto	465*	139,374 399(a)	
3.5-39	Isobutylbenzene		plant volatile turbine	552e 414	525 364(a)	
3.5-40	sec-Butylbenzene		auto	47,526	123*,139* 399(a)	4-15ppb
3.5-41	tert-Butylbenzene		diesel solvent	138 134	123*,139*	2-6ppb

TABLE 3.5 AROMATIC HYDROCARBONS: BENZENE AND DERIVATIVES
Emission and Detection

Species Number	Name	Chemical Formula	Emission		Detection	Ambient conc.
			Source	Ref.	Ref.	
3.5-42	1-Butenylbenzene				399(a),458(a)	
3.5-43	Pentylbenzene				399(a)	
3.5-44	Hexylbenzene		tobacco smoke	421	399(a),458(a)	
3.5-45	Octylbenzene	$(CH_2)_7CH_3$	tobacco smoke	421		
3.5-46	Nonylbenzene	$(CH_2)_8CH_3$			399(a)	
3.5-47	Decylbenzene	$(CH_2)_9CH_3$			399(a),458(a)	
3.5-48	Indane				200 364(a),458(a)	
3.5-49	Methylindane	CH_3	diesel tobacco smoke	138,176 421	86,525 399(a)	
3.5-50	Dimethylindane	$(CH_3)_2$	diesel	138,176	582 399(a)	

TABLE 3.5 AROMATIC HYDROCARBONS: BENZENE AND DERIVATIVES
Emission and Detection

Species Number	Name	Chemical Formula	Emission		Detection Ref.	Ambient conc.
			Source	Ref.		
3.5-51	Trimethylindane		diesel	138		
3.5-52	Propylindane				399(a)	
3.5-53	Indene		polymer mfr tobacco smoke	251 339(a)	200,582 364(a),458(a)	
3.5-54	1-Methylindene		tobacco smoke	339(a)	364(a),458(a)	
3.5-55	Dimethylindene				399(a)	
3.5-56	Ethylindene		tobacco smoke	339(a)	399(a),458(a)	
3.5-57	Benz[f]indene		coal comb. tobacco smoke	361(a) 406	361(a),362(a)	
3.5-58	Azulene		tobacco smoke	79(a),396(a)	209(a)	

TABLE 3.5 AROMATIC HYDROCARBONS: BENZENE AND DERIVATIVES
Emission and Detection

Species Number	Name	Chemical Formula	Emission		Detection	Ambient conc.
			Source	Ref.	Ref.	
3.5-59	Biphenyl		diesel phthal anhyd. mfr. plastics comb. tobacco smoke	309(a) 64* 181 339(a)	86,553* 362(a),363(a)	3.2-250pg/m^3
3.5-60	4-Methylbiphenyl		tobacco smoke	339(a)	86 361(a),362(a)	
3.5-61	Ethylbiphenyl				399(a),458(a)	
3.5-62	Ethylmethyl-biphenyl		tobacco smoke	130*(a)		
3.5-63	Diphenylmethane				364(a)	
3.5-64	Diphenyl ethane				399(a)	
3.5-65	Quaterphenyl				408(a)	
3.5-66	Benzo[c]tetraphene				363(a)	

TABLE 3.5 AROMATIC HYDROCARBONS: BENZENE AND DERIVATIVES
Emission and Detection

Species Number	Name	Chemical Formula	Emission		Detection	Ambient conc.
			Source	Ref.	Ref.	
3.5-67	Trinaphthene benzene		auto	311(a)		

TABLE 3.5 AROMATIC HYDROCARBONS: BENZENE AND DERIVATIVES

Reactions and Products

Species Number	Reactant	Chemical reactions					Remarks
		k	Ref.	Lifetime	Products	Ref.	
3.5-1	HO·	$(1.2\pm0.2)\times10^{-12}$	1122	1.9×10^{6}	[benzene ring with OH — phenol]	1134	
	O	$(1.6\pm0.1)\times10^{-14}$	1124		[methyl-substituted ring with OH]	1152	
	NO$_3$·	$<2.0\times10^{-17}$	1031				
	O$_2(^1\Delta)$	$\leqslant1.7\times10^{-17}$	1032				
3.5-2	HO·	$(6.4\pm0.6)\times10^{-12}$	1122	3.4×10^{5}	[methyl-substituted ring with OH]	1134	
	O	$(5.9\pm0.3)\times10^{-14}$	1124		[ring with CHO — benzaldehyde; OH]	1089	
	NO$_3$·	$<2.0\times10^{-17}$	1031				
3.5-3	HO·	$(1.4\pm0.2)\times10^{-11}$	1122	1.5×10^{5}	CHOCHO, [O=C–C(=O)–CHO dicarbonyl]	1102	
	O	$(1.8\pm0.2)\times10^{-13}$	1123		CHOCHO	1063	
	O$_2(^1\Delta)$	$\leqslant1.7\times10^{-17}$	1032				
	O$_3$						
3.5-4	HO·	$(2.4\pm0.3)\times10^{-11}$	1122	9.0×10^{4}	CHOCHO, [O=C–CH$_3$; CHO]	1102	
	O	$(3.5\pm0.4)\times10^{-13}$	1123				
3.5-5	HO·	$(1.5\pm0.2)\times10^{-11}$	1122	8.7×10^{5}	CHOCHO, [O=C–CH$_3$; CHO]	1102	
	O	$(1.8\pm0.2)\times10^{-13}$	1123				
3.5-6	HO·	$(3.3\pm0.5)\times10^{-11}$	1122	6.6×10^{4}	CHOCHO, [OCH$_3$-substituted ring]	1102	
	O	$(1.2\pm0.1)\times10^{-12}$	1123				
3.5-7	HO·	$(4.0\pm0.5)\times10^{-11}$	1122	5.4×10^{4}	[(CH$_3$)$_2$-substituted ring]	1152t	
	O	$(1.0\pm0.1)\times10^{-12}$	1123				

TABLE 3.5 AROMATIC HYDROCARBONS: BENZENE AND DERIVATIVES
Reactions and Products

Species Number	Reactant	Chemical reactions					Remarks
		k	Ref.	Lifetime	Products	Ref.	
3.5-8	HO· O	$(6.2\pm0.8)\times10^{-11}$ $(2.8\pm0.3)\times10^{-12}$	1122 1123	3.5×10^{4}			
3.5-9	HO·	1.2×10^{-11}	1167	1.9×10^{5}			
3.5-10	HO·	1.2×10^{-11}	1167	1.9×10^{5}			
3.5-11	HO·	1.2×10^{-11}	1167	1.9×10^{5}			
3.5-12	HO·	1.2×10^{-11}	1167	1.9×10^{5}			
3.5-13	HO·	$(8.0\pm1.7)\times10^{-12}$	1085	2.7×10^{5}			
3.5-14	HO·	$(1.4\pm0.3)\times10^{-11}$	1085	1.5×10^{5}			
3.5-15	HO·	$(2.0\pm0.4)\times10^{-11}$	1085	1.1×10^{5}			
3.5-16	HO·	$(1.3\pm0.3)\times10^{-11}$	1085	1.6×10^{5}			
3.5-21	O_3	3.0×10^{-17}	1144	3.3×10^{4}			
3.5-30	HO·	$(6.2\pm1.3)\times10^{-12}$	1085	3.5×10^{5}			
3.5-33	HO·	$(1.5\pm0.5)\times10^{-11}$	1033	1.4×10^{5}			

TABLE 3.5 AROMATIC HYDROCARBONS: BENZENE AND DERIVATIVES
Reactions and Products

Species Number	Chemical reactions						Remarks
	Reactant	k	Ref.	Lifetime	Products	Ref.	
3.5-36	HO·	$(6.2 \pm 1.3) \times 10^{-12}$	1085	3.5×10^5			

The same sequences can be initiated by oxygen atom addition to the aromatic ring or by H-atom abstraction by atomic oxygen. In the case of addition of either HO· or O, nitro- alkyl aromatics may be formed, as in

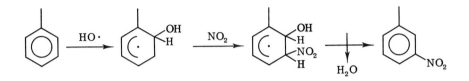

Of the numbered compounds in these sequences, I and IV have been detected in the ambient atmosphere (see Tables 5.11 and 4.3) and all are seen in smog chamber experiments (1089) . The peroxybenzoyl nitrate compounds (type V) are severe eye irritants, and may contribute to lachrymation even at very low concentrations in the atmosphere (1093).

3.6. AROMATIC COMPOUNDS: NAPHTHALENE AND DERIVATIVES

3.6.1. Identified Compounds

More than 30 naphthalenic compounds are known to be emitted into the atmosphere. Most of the compounds are alkyl derivatives of naphthalene and of tetralin (1,2,3,4-tetrahydronaphthalene), although several more complex substituents have also been detected. The naphthalenes are stable products of hydrocarbon combustion; the widest variety has thus far been detected in diesel exhaust and tobacco smoke.

3.6.2. Ambient Concentrations

The naphthalenes are present in urban atmospheres at concentrations of a few hundred ngm^{-3} . Naphthalene is the most abundant, having a concentration similar to that of the total of its methyl derivatives. No data are available concerning the atmospheric concentrations of any of the more complex naphthalenes.

3.6.3. Chemistry

Naphthalene chemistry bears obvious similarities both to that of the benzenes (Section 3.5) and of the polynuclear aromatic hydrocarbons (Section 3.7). The low vapor pressure of the naphthalenes makes it probable that their chemistry will primarily occur within the liquid aerosol phase. Reaction with the hydroxyl radical is the most likely process for the parent compound:

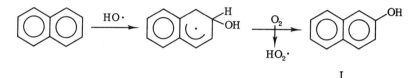

The alkyl naphthalenes will form naphthols also, but can in addition undergo alkyl hydrogen abstraction to give other classes of products. For 1-methyl naphthalene, for example,

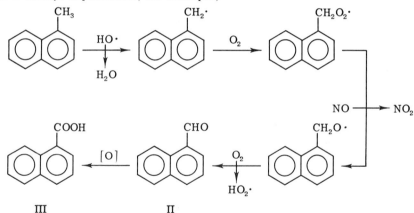

Compounds I, II, and III have all been detected in ambient aerosols.

3.7. POLYNUCLEAR AROMATIC HYDROCARBONS

3.7.1. Identified Compounds

A numerous family of polynuclear aromatic hydrocarbons (PNAHs) is present in the lower atmosphere. The size range is from fluorene ($C_{13}H_{11}$) to the seven-ring compound coronene ($C_{24}H_{12}$) . Virtually every isomer that can be constructed from three, four, five, six, or seven benzene rings is known to be emitted into the air, as are many similar compounds containing five-member ring structures or alkyl groups. The total number of species is more than 130.

TABLE 3.6 AROMATIC HYDROCARBONS:NAPHTHALENE AND DERIVATIVES
Emission and Detection

Species Number	Name	Chemical Formula	Emission		Detection	Ambient conc.
			Source	Ref.	Ref.	
3.6-1	Naphthalene		auto	537	86,200	.03-300ng/m³
			auto	311(a)	209(a),553*(a)	
			brewing	144		
			diesel	176		
			diesel	309(a)		
			phthal anhyd. mfr.	64*		
			polymer comb.	304,332		
			refuse comb.	26(a)		
			river water odor	202		
			tobacco smoke	406,421		
			tobacco smoke	79(a),339(a)		
3.6-2	1-Methylnaphalene		auto	537	86	10-170ng/m³
			brewing	144	209(a),214*(a)	
			coal tar mfr.	338(a)		
			diesel	176		
			diesel	309(a)		
			landfill	365		
			tobacco smoke	406,421		
			tobacco smoke	339(a)		
			vulcanization	308		
3.6-3	2-Methylnaphthalene		landfill	365	86	
			tobacco smoke	406,421	399(a),458(a)	
			tobacco smoke	339(a),396(a)		
3.6-4	1,2-Dimethyl-naphthalene		tobacco smoke	339(a)		
3.6-5	1,4-Dimethyl-naphthalene				200	

TABLE 3.6 AROMATIC HYDROCARBONS:NAPHTHALENE AND DERIVATIVES
Emission and Detection

Species Number	Name	Chemical Formula	Emission		Detection Ref.	Ambient conc.
			Source	Ref.		
3.6-6	1,6-Dimethyl-naphthalene				86,200 364(a),458(a)	
3.6-7	1,8-Dimethyl-naphthalene		tobacco smoke	339(a),396(a)	86	
3.6-8	2,3-Dimethyl-naphthalene		coal tar mfr.	338(a)	200 589(a)	
3.6-9	2,6-Dimethyl-naphthalene		vulcanization	308	86 214*(a)	2-70ng/m^3
3.6-10	Trimethyl-naphthalene	(CH$_3$)$_3$ / (CH$_3$)$_4$	diesel tobacco smoke	309(a) 339(a)	214*(a),589(a)	3-80ng/m^3
3.6-11	Tetramethyl-naphthalene	(CH$_3$)$_5$	tobacco smoke	339(a)		
3.6-12	Pentamethyl-naphthalene		tobacco smoke	339(a)		
3.6-13	1-Ethyl-naphthalene		diesel tobacco smoke	309(a) 339(a)	200 399(a)	
3.6-14	2-Ethyl-naphthalene		vulcanization	308		

TABLE 3.6 AROMATIC HYDROCARBONS:NAPHTHALENE AND DERIVATIVES
Emission and Detection

Species Number	Name	Chemical Formula	Emission Source	Emission Ref.	Detection Ref.	Ambient conc.
3.6-15	2-Isopropyl-naphthalene		diesel	309(a)		
3.6-16	Undecyl-naphthalene	$(CH_2)_{10}CH_3$	auto	311(a)		
3.6-17	1,2-Dihydrotrimethyl-naphthalene	$(CH_3)_3$	diesel	309(a)		
3.6-18	Tetralin		diesel river water odor	138 202	364(a)	
3.6-19	Methyltetralin	CH_3	diesel tobacco smoke	138,176 421		
3.6-20	Dimethyltetralin	$(CH_3)_2$	diesel diesel	138 309(a)		
3.6-21	Trimethyltetralin	$(CH_3)_3$	diesel	138		
3.6-22	Decahydro-naphthalene		diesel	309(a)	200	

126

TABLE 3.6 AROMATIC HYDROCARBONS:NAPHTHALENE AND DERIVATIVES
Emission and Detection

Species Number	Name	Chemical Formula	Emission		Detection	Ambient conc.
			Source	Ref.	Ref.	
3.6-23	Acenaphthalene		diesel	138	86,200	
			diesel	309(a)	209(a),214(a)	
			tobacco smoke	396(a)		
3.6-24	Methyl-acenapthalene		tobacco smoke	130*(a)		
3.6-25	Dimethylcyclopent-acenaphthalene		auto	311(a)		
3.6-26	Benzace-naphthalene		tobacco smoke	130*(a)		
3.6-27	Diphenyl-acenaphthalene		auto	311(a)	408(a)	
3.6-28	Acenaphthylene		diesel	309(a)	209(a),528(a)	
			refuse comb.	26(a)		
			tobacco smoke	79(a),339(a)		
3.6-29	1-Methyl-acenaphthylene		tobacco smoke	406		
3.6-30	Benzace-naphthylene				408(a)	
3.6-31	β,β-Binaphthyl		diesel	309(a)	362(a),363(a)	

TABLE 3.6 AROMATIC HYDROCARBONS:NAPHTHALENE AND DERIVATIVES
Emission and Detection

| Species Number | Name | Chemical Formula | Emission | | Detection | Ambient |
			Source	Ref.	Ref.	conc.
3.6-32	Methyl-β,β-binaphthyl				362(a)	

Polycyclic aromatic molecular structures are formed whenever organic substances are exposed to high temperatures (1138,1139). Two sources account for most of the atmospheric emissions of PNAHs: internal combustion engines and coal burners. The compounds are also present to a high degree in tobacco smoke. Since tobacco smoke has been extensively analyzed, its compounds are well represented in Table 3.7.

The PNAHs have very low vapor pressures and are not known to exist in the atmosphere in the gas phase.

3.7.2. Ambient Concentrations

The total concentrations of PNAHs in ambient aerosols are typically several hundred ng m^{-3} (189). The distribution of individual compounds within this total amount is not well established and is probably not uniform. As would be expected (1138), the stable compounds made up of tight clusters of benzene rings appear to be the most abundant. The following PNAHs have been identified in the ambient aerosol at concentrations $\geqslant 100$ ng m^{-3} :

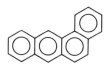

Benz[*a*]anthracene

Benzo[*k*]fluoranthene

Pyrene

Benzo[*ghi*]perylene

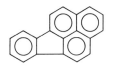

Benzo[*cd*]fluoranthene

Benzo[*cd*]pyrene

3.7.3. Chemistry

PNAHs are moderately reactive in the environment(1260), and their photooxidation in wet aerosol solution is expected to lead to formation of endoperoxides (424,1090). Thus, for anthracene,

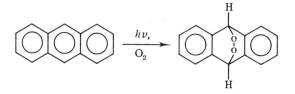

If the bridgehead carbon atoms are also bonded to hydrogen atoms, conversion to quinones occurs rapidly (424,1261):

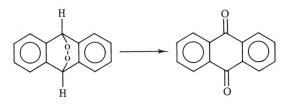

Ketone formation can also take place by oxygen addition to a saturated C−H bond (424). For fluorene,

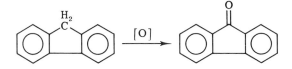

Eleven mono- or diketones of PNAH compounds have been detected in atmospheric aerosols, thus providing supporting evidence for the oxidation process presented here (Table 3.7-2).

A more complete discussion of the action of various oxidants on PNAHs can be found in reference 424.

TABLE 3.7-2. POLYNUCLEAR AROMATIC HYDROCARBONS WHOSE KETONES OR QUINONES HAVE BEEN DETECTED IN AMBIENT AEROSOLS

PNAH	Ketone or Quinone

Fluorene

Fluorenone

Anthracene

9,10-Anthraquinone

Phenalene

Phenalene-9-one

7H-Benz[*de*]anthracene

Benz[*de*]anthracene-7-one

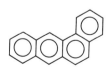

Benz[*a*]anthracene

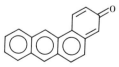

Benz[*a*]anthrone

**TABLE 3.7-2. POLYNUCLEAR AROMATIC HYDROCARBONS
WHOSE KETONES OR QUINONES HAVE BEEN DETECTED
IN AMBIENT AEROSOLS**

PNAH	Ketone or Quinone

Triphenylene

Triphenylene-1-one

Benzo[*a*]pyrene

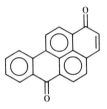

Benzo[*a*]pyrene-1,6-quinone

Benzo[*a*]pyrene

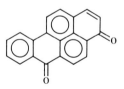

Benzo[*a*]pyrene-3,6-quinone

Benzo[*a*]pyrene

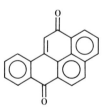

Benzo[*a*]pyrene-6,12-quinone

TABLE 3.7-2. POLYNUCLEAR AROMATIC HYDROCARBONS WHOSE KETONES OR QUINONES HAVE BEEN DETECTED IN AMBIENT AEROSOLS

PNAH	Ketone or Quinone
 Dibenzo[cd,jk]pyrene	Dibenzo[cd,jk]pyrene-6,12-quinone
Dibenzo[b,def]chrysene	Dibenzo[b,def]chrysene-7,14-quinone

TABLE 3.7 POLYNUCLEAR AROMATIC HYDROCARBONS
Emission and Detection

Species Number	Name	Chemical Formula	Emission		Detection	Ambient conc.
			Source	Ref.	Ref.	
3.7-1	Fluorene		auto coal comb. tobacco smoke	311(a) 361(a) 79(a),396(a)	86,200 209(a),362(a)	
3.7-2	Dihydrofluorene		coal comb.	361(a)	362(a)	
3.7-3	1-Methylfluorene		coal comb. tobacco smoke	361(a) 130*(a),396(a)	362(a)	
3.7-4	2-Methylfluorene		coal comb. refuse comb. tobacco smoke	361(a) 26(a) 130*(a)	362(a)	
3.7-5	9-Methylfluorene		coal comb. tobacco smoke	361(a) 396(a)	362(a)	
3.7-6	Dimethylfluorene		diesel	309(a)		
3.7-7	Trimethylfluorene		diesel	309(a)		

TABLE 3.7 POLYNUCLEAR AROMATIC HYDROCARBONS
Emission and Detection

Species Number	Name	Chemical Formula	Emission			Detection Ref.	Ambient conc.
			Source	Ref.			
3.7-8	Ethylfluorene	(structure with C₂H₅)	diesel	309(a)			
3.7-9	Isoamylfluorene	(structure)	auto	311(a)			
3.7-10	Benzo[a]fluorene	(structure)	coal comb. tobacco smoke	361(a) 130*(a),396(a)		209(a),343(a)	
3.7-11	Dihydrobenzo[a]-fluorene	(structure)	coal comb.	361(a)			
3.7-12	9-Methylbenzo[a]-fluorene	(structure)	tobacco smoke	396(a)			
3.7-13	Benzo[b]fluorene	(structure)	auto coal comb. tobacco smoke	98(a),284(a) 361(a) 130*(a),396(a)		209(a),408(a)	
3.7-14	Benzo[c]fluorene	(structure)	coal comb. tobacco smoke	361(a) 130*(a)		209(a),363(a)	

TABLE 3.7 POLYNUCLEAR AROMATIC HYDROCARBONS
Emission and Detection

Species Number	Name	Chemical Formula	Emission		Detection	Ambient conc.
			Source	Ref.	Ref.	
3.7-15	Dihydrobenzo[c]-fluorene				362(a),363(a)	
3.7-16	Dibenzo[a,l]-fluorene		tobacco smoke	396(a)		
3.7-17	Naphtho[1,2-a]-fluorene		tobacco smoke	396(a)		
3.7-18	Phenanthrene		aluminum mfr. auto coal comb. petroleum mfr. refuse comb. tobacco smoke	551*(a) 284(a),285(a) 56(a),285(a) 285(a),338*(a) 26(a),285(a) 79(a),218(a)	553* 2(a),553*(a)	4.9-340ng/m³
3.7-19	Dihydro-phenanthrene		coal comb.	361(a)	362(a)	
3.7-20	Tetrahydro-phenanthrene		auto	311(a)		

136

TABLE 3.7 POLYNUCLEAR AROMATIC HYDROCARBONS
Emission and Detection

| Species Number | Name | Chemical Formula | Emission | | Detection | Ambient conc. |
			Source	Ref.	Ref.	
3.7-21	Octahydro-phenanthrene		auto coal comb.	311(a) 361(a)		
3.7-22	Perhydro-phenanthrene		diesel	309(a)		
3.7-23	1-Methyl-phenanthrene		coal comb. coal tar mfr. tobacco smoke	361(a) 338(a) 130*(a)	362(a),363(a)	
3.7-24	2-Methyl-phenanthrene		tobacco smoke	130*(a)	408(a)	
3.7-25	3-Methyl-phenanthrene		tobacco smoke	130*(a)	408(a)	
3.7-26	9-Methyl-phenanthrene		coal tar mfr. tobacco smoke	338(a) 130*(a),396(a)	408(a)	
3.7-27	2,5-Dimethyl-phenanthrene		diesel tobacco smoke	309(a) 396(a)		

TABLE 3.7 POLYNUCLEAR AROMATIC HYDROCARBONS
Emission and Detection

Species Number	Name	Chemical Formula	Emission			Detection Ref.	Ambient conc.
			Source	Ref.			
3.7-28	4,5-Dimethylperhydro-phenanthrene		diesel	309(a)			
3.7-29	Ethyl-phenanthrene		coal comb.	361(a)		363(a)	
3.7-30	4-H-Cyclopenta-[def] phenanthrene		tobacco smoke	130*(a)		408(a)	
3.7-31	Methyl-4-H-cyclopenta[def]-phenanthrene		tobacco smoke	130*(a)		408(a)	
3.7-32	Ethyl-4-H-cyclopenta[def]-phenanthrene		tobacco smoke	130*(a)		408(a)	
3.7-33	Ethylmethyl-4-H-cyclopenta[def]-phenanthrene					408(a)	

TABLE 3.7 POLYNUCLEAR AROMATIC HYDROCARBONS
Emission and Detection

| Species Number | Name | Chemical Formula | Emission | | Detection | Ambient conc. |
			Source	Ref.	Ref.	
3.7-34	Benzo[c]-phenanthrene		asphalt mfr. coal comb. tobacco smoke	75(a),338(a) 361(a) 396(a)	213(a),408(a)	
3.7-35	Dihydrobenzo[c]-phenanthrene				362(a),363(a)	
3.7-36	Methylbenzo[c]-phenanthrene				408(a)	
3.7-37	Dimethylbenzo-phenanthrene		auto	311(a)		
3.7-38	Benzo[l]phenanthrene		auto	87(a),312*(a)		
3.7-39	Methylbenzo[l]-phenanthrene		coal comb.	361(a)		

TABLE 3.7 POLYNUCLEAR AROMATIC HYDROCARBONS
Emission and Detection

Species Number	Name	Chemical Formula	Emission		Detection	Ambient conc.
			Source	Ref.	Ref.	
3.7-40	Dibenzo[b,h]-phenanthrene		auto diesel	209(a) 209(a)		
3.7-41	Anthracene		aluminum mfr. asphalt mfr. auto coal comb. petroleum mfr. refuse comb. tobacco smoke	551*(a) 75*(a),285(a) 284(a),285(a) 56(a),285(a) 285(a) 285(a) 79(a),218(a)	41(a),209(a)	
3.7-42	Dihydroanthacene		coal comb.	361(a)	362(a)	
3.7-43	Octahydro-anthracene		coal comb.	361(a)		
3.7-44	1-Methylanthracene		aluminum mfr. coal comb. tobacco smoke	551*(a) 361(a) 130*(a)	362(a),363(a)	
3.7-45	2-Methylanthracene		coal tar mfr. tobacco smoke	338*(a) 130*(a),396(a)	408(a)	

140

TABLE 3.7 POLYNUCLEAR AROMATIC HYDROCARBONS
Emission and Detection

| Species Number | Name | Chemical Formula | Emission | | Detection | Ambient conc. |
			Source	Ref.	Ref.	
3.7-46	Ethylanthracene		coal comb.	361(a)	363(a)	
3.7-47	Benz[a]-anthracene		aluminum mfr. auto coal comb. diesel tobacco smoke	551*(a) 87(a),312(a) 56(a),361(a) 433*(a) 79(a),130*(a)	98(a),188*(a)	1.1-250ng/m³
3.7-48	Dihydrobenz[a]-anthracene				363(a)	
3.7-49	2-Methylbenz[a]-anthracene		coal comb. tobacco smoke	361(a) 130*(a)	362(a),363(a)	
3.7-50	3-Methylbenz[a]-anthracene		auto tobacco smoke	311(a) 130*(a),396(a)		
3.7-51	4-Methylbenz[a]-anthracene		tobacco smoke	130*(a)		
3.7-52	5-Methylbenz[a]-anthracene		tobacco smoke	130*(a)		

TABLE 3.7 POLYNUCLEAR AROMATIC HYDROCARBONS
Emission and Detection

Species Number	Name	Chemical Formula	Emission		Detection	Ambient conc.
			Source	Ref.	Ref.	
3.7-53	6-Methylbenz[a]-anthracene		tobacco smoke	130*(a)		
3.7-54	8-Methylbenz[a]-anthracene		tobacco smoke	130*(a)		
3.7-55	9-Methylbenz[a]-anthracene		tobacco smoke	130*(a)		
3.7-56	10-Methylbenz[a]-anthracene		tobacco smoke	130*(a)		
3.7-57	7,12-Dimethyl-benz[a]anthracene		asphalt mfr.	75(a)		
3.7-58	9,10-Dimethylbenz[a]-anthracene		coal tar mfr. tobacco smoke	338*(a) 396(a)		
3.7-59	5,6-Cyclopenteno-benz[a]anthracene		tobacco smoke	396(a)		

TABLE 3.7 POLYNUCLEAR AROMATIC HYDROCARBONS
Emission and Detection

Species Number	Name	Chemical Formula	Emission		Detection	Ambient conc.
			Source	Ref.	Ref.	
3.7-60	6,7-Cyclopenteno-benz[a]anthracene		tobacco smoke	396(a)	553* 362(a),553*(a)	30-806pg/m^3
3.7-61	Dibenz[a,c]-anthracene					
3.7-62	Dibenz[a,h]-anthracene		auto coal comb. tobacco smoke	312*(a) 361(a) 396(a)	213(a),362(a)	
3.7-63	Methyldibenz[a,h]-anthracene		coal comb.	361(a)	362(a),408(a)	

143

TABLE 3.7 POLYNUCLEAR AROMATIC HYDROCARBONS
Emission and Detection

Species Number	Name	Chemical Formula	Emission		Detection		Ambient conc.
			Source	Ref.	Ref.		
3.7-64	Fluoranthene		aluminum mfr. asphalt mfr.	551*(a) 285*(a)	553* 41(a),553*(a)		.16-4.1ng/m^3
			auto	311(a),312*(a)			
			coal comb.	56(a),285(a)			
			diesel	433*(a)			
			petroleum mfr.	285(a),338*(a)			
			refuse comb.	285(a),341*(a)			
			tobacco smoke	79(a),130*(a)			
3.7-65	Dihydrofluor anthene		coal comb.	361(a)	362(a)		
3.7-66	Octahydrofluor anthene		coal comb.	361(a)	361(a),362(a)		
3.7-67	1-Methylfluor- anthene		tobacco smoke	130*(a)	408(a)		
3.7-68	2-Methylfluor- anthene		aluminum mfr. tobacco smoke	551*(a) 130*(a)	408(a)		
3.7-69	3-Methylfluor- anthene		coal comb. tobacco smoke	361(a) 130*(a)	362(a),363(a)		

TABLE 3.7 POLYNUCLEAR AROMATIC HYDROCARBONS
Emission and Detection

Species Number	Name	Chemical Formula	Emission		Detection	Ambient conc.
			Source	Ref.	Ref.	
3.7-70	7-Methylfluor-anthene		tobacco smoke	130*(a)	408 (a)	
3.7-71	8-Methylfluoranthene		tobacco smoke	130*(a)	408 (a)	
3.7-72	Dimethyl fluor-anthene		tobacco smoke	396 (a)		
3.7-73	Benzo[a]fluor-anthene		tobacco smoke	406		
3.7-74	Benzo[b]fluor-anthene		aluminum mfr. auto diesel refuse comb.	551*(a) 209(a),312*(a) 209(a),433*(a) 26(a)	98(a),189(a)	
3.7-75	Benzo[cd]fluor-anthene		tobacco smoke	396 (a)	179*(a)	240ng/m^3

TABLE 3.7 POLYNUCLEAR AROMATIC HYDROCARBONS
Emission and Detection

Species Number	Name	Chemical Formula	Emission		Detection	Ambient conc.
			Source	Ref.	Ref.	
3.7-76	Benzo[ghi]fluoranthene		auto coal comb.	87(a),312*(a) 361(a)	209(a),362(a)	
3.7-77	Methylbenzo[ghi]fluoranthene				408(a)	
3.7-78	Dihydromethyl benzo[ghi]fluoranthene				362(a),570(a)	
3.7-79	Benzo[j]fluoranthene		auto coal comb. tobacco smoke	87(a),312*(a) 361(a) 130*(a),396(a)	98(a),362(a)	
3.7-80	Benzo[k]fluoranthene		auto diesel tobacco smoke	87(a),312*(a) 209(a),433*(a) 130*(a),396(a)	98(a),188*(a)	0.24-158ng/m^3

146

TABLE 3.7 POLYNUCLEAR AROMATIC HYDROCARBONS
Emission and Detection

Species Number	Name	Chemical Formula	Emission		Detection	Ambient conc.
			Source	Ref.	Ref.	
3.7-81	Benzol[mno]-fluoranthene		tobacco smoke	396(a)	179(a),182(a)	
3.7-82	Methylbenzol[mno]-fluoranthene				362(a)	
3.7-83	Indenol1,2,3-cd-fluoranthene		auto	312*(a)		
3.7-84	Methyldibenzol[b,k]-fluoranthene				363(a)	
3.7-85	Acephenan-thrylene		tobacco smoke	406(a)		

147

TABLE 3.7 POLYNUCLEAR AROMATIC HYDROCARBONS
Emission and Detection

Species Number	Name	Chemical Formula	Emission		Detection	Ambient conc.
			Source	Ref.	Ref.	
3.7-86	Pyrene		aluminum mfr. asphalt mfr. auto coal comb. diesel petroleum mfr. refuse comb. tobacco smoke	551*(a) 75*(a),285(a) 87(a),311(a) 56(a),285(a) 209(a),433*(a) 285(a) 26(a),285(a) 79(a),218(a)	41(a),209*(a)	.09-167ng/m^3
3.7-87	Dihydro-pyrene		coal comb.	361(a)	362(a)	
3.7-88	Octahydro-pyrene				361(a),362(a)	
3.7-89	1-Methylpyrene		aluminum mfr. coal comb. tobacco smoke	551*(a) 361(a) 79(a),130*(a)	209(a),362(a)	
3.7-90	2-Methylpyrene		tobacco smoke	130*(a),406(a)	408(a)	
3.7-91	4-Methylpyrene		tobacco smoke	130*(a),396(a)	408(a)	

148

TABLE 3.7 POLYNUCLEAR AROMATIC HYDROCARBONS
Emission and Detection

| Species Number | Name | Chemical Formula | Emission | | Detection | Ambient |
			Source	Ref.	Ref.	conc.
3.7-92	2,7-Dimethyl-pyrene				209(a)	
3.7-93	o-Phenylene pyrene		coal comb.	361(a)	363(a),570(a)	
3.7-94	Cyclopenta[cd]-pyrene		auto coal comb. diesel	340(a),556(a) 340(a) 340(a)		
3.7-95	3,4-Dihydro-cyclopenta[cd]-pyrene		coal comb.	340(a)		

149

TABLE 3.7 POLYNUCLEAR AROMATIC HYDROCARBONS
Emission and Detection

Species Number	Name	Chemical Formula	Emission		Detection	Ambient conc.
			Source	Ref.	Ref.	
3.7-96	Benzo[a]pyrene		aluminum mfr.	551(a),285(a)	553*	.004-80ng/m³
			asphalt mfr.	75(a),285(a)	41(a),113*(a)	
			auto	87(a),311(a)		
			coal comb.	56(a),285(a)		
			coking	145*(a)		
			diesel	209(a),433*(a)		
			petroleum mfr.	285(a)		
			refuse comb.	26(a),285(a)		
			rubber abrasion	424(a)		
			steel molds	201(a),277(a)		
			tobacco smoke	4(a),130*(a)		
3.7-97	Methylbenzo[a]pyrene		auto	87(a),556(a)		
			coal comb.	361(a)		
3.7-98	Dimethylbenzo[a]pyrene		coal comb.	361(a)		
3.7-99	Benzo[cd]pyrene		tobacco smoke	396(a)	113*(a),179*(a)	133-415ng/m³
3.7-100	3,4-Dihydrobenzo[cd]pyrene		tobacco smoke	396(a)		

TABLE 3.7 POLYNUCLEAR AROMATIC HYDROCARBONS
Emission and Detection

Species Number	Name	Chemical Formula	Emission		Detection	Ambient conc.
			Source	Ref.	Ref.	
3.7-101	Benzo[e]pyrene		aluminum mfr. asphalt mfr. auto coal comb. diesel petroleum mfr. refuse comb. tobacco smoke	551*(a) 75(a),285(a) 87(a),312*(a) 56(a),285(a) 433*(a) 285(a) 285(a) 130*(a)	98(a),209*(a)	0.54-23ng/m^3
3.7-102	Methylbenzo[e]-pyrene					
3.7-103	Dibenzo[a,b]-pyrene		auto	284(a)		
3.7-104	Dibenzo[a,e]-pyrene		auto	209(a)	189(a),375(a)	
3.7-105	Methyldibenzo[a,e]-pyrene				363(a)	

151

TABLE 3.7 POLYNUCLEAR AROMATIC HYDROCARBONS
Emission and Detection

Species Number	Name	Chemical Formula	Emission		Detection Ref.	Ambient conc.
			Source	Ref.		
3.7-106	Dibenzo[cd,jk]-pyrene		auto coal comb. diesel petroleum mfr. refuse comb. tobacco smoke	209(a),312*(a) 56(a),285(a) 209(a) 285(a) 285(a) 79(a),130*(a)	98(a),189(a)	
3.7-107	Dibenzo[cd,el]-pyrene		tobacco smoke	396(a)	375(a)	
3.7-108	Dibenzo[e,l]-pyrene				98(a)	
3.7-109	Indeno[1,2,3-cd]-pyrene		auto diesel refuse comb.	312*(a),360(a) 433*(a) 341*(a)	98(a)	
3.7-110	Naphtho[1,2,3-cd]-pyrene		tobacco smoke	396(a)		

TABLE 3.7 POLYNUCLEAR AROMATIC HYDROCARBONS
Emission and Detection

Species Number	Name	Chemical Formula	Emission		Detection Ref.	Ambient conc.
			Source	Ref.		
3.7-111	Chrysene		aluminum mfr. auto diesel refuse comb. tobacco smoke	551*(a) 87(a),311(a) 433*(a) 26(a) 130*(a),396(a)	188*(a),373*(a)	1.2-4.8ng/m^3
3.7-112	Dihydrochrysene				363(a)	
3.7-113	Hexahydrochrysene				363(a),570(a)	
3.7-114	1-Methylchrysene		coal comb. tobacco smoke	361(a) 130*(a),396(a)	362(a),363(a)	
3.7-115	2-Methylchrysene		tobacco smoke	130*(a)	408(a)	
3.7-116	3-Methylchrysene		tobacco smoke	130*(a)	408(a)	
3.7-117	5-Methylchrysene		tobacco smoke	130*(a)		

TABLE 3.7 POLYNUCLEAR AROMATIC HYDROCARBONS
Emission and Detection

Species Number	Name	Chemical Formula	Emission		Detection	Ambient conc.
			Source	Ref.	Ref.	
3.7-118	6-Methylchrysene		tobacco smoke	130*(a)	408(a)	
3.7-119	Dimethylchrysene	(CH₃)₂	tobacco smoke	396(a)	362(a),570(a)	
3.7-120	Dibenzo[b,def]-chrysene		asphalt mfr.	75(a)	98(a),189(a)	
3.7-121	3-Methyl-cholanthrene		asphalt mfr.	75(a)	362(a)	
3.7-122	Naphthacene		auto	209(a)		
3.7-123	Benzo[a]-naphthacene		tobacco smoke	396(a)		

TABLE 3.7 POLYNUCLEAR AROMATIC HYDROCARBONS
Emission and Detection

Species Number	Name	Chemical Formula	Emission		Detection	Ambient conc.
			Source	Ref.	Ref.	
3.7-124	Dibenzo[a,c]-naphthacene		tobacco smoke	396(a)		
3.7-125	Dibenzo[a,j]-naphthacene		tobacco smoke	396(a)		
3.7-126	Dibenzo[a,l]-naphthacene		auto diesel	209(a) 209(a)	209(a)	
3.7-127	Dibenzo[de,qr]-naphthacene		tobacco smoke	396(a)		
3.7-128	Naphtho[2,1,8-qra]-naphthacene				189(a)	
3.7-129	Picene				363(a)	

TABLE 3.7 POLYNUCLEAR AROMATIC HYDROCARBONS
Emission and Detection

Species Number	Name	Chemical Formula	Emission		Detection	Ambient conc.
			Source	Ref.	Ref.	
3.7-130	Perylene		auto coal comb. petroleum mfr. refuse comb. tobacco smoke	311(a),312*(a) 56(a),285(a) 285(a) 285(a),341(a) 130*(a),396(a)	98(a),209*(a)	.034-31ng/m^3
3.7-131	Benzo[ghi]-perylene		asphalt mfr. auto coal comb. diesel petroleum mfr. refuse comb. tobacco smoke	285(a) 87(a),311(a) 56(a),285(a) 209(a) 285(a) 285(a),341(a) 130*(a),396(a)	98(a),188*(a)	2.0-125ng/m^3
3.7-132	Methylbenzo[ghi]-perylene		auto coal comb.	556(a) 361(a)		
3.7-133	Dibenzo[b,pqr]-perylene		auto	209(a)		

TABLE 3.7 POLYNUCLEAR AROMATIC HYDROCARBONS
Emission and Detection

Species Number	Name	Chemical Formula	Emission		Detection Ref.	Ambient conc.
			Source	Ref.		
3.7-134	Benzo[rsd]-pentaphene		asphalt mfr.	75(a)	98(a),189(a)	
3.7-135	Pentacene				375(a)	
3.7-136	Coronene		auto coal comb. petroleum mfr. refuse comb. tobacco smoke	311(a),312*(a) 56(a),285(a) 285(a) 285(a),341(a) 79(a),396(a)	98(a),188*(a)	0.8-27ng/m^3

4

CARBONYL COMPOUNDS

4.0. INTRODUCTION

Hundreds of organic compounds are included in the trace gases present in the lower atmosphere. Of these, only the aliphatic aldehydes and ketones absorb the available solar radiation ($\lambda > 290$ nm) and dissociate to produce free radical fragments (1231). This property gives them a central role in many atmospheric chemical processes, since less than a dozen other atmospheric compounds (primarily the oxides and acids of nitrogen) photodissociate at tropospheric wavelengths. The molecular fragments produced by photodissociation control most atmospheric chemical processes; in the absence of the carbonyl compounds, atmospheric chemistry would be much simpler, better understood, and much less interesting.

Aldehydes and ketones are emitted from a wide variety of anthropogenic and natural sources. In addition, they are the first stable products in the atmospheric oxidation reactions of hydrocarbons. Typical chains for such processes are

$$RCH_3 \xrightarrow{HO\cdot} RCH_2\cdot \xrightarrow{O_2} RCH_2O_2\cdot \xrightarrow{NO} RCH_2O\cdot \xrightarrow{O_2} RCHO$$

$$RCH_2R' \xrightarrow{HO\cdot} R\dot{C}HR' \xrightarrow{O_2} RC(\dot{O}_2)HR' \xrightarrow{NO} RC(\dot{O})HR' \xrightarrow{O_2} RC(O)R'$$

The nonaliphatic carbonyl compounds do not photodissociate in the troposphere; their atmospheric chemistry resembles that of esters and alcohols in being primarily that of increasing oxidation of the initial compounds rather than that of the production of reactive fragments.

4.1. ALIPHATIC ALDEHYDES

4.1.1. Identified Compounds

The aliphatic aldehydes emitted into or detected in the air comprise a group of more than twenty five compounds. The n-aldehydes from C_1 to C_{12} plus C_{14} are included, as are a few of the common isomers. Nine difunctional compounds of this group have been detected in aerosols: the C_5, C_6, and C_7 dialdehydes, hydroxy aldehydes, and aldehydic acids.

The aliphatic aldehydes are common products of combustion and many have been found in motor vehicle exhaust. In addition, the C_1 to C_4 compounds are often the most common oxygenated organic species in a great variety of industrial processes. Aldehydes are common products of a variety of microbial and vegetative processes. A low olfactory threshold is a common aldehydic property (1220,1257) and many of the aldehydes are thus readily detectable in low concentrations. Those with low carbon numbers are acrid compounds that are significant eye irritants; those with high carbon numbers are generally regarded as more pleasant and are sometimes used as perfume and flavoring constituents.

4.1.2. Ambient Concentrations

Despite their apparent ubiquity in the lower atmosphere, the ambient concentrations of the aliphatic aldehydes are small. Formaldehyde, the most common, has occasionally been detected at concentrations above 100 ppb, but urban concentrations of 10 ppb (1147) and nonurban concentrations of 1 ppb (23) appear to be more common. Ethanal concentrations, when measured, have been in the range of $1-10$ ppb. Although propanal, *n*-butanal, isobutanal, and octanal have also been detected in ambient air, no concentration studies have been reported.

4.1.3. Chemistry

The aliphatic aldehydes are important compounds in atmospheric chemistry for two distinct reasons. First, they (especially formaldehyde) are common products of organic chemical reaction chains. Second, the carbonyl group absorbs solar photons at near ultraviolet wavelengths and the activated compound readily dissociates. Although stable compounds can result from formaldehyde, the free radical flux from photolysis of the alkyl aldehydes is substantial. Reactions such as the following occur:

$$HCHO \xrightarrow{h\nu} H\cdot + CHO\cdot$$

$$HCHO \xrightarrow{h\nu} H_2 + CO$$

$$CH_3CHO \xrightarrow{h\nu} CH_3\cdot + CHO\cdot$$

$$CH_3CH_2CHO \xrightarrow{h\nu} CH_3CH_2\cdot + CHO\cdot.$$

The hydrogen atom arising from formaldehyde photolysis readily forms the hydroperoxyl radical by reaction with oxygen

$$H\cdot + O_2 \xrightarrow{M} HO_2\cdot.$$

This reaction is the principal source of odd hydrogen $(HO\cdot, HO_2\cdot)$ radicals in the urban atmosphere (1007). The exact source strength is uncertain, however, because of uncertainty in the branching ratio in the atmosphere for the two formaldehyde dissociation processes (1232). The alkyl radicals $(C_xH_{2x+1}\cdot)$ react promptly to form alkylperoxy radicals

$$C_xH_{2x+1}\cdot + O_2 \xrightarrow{M} C_xH_{2x+1}O_2\cdot$$

which are important atmospheric oxidizing species.

The fate of the formyl $(CHO\cdot)$ radical in the atmosphere is presently somewhat uncertain. The reaction with O_2 to form an odd hydrogen radical

$$CHO\cdot + O_2 \xrightarrow{M} CO + HO_2\cdot$$

is apparently favored (1170) over three-body addition (but see 1148):

$$CHO\cdot + O_2 \xrightarrow{M} CHO(O_2)\cdot.$$

In the latter case, if the peroxyl radical is formed it will probably participate in the following reactions:

$$CHO(O_2)\cdot + NO \rightarrow CHO(O)\cdot + NO_2$$

$$CHO(O)\cdot + O_2 \rightarrow CO_2 + HO_2\cdot.$$

In either case, the formyl radical eventually produces an $HO_2\cdot$ radical. The first sequence is a chemical source of carbon monoxide, and the latter a source of carbon dioxide. Resolution of the reaction chains of the formyl radical thus has direct impact on studies of the chemical cycles of the oxides of carbon in the atmosphere.

TABLE 4.1 ALIPHATIC ALDEHYDES
Emission and Detection

| Species Number | Name | Chemical Formula | Emission | | Detection | Ambient |
			Source	Ref.	Ref.	conc.
4.1-1	Formaldehyde	HCHO	animal waste	25,80	23*,89	1-160ppb
			auto	15*,217*		
			building resin	9*,270*		
			casting resin	97*		
			charcoal mfr.	58*		
			coffee mfr.	133*		
			diesel	1,330		
			forest fire	354		
			HCHO mfr.	535*		
			lacquer mfr.	427,428		
			litho. coater	133*,535*		
			microbes	302		
			petroleum mfr.	221*,390		
			phthalic acid mfr.	535*		
			plant volatile	552e		
			plastics comb.	354		
			printing	535*		
			refuse comb.	571		
			spray painter	133*		
			tobacco smoke	4,554*		
			turbine	594		

TABLE 4.1 ALIPHATIC ALDEHYDES
Emission and Detection

Species Number	Name	Chemical Formula	Emission		Detection	Ambient conc.
			Source	Ref.	Ref.	
4.1-2	Acetaldehyde	CH_3CHO	animal waste	18,25	171*,550*	1.5-9.6ppb
			auto	15*,217*	399(a)	
			casting resin	97*		
			charcoal mfr.	58*		
			coffee mfr.	133*		
			diesel	1,176		
			fish meal mfr.	110,255*		
			forest fire	183,354		
			HCHO mfr.	535*		
			insects	471		
			litho. coater	133*,535		
			microbes	302		
			petroleum mfr.	111,221*		
			phthalic acid mfr.	535*		
			plant volatile	552e		
			plastics comb.	354		
			printing	535*		
			refuse comb.	439		
			sewage tmt.	422*		
			spray painter	133*		
			tobacco smoke	446,448		
			tobacco smoke	396(a),554(a)		
			turbine	273,414		
			volcano	234		
4.1-3	Hydroxyethanal	$HOCH_2CHO$	turbine	414		
4.1-4	Glyoxylic acid	$HOOCCHO$	tobacco smoke	396(a)		

162

TABLE 4.1 ALIPHATIC ALDEHYDES
Emission and Detection

| Species Number | Name | Chemical Formula | Emission | | Detection | Ambient |
			Source	Ref.	Ref.	conc.
4.1-5	Propanal	CHO	animal waste	25,80	358,380	
			auto	15*,217*		
			coffee mfr.	133		
			diesel	176,381		
			fish meal mfr.	254*		
			forest fire	376,535*		
			insects	471		
			litho. coater	33,535*		
			microbes	302		
			petroleum mfr.	221*,390		
			plant volatile	552e		
			rendering plant	102		
			sewage tmt.	422*		
			spray painter	133		
			tobacco smoke	396(a)		
			turbine	273,414		
			volcano	234		
4.1-6	2-Oxopropanal	CHO	tobacco smoke	396		

TABLE 4.1 ALIPHATIC ALDEHYDES
Emission and Detection

| Species Number | Name | Chemical Formula | Emission | | Detection | Ambient |
			Source	Ref.	Ref.	conc.
4.1-7	*n*-Butanal	⌬CHO	animal waste	18	165,582	
			auto	132,284		
			coffee mfr.	133		
			diesel	176,330		
			fish meal mfr.	254*		
			forest fire	376		
			lacquer mfr.	427		
			litho. coater	133		
			microbes	302		
			petroleum mfr.	390		
			plant volatile	552e		
			spray painter	133		
			tobacco smoke	396(a)		
			turbine	414		
4.1-8	Isobutanal	⌬CHO	animal waste	25,80	162,582	
			auto	55,132		
			fish meal mfr.	254*		
			forest fire	376		
			industrial	133		
			insects	471		
			microbes	302		
			plant volatile	552e		
			tobacco smoke	396(a)		
			turbine	414		
4.1-9	2-Methylbutanal	⌬CHO	microbes	302		
4.1-10	2-Ethylbutanal	⌬CHO	auto	519		

TABLE 4.1 ALIPHATIC ALDEHYDES
Emission and Detection

Species Number	Name	Chemical Formula	Emission		Detection	Ambient conc.
			Source	Ref.	Ref.	
4.1-11	*n*-Pentanal	CHO	animal waste	25,80		
			auto	132,519		
			diesel	176,381		
			microbes	302		
			plant volatile	552e		
			turbine	414		
4.1-12	Isopentanal	CHO	auto	67*,132		
			microbes	302		
			plant volatile	552e		
			turbine	414		
4.1-13	Neopentanal	CHO	auto	284,313		
			turbine	414		
4.1-14	1,5-Pentanedial	CHO CHO			214*(a)	$30\text{-}300\text{ng}/\text{m}^3$
4.1-15	5-Hydroxypentanal	HO CHO			214*(a)	$100\text{-}310\mu\text{g}/\text{m}^3$
4.1-16	5-Oxopentanoic acid	O COOH			214(a)*,532(a)	$.40\text{-}1.36\mu\text{g}/\text{m}^3$
4.1-17	Hexanal	CHO	diesel	176,330		
			microbes	302		
			plant volatile	503e,552e		
			turbine	273,414		
4.1-18	1,6-Hexanedial	CHO CHO			214*(a)	$240\text{ng}/\text{m}^3$
4.1-19	6-Hydroxyhexanal	HO CHO			214*(a)	$30\text{-}400\text{ng}/\text{m}^3$

TABLE 4.1 ALIPHATIC ALDEHYDES
Emission and Detection

Species Number	Name	Chemical Formula	Emission		Detection Ref.	Ambient conc.
			Source	Ref.		
4.1-20	Galactose	(structure)	tobacco smoke	542(a)	214*(a),532(a)	.06-2.54μg/m^3
4.1-21	6-Oxohexanoic acid	(structure) COOH				
4.1-22	Heptanal	(structure) CHO	animal waste diesel microbes plant volatile turbine	25,80 176,381 302 514e,552e 414		
4.1-23	1,7-Heptanedial	(structure) CHO			214*(a)	70-540ng/m^3
4.1-24	7-Hydroxyheptanal	(structure) CHO			214*(a)	90-180ng/m^3
4.1-25	7-Oxoheptanoic acid	(structure) COOH			214*(a),532(a)	50-820ng/m^3
4.1-26	Octanal	$CH_3(CH_2)_6CHO$	animal waste diesel microbes plant volatile	25,80 176,381 302 503e,552e	86	
4.1-27	Nonanal	$CH_3(CH_2)_7CHO$	microbes plant volatile	302 503e,552e		
4.1-28	Decanal	$CH_3(CH_2)_8CHO$	animal waste auto plant volatile	25,80 67* 504e,552e		
4.1-29	Undecanal	$CH_3(CH_2)_9CHO$	plant volatile	504e,552e		

TABLE 4.1 ALIPHATIC ALDEHYDES
Emission and Detection

Species Number	Name	Chemical Formula	Emission		Detection Ref.	Ambient conc.
			Source	Ref.		
4.1-30	Dodecanal	$CH_3(CH_2)_{10}CHO$	plant volatile	504e,531		
4.1-31	Tetradecanal	$CH_3(CH_2)_{12}CHO$	plant volatile	504e		

TABLE 4.1 ALIPHATIC ALDEHYDES
Reactions and Products

Species Number	Reactant	Chemical reactions			Products	Ref.	Remarks
		k	Ref.	Lifetime			
4.1-1	$h\nu$	8.0×10^{-5}	1125	1.3×10^4	H_2CO	1232	B
	$HO\cdot$	1.3×10^{-11}	1170				
	$HO_2\cdot$	4.9×10^{-15}	1152				
	O	1.5×10^{-13}	1170		$HO\cdot$	1170	
	O_3	$\leqslant 2.1\times10^{-24}$	1171				
4.1-2	$h\nu$	2.3×10^{-5}	1043	4.4×10^4			B
	$HO\cdot$	$(1.5\pm0.2)\times10^{-11}$	1168				
	$NO_3\cdot$	$(1.2\pm0.3)\times10^{-15}$	1046		HNO_3	1046	
	O	3.1×10^{-13}	1172		$CH_3C(O)O_2NO_2$	1233	
	NO_2	2.0×10^{-25}	1169				
4.1-5	$h\nu$	3.2×10^{-5}	1022	3.1×10^4	$CH_3CH_2O_2H, CO$	1233	B
	$HO\cdot$	$(4.8\pm0.5)\times10^{-11}$	1163				
	O	$(2.4\pm0.8)\times10^{-13}$	1044		$CH_3CH_2C(O)O_2NO_2$	1233	
4.1-6							B
4.1-7	$h\nu$	2.7×10^{-5}	1022	3.7×10^4	$H_2C{=}CH_2, CH_3CHO$	1231	B
	O	$(9.5\pm0.2)\times10^{-13}$	1251				
4.1-8	O	$(1.2\pm0.1)\times10^{-12}$	1251	3.3×10^7			
4.1-9	$h\nu$				$H_2C{=}CH_2$, $\diagup\!\diagdown$CHO	1231	
4.1-11	$h\nu$				CH_3CHO, $\diagdown\!\diagup$	1231	B

168

TABLE 4.1 ALIPHATIC ALDEHYDES
Reactions and Products

Species Number	Chemical reactions						Remarks
	Reactant	Ref.	k	Lifetime	Products	Ref.	
4.1-12	hν				CH_3CHO, ⬱	1231	
4.1-14							B
4.1-16							B
4.1-21							B

4.2. OLEFINIC ALDEHYDES

4.2.1. Identified Compounds

This group is comprised of 15 compounds of widely varying molecular size, source diversity, and atmospheric importance. Acrolein is an important industrial chemical and is produced by a variety of combustion processes; it is apparently a common, although seldom measured, urban atmospheric constituent, and is one of the more potent lachrymators in photochemical smog (439). Combustion is also responsible for several other small olefinic aldehydes. Some of the larger compounds are terpene derivatives emitted by vegetation.

A subgroup of four compounds [4.2-6 to 4.2-8, 4.2-11] have no known physical source. Their structures are typical of those expected from atmospheric reactions of oxygenated organic compounds, however (see Chapter 5).

4.2.2. Ambient Concentrations

Acrolein is the only olefinic aldehyde whose concentration in the atmosphere has been determined. The measured values were several ppb, or about 15% that of formaldehyde. This is sufficient to render acrolein a significant, if not major, atmospheric oxygenated organic compound.

4.2.3. Chemistry

The rate constants for the reactions of atomic oxygen with acrolein and crotonaldehyde have been measured and found to be large. A product analysis suggests that addition occurs at the double bond. Addition would also be expected for reaction with the hydroxyl radical. Using acrolein chemistry to typify that of the olefinic aldehydes, sequences of the following type are anticipated:

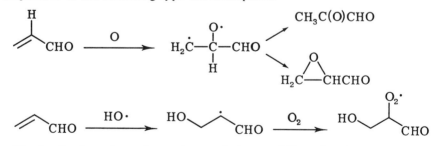

with the final products in each case being multifunctional oxygenated compounds.

The reactions of ozone with olefinic aldehydes have not been studied, but should proceed in a manner similar to ozone−olefin sequences (Section 3.2.3.).

TABLE 4.2 OLEFINIC ALDEHYDES
Emission and Detection

Species Number	Name	Chemical Formula	Emission Source	Emission Ref.	Detection Ref.	Ambient conc.
4.2-1	Acrolein	CHO	auto	15*,243*	5*,411	1-13ppb
			coffee mfr.	133*		
			diesel	314*,432*		
			fish oil mfr.	110,241*		
			forest fire	354,376		
			lacquer mfr.	8,427		
			litho. coater	58*,133*		
			plastics comb.	46		
			spray painting	133*		
			syn. rubber mfr.	58*		
			tobacco smoke	4,448		
			tobacco smoke	396(a),554(a)		
			turbine	273,414		
4.2-2	Methacrolein	CHO	auto	67*,519		
			volcano	234		
4.2-3	Crotonaldehyde	CHO	auto	15*,217*		
			diesel	1,314*		
			tobacco smoke	396(a),554(a)		
			turbine	414		
			volcano	234		
4.2-4	Tiglaldehyde	CHO	auto	217*		
4.2-5	Mesityl oxide	CHO	solvent	134		
4.2-6	2-Formyl-4-hydroxy-2-butenoic acid	HO—COOH CHO			532(a)	

TABLE 4.2 OLEFINIC ALDEHYDES
Emission and Detection

Species Number	Name	Chemical Formula	Emission		Detection	Ambient conc.
			Source	Ref.	Ref.	
4.2-7	2-Methyl-4-penten-1,5-dial				532(a)	
4.2-8	2-Formyl-5-hydroxy-2-pentenoic acid				532(a)	
4.2-9	1-Hexen-2-al		plant volatile	512e		
4.2-10	*trans*-2-Hexenal		microbes	302		
			plant volatile	552e		
4.2-11	1-Oxo-2-methyl-6-hydroxy-2,4-hexanediene				532(a)	
4.2-12	Citral		plant volatile	498,505e		
4.2-13	Citronellal		plant volatile	498,506e		
4.2-14	Hydroxycitronellal		plant volatile	498		
4.2-15	2,6-Nonadien-1-al		plant volatile	531		

TABLE 4.2 OLEFINIC ALDEHYDES
Reactions and Products

| Species Number | Chemical reactions | | | | | | Remarks |
	Reactant	k	Ref.	Lifetime	Products	Ref.	
4.2-1	O	$(2.7\pm0.9)\times10^{-13}$	1044	1.5×10^{8}	$C_2H_2, H_2C{=}CH_2$	1044	B
4.2-2							B
4.2-3	O	8.3×10^{-13}	1045	4.8×10^{7}			B

4.3. CYCLIC ALDEHYDES

4.3.1. Identified Compounds
Four of the five compounds in this group are furan derivatives. Furfural, probably the most common, is emitted from both anthropogenic and natural sources, as is its methyl derivative. No source strength data have been acquired for any of these compounds.

4.3.2. Ambient Concentrations
No data are available concerning the atmospheric concentrations of these compounds.

4.3.3. Chemistry
The chemistry of the cyclic aldehydes has not been studied in the laboratory. It is reasonable to anticipate that it will have reaction analogues with the cyclic hydrocarbons, however, in addition to reactions involving the formyl group. Examples of likely processes are

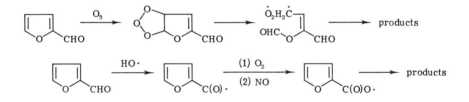

In either case, ring cleavage is likely and the eventual products will be the multifunctional straight chain compounds often found in atmospheric aerosols (1063).

TABLE 4.3 CYCLIC ALDEHYDES
Emission and Detection

Species Number	Name	Chemical Formula	Emission		Detection Ref.	Ambient conc.
			Source	Ref.		
4.3-1	Furfural		brewing	144		
			diesel	137,176		
			forest fire	354		
			microbes	302		
			plant volatile	552e		
			tobacco smoke	396,554		
			tobacco smoke	579(a)		
4.3-2	5-Methylfurfural		brewing	144		
			diesel	137		
			forest fire	354		
4.3-3	Furfural alcohol		diesel	137		
4.3-4	5-Hydroxymethyl-furfural		tobacco smoke	554(a)		
4.3-5	Safranal		plant volatile	531		

175

4.4. AROMATIC ALDEHYDES

4.4.1. Identified Compounds

Many aldehyde derivatives of the alkyl benzenes and naphthalenes are known to be emitted during combustion processes; Table 4.4 lists more than a dozen such compounds. Aromatic aldehydes are often common volatile compounds in vegetation; more than a dozen vegetative emittants are listed. Although atmospheric measurements are sparse, it appears that aromatic aldehydes may thus normally be present in both urban and remote areas.

4.4.2. Ambient Concentrations

4-Methylbenzaldehyde is the only aromatic aldehyde whose ambient concentration has been determined; the values derived were much less than 1 ppb.

4.4.3. Chemistry

The products of the reactions of atomic oxygen with several of the benzaldehydes in the presence of NO_2 have been found to include peroxybenzoylnitrates. These compounds presumably form by the following chemical sequence:

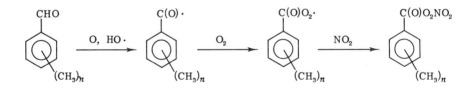

A similar chain would be anticipated with the hydroxyl radical as the abstracting species in place of atomic oxygen. Addition to the ring is also possible, however; the result would be the family of hydroxybenzaldehydes.

Many of the vegetative emittants are more highly substituted than are the anthropogenic aromatic aldehydes. Their reactions will parallel those described above, with appropriate adjustments for steric and functional group factors and bond energies.

TABLE 4.4 AROMATIC ALDEHYDES
Emission and Detection

Species Number	Name	Chemical Formula	Emission		Detection		Ambient conc.
			Source	Ref.	Ref.		
4.4-1	Benzaldehyde	CHO (benzene ring)	auto diesel plant volatile refuse comb. tobacco smoke turbine	67*,217* 1,176 531,552e 26(a) 396(a) 273,414	86,358 532(a)		
4.4-2	2-Methylbenz-aldehyde	CHO (methylbenzene ring)	auto diesel	48,217* 137	86,525		
4.4-3	3-Methylbenz-aldehyde	CHO (methylbenzene ring)	auto	48,537	86,525		
4.4-4	4-Methylbenz-aldehyde	CHO (methylbenzene ring)	auto turbine	48,537 414	358,455*		50-280ppt
4.4-5	Dimethylbenz-aldehyde	CHO $(CH_3)_2$ (benzene ring)	auto diesel	537 138	86		
4.4-6	Ethylbenz-aldehyde	CHO C_2H_5 (benzene ring)	auto diesel	217* 176			
4.4-7	Ethyltetramethyl-benzaldehyde	OHC C_2H_5 $(CH_3)_4$ (benzene ring)	diesel	138			
4.4-8	Propylbenz-aldehyde	CHO C_3H_8 (benzene ring)	diesel	138			
4.4-9	Cuminaldehyde	CHO (isopropylbenzene ring)	plant volatile	507e,531			
4.4-10	Dihydrocumin-aldehyde	CHO (isopropyl cyclohexene ring)	plant volatile	531			

TABLE 4.4 AROMATIC ALDEHYDES
Emission and Detection

Species Number	Name	Chemical Formula	Emission		Detection	Ambient conc.
			Source	Ref.	Ref.	
4.4-11	Butylbenz-aldehyde		diesel	138		
4.4-12	Phenylacet-aldehyde				358	
4.4-13	β-Phenylpropanal		plant volatile	510e		
4.4-14	Cinnamaldehyde		diesel plant volatile	138 326,533		
4.4-15	Salicylaldehyde		auto	217*		
4.4-16	p-Hydroxybenz-aldehyde		plant volatile	552e		
4.4-17	Hydroxyvallyl-benzaldehyde		diesel	138		
4.4-18	Anisaldehyde		diesel microbes plant volatile	138 302 510e,533		
4.4-19	o-Methoxycinnam-aldehyde		plant volatile	514e		

TABLE 4.4 AROMATIC ALDEHYDES
Emission and Detection

Species Number	Name	Chemical Formula	Emission Source	Emission Ref.	Detection Ref.	Ambient conc.
4.4-20	Phthalaldehyde	(structure)	diesel	138		
4.4-21	Terephthalaldehyde	(structure)	auto	311(a)		
4.4-22	4-Formylbenzoic acid	(structure)	auto	309(a)		
4.4-23	Piperonal	(structure)	diesel / plant volatile	138 / 510e		
4.4-24	Vanillin	(structure)	diesel / plant volatile / wood comb. / wood pulping	138 / 501,531 / 344*(a) / 219,267		
4.4-25	Veratraldehyde	(structure)	plant volatile	514e		
4.4-26	Coniferlaldehyde	(structure)	plant volatile / wood comb.	531 / 344(a)		
4.4-27	Syringicaldehyde	(structure)	wood comb.	344(a)		
4.4-28	Sinapylaldehyde	(structure)	wood comb.	344(a)		

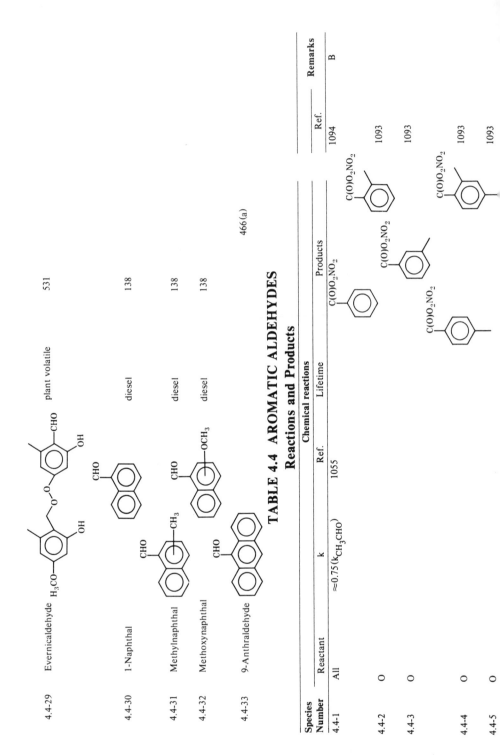

4.4-29 Evernicaldehyde — plant volatile — 531

4.4-30 1-Naphthal — diesel — 138

4.4-31 Methylnaphthal — diesel — 138

4.4-32 Methoxynaphthal — diesel — 138

4.4-33 9-Anthraldehyde — 466(a)

TABLE 4.4 AROMATIC ALDEHYDES
Reactions and Products

Species Number	Reactant	Chemical reactions					Remarks
		k	Ref.	Lifetime	Products	Ref.	
4.4-1	All	$\approx 0.75(k_{CH_3CHO})$	1055		$C(O)O_2NO_2$	1094	B
4.4-2	O				$C(O)O_2NO_2$	1093	
4.4-3	O				$C(O)O_2NO_2$	1093	
4.4-4	O				$C(O)O_2NO_2$	1093	
4.4-5	O				$C(O)O_2NO_2$	1093	

4.5. ALIPHATIC KETONES

4.5.1. Identified Compounds

The aliphatic ketones are represented in atmospheric emissions by more than 25 different compounds. This includes the methyl ketones from C_3 to C_8 plus C_{11} and C_{16}, the ethyl ketones from C_5 to C_9, and dipropyl ketone. Several hydroxy and alkyl derivatives are also known. The principal atmospheric sources of the compounds are combustion (motor vehicles, forest fires, etc.), solvent evaporation, and plant volatilization, although numerous secondary sources have been identified.

4.5.2. Ambient Concentrations

Only four of the aliphatic ketones have been detected thus far in ambient air: acetone, butanone, biacetyl, and 4-methyl-2-pentanone. All have been found in the gas phase. Concentrations have been measured for the first two; they are both present in urban air at levels of several ppb.

4.5.3. Chemistry

The primary fate of aliphatic ketones in the troposphere is photodissociation. The cleavage occurs at the α position (1095, 1107), the weaker of the two bonds being favored (1231). Both radicals are expected to add molecular oxygen.

$$RC(O)R' \xrightarrow{h\nu} RC(O)\cdot + R'\cdot$$

$$RC(O)\cdot \xrightarrow{O_2} RC(O)O_2\cdot$$

$$R'\cdot \xrightarrow{O_2} R'O_2\cdot.$$

The subsequent chemistry of the peroxy radicals has been discussed in Section 3.1.3.

A secondary reaction chain for the aliphatic ketones involves initial hydrogen abstraction by either HO· or O, followed by peroxy formation and scission to produce lower aldehydes and oxides of carbon (439). For butanone, for example,

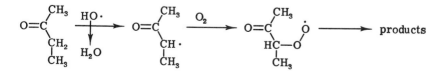

TABLE 4.5 ALIPHATIC KETONES
Emission and Detection

Species Number	Name	Chemical Formula	Emission		Detection	Ambient conc.
			Source	Ref.	Ref.	
4.5-1	Acetone		animal waste	436*	33*,306*	0.08-6.8ppb
			auto	15*,217*		
			chemical mfr.	566*		
			diesel	176		
			fish oil mfr.	110		
			forest fire	354,376		
			insects	471		
			microbes	302		
			petroleum stor.	58*		
			phthalic acid mfr.	535*		
			plant volatile	552e		
			plastics comb.	354		
			printing	535*,543*		
			refuse comb.	439		
			refuse comb.	26(a)		
			solvent	88,134		
			tobacco smoke	446,453		
			tobacco smoke	396(a)		
			turbine	414		
			volcano	234		
			wood pulping	19,267		
4.5-2	Hydroxypropanone		diesel	1		
4.5-3	Pyruvic acid		tobacco smoke	396(a)		

TABLE 4.5 ALIPHATIC KETONES
Emission and Detection

Species Number	Name	Chemical Formula	Emission Source	Emission Ref.	Detection Ref.	Ambient conc.
4.5-4	Butanone	(structure)	auto	15*,217*	358,469	
			forest fire	354,535*		
			industrial	133		
			insects	471		
			printing	535*,543*		
			solvent	88,134		
			tobacco smoke	396,446		
			turbine	414		
			volcano	234		
			wood pulping	19,267		
4.5-5	Reductic acid	(structure with HO, COOH)	tobacco smoke	396(a)		
4.5-6	Methylbutanone	(structure)	auto	284		
			wood pulping	19,267		
4.5-7	Dimethylbutanone	(structure)	auto	284		
			solvent	134		
4.5-8	3-Hydroxy-butanone	(structure with OH)	animal waste	25,438		
			microbes	302		
4.5-9	Biacetyl	(structure)	animal waste	25	352,358	
			forest fire	439		
			plant volatile	552e		
			tobacco smoke	396		
4.5-10	2-Pentanone	(structure)	auto	271	582	
			forest fire	354		
			solvent	134		
			turbine	414		
			wood pulping	19		

TABLE 4.5 ALIPHATIC KETONES
Emission and Detection

Species Number	Name	Chemical Formula	Emission Source	Emission Ref.	Detection Ref.	Ambient conc.
4.5-11	4-Methyl-2-pentanone		auto landfill plant volatile solvent turbine wood pulping	284 365 552e 315,442 414 267,545	358	
4.5-12	4,4-Dimethyl-2-pentanone		turbine	414		
4.5-13	Diacetone alcohol		solvent	88,134		
4.5-14	4-Hydroxy-2-pentanone		diesel	138		
4.5-15	Levulinic acid		tobacco smoke	396(a)		
4.5-16	2,3-Pentanedione		forest fire tobacco smoke	354 396		
4.5-17	2-Oxopentanedioic acid		tobacco smoke	396(a)		
4.5-18	3-Pentanone		auto chemical mfr. tobacco smoke wood pulping	284 566* 396 267,545	566*	2ppb
4.5-19	2-Methyl-3-pentanone		chemical mfr. turbine	566* 414		
4.5-20	2-Hexanone		solvent	134		

TABLE 4.5 ALIPHATIC KETONES
Emission and Detection

Species Number	Name	Chemical Formula	Emission		Detection Ref.	Ambient conc.
			Source	Ref.		
4.5-21	3-Hexanone		turbine	414		
4.5-22	2,2,5-Trimethyl-hexane-3,4-dione		turbine	414		
4.5-23	2-Heptanone		plant volatile	514e,552e		
			wood pulping	545		
4.5-24	Methyl-2-heptanone		plant volatile	497		
4.5-25	3-Heptanone		diesel	414	566*	2ppb
4.5-26	4-Heptanone		diesel	138		
			solvent	134		
			tobacco smoke	396		
4.5-27	2,6-Dimethyl-4-heptanone		solvent	134		
4.5-28	2-Octanone		plant volatile	514e,531		
4.5-29	3-Octanone		microbes	302		
			plant volatile	509e,552e		
4.5-30	2-Nonanone		plant volatile	552e		
4.5-31	3-Nonanone				358	
4.5-32	2-Undecanone	$CH_3C(O)(CH_2)_8CH_3$	plant volatile	512e,531		

TABLE 4.5 ALIPHATIC KETONES
Emission and Detection

Species Number	Name	Chemical Formula	Emission		Detection	Ambient conc.
			Source	Ref.	Ref.	
4.5-33	2-Hexadecanone	$CH_3C(O)(CH_2)_{13}CH_3$	tobacco smoke	396		
4.5-34	Acetic anhydride		industrial	60		

TABLE 4.5 ALIPHATIC KETONES
Reactions and Products

Species Number	Reactant	Chemical reactions					Remarks
		k	Ref.	Lifetime	Products	Ref.	
4.5-1	hν	1.4×10^{-5}	1022	7.1×10^4	$CH_3C(O)O_2NO_2$	1109	B
	O	$(6.7\pm1.4)\times10^{-16}$	1047				
4.5-4	hν	1.4×10^{-5}	1022	7.1×10^4			B
	HO·	$(3.3\pm1.0)\times10^{-12}$	1033				
4.5-9	hν	1.7×10^{-4}	1022	5.9×10^3	$CH_3C(O)O_2NO_2$	1022	B
4.5-10							B
4.5-11	HO·	$(1.5\pm0.5)\times10^{-11}$	1033	1.4×10^5			
4.5-18							B
4.5-27	HO·	$(2.5\pm0.8)\times10^{-11}$	1033	8.7×10^4			

TABLE 4.6 OLEFINIC KETONES
Emission and Detection

Species Number	Name	Chemical Formula	Emission		Detection	Ambient conc.
			Source	Ref.	Ref.	
4.6-1	3-Butene-2-one		auto turbine	271,284 414	358t	
4.6-2	3-Methyl-3-butene-2-one		auto	271		
4.6-3	4-Methyl-3-pentene-2-one		auto	271		
4.6-4	5-Hexen-2-one		turbine	414		
4.6-5	6-Methyl-5-hepten-2-one		plant volatile	514e		
4.6-6	6,9-Pentadecadien-2-one		plant volatile	552e		

TABLE 4.6 OLEFINIC KETONES
Reactions and Products

Species Number	Reactant	Chemical reactions					Remarks
		k	Ref.	Lifetime	Products	Ref.	
4.6-1							B

The products will then follow the aliphatic aldehyde chains described in Section 4.1.3 and form peroxy radicals different in structure from but similar in effect to the peroxy radicals produced by aliphatic ketone photolysis.

4.6. OLEFINIC KETONES

4.6.1. Identified Compounds

Olefinic ketones have not commonly been found in the atmosphere. Of the six known compounds, four are products of internal combustion engines and two are emitted from vegetation.

4.6.2. Ambient Concentrations

No data are available concerning the atmospheric concentrations of these compounds.

4.6.3. Chemistry

Ketones containing one or more olefinic groups do not photodissociate in the troposphere (1231). No studies of their reactions with other atmospheric species have been reported, but by analogy with the olefinic hydrocarbons, the principal chemical loss mechanism is expected to be hydroxyl addition to a double bond. Prompt addition of molecular oxygen will follow, and reactions with NO and SO_2 will eventually produce multifunctional compounds of low vapor pressure. Thus, for 3-butene-2-one, the initial steps would be

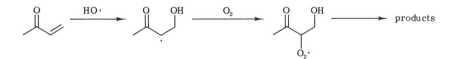

4.7. CYCLIC KETONES

4.7.1. Identified Compounds

Nearly 40 cyclic ketones are emitted into the atmosphere, mostly by vegetation. The compounds have pleasant odors and are widely used as fragrance materials. A few cyclic ketones are emitted from internal combustion engines. Several of these compounds have industrial uses (and hence sources).

4.7.2. Ambient Concentrations

No data are available concerning the atmospheric concentrations of these compounds.

4.7.3. Chemistry

Despite extensive studies of the photolysis of cyclic ketones in solution, gas phase studies at atmospheric temperatures are rare. The photoprocesses appear to be either collisional relaxation to the original compound or an isomer, or expulsion of CO with diradical formation (1231). In either case there is no evidence that the cyclic ketones are significant participants in atmospheric chemical processes. Hydrogen abstraction from the ring is the most probable chemical loss process; the sequence should parallel those of the cyclic hydrocarbons (Section 3.4.3.).

4.8. AROMATIC KETONES AND QUINONES

4.8.1. Identified Compounds

Aromatic ketones appear to be ubiquitous constituents of the atmosphere. More than 60 compounds in this group have known sources, the largest subgroup being derivatives of the indane produced by diesel engine combustion. Internal combustion engines are also the source (either directly or through oxidation) of the keto derivatives of polynuclear aromatic hydrocarbons (see Section 3.7.3.). A few aromatic ketones are emitted by vegetation.

4.8.2. Ambient Concentrations

Only acetophenone and 4-phenylbutanone have been detected in the gas phase in the ambient atmosphere; no concentration data are available. The polynuclear aromatic ketones have been found on several occasions as atmospheric aerosol constituents at very low concentrations (no higher than perhaps 20 ng/m^3 for all the compounds

TABLE 4.7 CYCLIC KETONES
Emission and Detection

Species Number	Name	Chemical Formula	Emission Source	Emission Ref.	Detection Ref.	Ambient conc.
4.7-1	Cyclopentanone		turbine	414		
4.7-2	1,2-Cyclopentanedione		plant volatile	552e		
4.7-3	Cyclohexanone		building resin solvent	9* 88,134		
4.7-4	1,3-Dimethyl-cyclohexan-5-one		diesel	138		
4.7-5	1,3,3-Trimethyl-cyclohexan-2-one		plant volatile	515e		
4.7-6	Menthone		plant volatile	516e,531		
4.7-7	Isomenthone		plant volatile	515e		
4.7-8	Fenchone		wood pulping	267,545		
4.7-9	Camphor		plant volatile	497e,514e		
4.7-10	Thujone		plant volatile	510e,514e		
4.7-11	Pulegone		plant volatile	510e,531		
4.7-12	Isophorone		solvent	88,134		

TABLE 4.7 CYCLIC KETONES
Emission and Detection

Species Number	Name	Chemical Formula	Emission		Detection	Ambient conc.
			Source	Ref.	Ref.	
4.7-13	Piperitone		plant volatile	531		
4.7-14	Perinaphthanone				188(a)	
4.7-15	Carvone		plant volatile	506e, 531		
4.7-16	Methylcyclo-hexanedione		diesel	138		
4.7-17	α-Ionone		plant volatile trees	513e, 531 440		
4.7-18	β-Ionone		plant volatile trees	513e, 531 440		
4.7-19	α-Irone		plant volatile trees	513e, 514e 440		
4.7-20	2-Methyl-cyclooctanone		turbine	414		
4.7-21	Cycloheptadec-9-ene-1-one	$C(O)(CH_2)_7-CH=CH(CH_2)_6CH_2$	animal secretion	510e		

TABLE 4.7 CYCLIC KETONES
Emission and Detection

| Species Number | Name | Chemical Formula | Emission | | Detection | Ambient |
			Source	Ref.	Ref.	conc.
4.7-22	Cedrene		plant volatile	510e		
4.7-23	β,β-Dimethyl-propiolactone		turbine	414		
4.7-24	γ-Butyrolactone		plant volatile	552e		
4.7-25	γ-Valerolactone		diesel	1		
4.7-26	γ-Hexaloctone		turbine	414		
4.7-27	γ-Octalactone		plant volatile	552e		
4.7-28	γ-Nonalactone	$CH_3(CH_2)_4$	plant volatile	514e		
4.7-29	γ-Undecalactone	$CH_3(CH_2)_6$	plant volatile	514e		
4.7-30	Maleic anhydride		maleic anhyd. mfr. phthal anhyd. mfr.	351 58*,64*		
4.7-31	δ-Valerolactone		diesel	138		
4.7-32	δ-Octalactone		plant volatile	552e		
4.7-33	Maltol		plant volatile	514e		
4.7-34	Cyclopentade-calactone	$H_2C(CH_2)_{13}C=O$	plant volatile	514e		
4.7-35	ω-Hexadecenlactone	$HC=CH(CH_2)_{13}C=O$	plant volatile	514e		

194

put together).

4.8.3. Chemistry

The aromatic ketones absorb light readily at tropospheric wavelengths. For compounds whose carbonyl carbon is a ring constituent, collisional deactivation is the most likely fate. If the carbonyl carbon is a ligand group member, cleavage can occur by pathways similar to those of the aliphatic ketones.

As with the substituted benzenes, interaction with hydroxyl radicals should result in HO· addition to the aromatic ring or hydrogen abstraction from the ligand group. These processes may be illustrated by those expected for acetophenone:

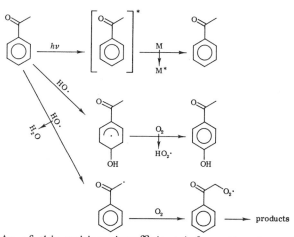

As of this writing, insufficient information is available to specify the branching ratios for these options although in any case it appears unlikely that the aromatic ketones play an important role in gas phase atmospheric chemistry. Because of their frequent presence in aerosols, however, their participation in aerosol chemistry may be more important and should be studied in greater detail than has thus far been the case.

TABLE 4.8 AROMATIC KETONES AND QUINONES
Emission and Detection

Species Number	Name	Chemical Formula	Emission		Detection	Ambient conc.
			Source	Ref.	Ref.	
4.8-1	1,4-Benzoquinone		tobacco smoke	579(a)	213*(a)	<15-80ng/m^3
4.8-2	Methyl-1,4-benzoquinone		tobacco smoke	579(a)		
4.8-3	2,3-Dimethyl-1,4-benzoquinone		tobacco smoke	579(a)		
4.8-4	2,5-Dimethyl-1,4-benzoquinone		tobacco smoke	579(a)		
4.8-5	Trimethyl-1,4-benzoquinone		tobacco smoke	579(a)		
4.8-6	Tetramethyl-1,4-benzoquinone		tobacco smoke	579(a)		

TABLE 4.8 AROMATIC KETONES AND QUINONES
Emission and Detection

| Species Number | Name | Chemical Formula | Emission | | Detection | Ambient |
			Source	Ref.	Ref.	conc.
4.8-7	Acetophenone		auto	217*	86,202	
			diesel	138		
			plant volatile	503,552e		
			river water odor	202		
4.8-8	*p*-Methyl acetophenone		plant volatile	510e		
4.8-9	Dimethyl acetophenone		diesel	138		
4.8-10	*m*-Hydroxy-acetophenone		tobacco smoke	396(a)		
4.8-11	*p*-Hydroxy-acetophenone		diesel	138		
			tobacco smoke	396(a)		
4.8-12	*p*-Methoxy-acetophenone		plant volatile	510e		

TABLE 4.8 AROMATIC KETONES AND QUINONES
Emission and Detection

Species Number	Name	Chemical Formula	Emission		Detection	Ambient
			Source	Ref.	Ref.	conc.
4.8-13	Dihydroxy-acetophenone		diesel	138		
4.8-14	Acetovanillone		wood pulping	267,545		
4.8-15	Benzoyl peroxide		polymer mfr.	407*		
4.8-16	Phenylvinylketone		diesel	138		
4.8-17	4-Phenyl-2-butanone				358	
4.8-18	4-(p-Methoxyphenyl)-butan-2-one		plant volatile	510e		
4.8-19	Benzophenone				591	

TABLE 4.8 AROMATIC KETONES AND QUINONES
Emission and Detection

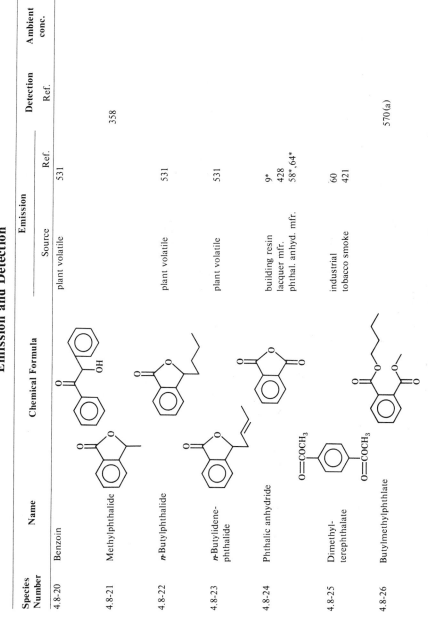

Species Number	Name	Chemical Formula	Emission		Detection	Ambient conc.
			Source	Ref.	Ref.	
4.8-20	Benzoin		plant volatile	531		
4.8-21	Methylphthalide				358	
4.8-22	*n*-Butylphthalide		plant volatile	531		
4.8-23	*n*-Butylidene-phthalide		plant volatile	531		
4.8-24	Phthalic anhydride		building resin lacquer mfr. phthal. anhyd. mfr.	9* 428 58*,64*		
4.8-25	Dimethyl-terephthalate		industrial tobacco smoke	60 421		
4.8-26	Butylmethylphthlate				570(a)	

199

TABLE 4.8 AROMATIC KETONES AND QUINONES
Emission and Detection

| Species Number | Name | Chemical Formula | Emission | | Detection | Ambient conc. |
			Source	Ref.	Ref.	
4.8-27	Diethylphthalate		auto polymer films tobacco smoke	311(a) 84 421	363(a),443(a)	
4.8-28	Ethylbutylphthlate				570(a)	
4.8-29	Dibutylphthalate		auto PVC mfr. refuse comb. water works emiss.	311(a) 521* 26(a) 248(a)	98(a),363(a)	
4.8-30	Diisobutyl- phthalate		auto	311(a)	363(a)	
4.8-31	Di-sec-butyl- phthalate				363(a)	
4.8-32	Butylphthalyl- butylglycolate		water works emiss.	248(a)		

TABLE 4.8 AROMATIC KETONES AND QUINONES
Emission and Detection

Species Number	Name	Chemical Formula	Emission		Detection	Ambient conc.
			Source	Ref.	Ref.	
4.8-33	Di(2-ethylhexyl)-phthalate		industrial water works emiss.	60 248(a)	363(a),443(a)	
4.8-34	Dioctalphthalate		auto PVC mfr.	311(a) 92,521*	363(a)	
4.8-35	n-Octal-n-decylphthalate		industrial	60		
4.8-36	Diisodecyl-phthalate		industrial PVC mfr.	60 521*		
4.8-37	Benzylphthalate		PVC mfr.	521*		
4.8-38	Butylbenzyl-phthalate		auto	311(a)	363(a),443(a)	

TABLE 4.8 AROMATIC KETONES AND QUINONES
Emission and Detection

Species Number	Name	Chemical Formula	Emission		Detection	Ambient conc.
			Source	Ref.	Ref.	
4.8-39	1-Indanone		diesel	138	364(a)	
4.8-40	Dimethyl-1-indanone		diesel	138		
4.8-41	3,3-Dimethyl-5-*tert*-butyl-indanone				364(a)	
4.8-42	Hydroxyindanone		diesel	138, 227		
4.8-43	Methylhydroxy indanone		diesel	138		
4.8-44	Trimethylhydroxy-indanone		diesel	138		
4.8-45	Tetramethyl-hydroxyindanone		diesel	138		
4.8-46	Methoxyindanone		diesel	227		

202

TABLE 4.8 AROMATIC KETONES AND QUINONES
Emission and Detection

Species Number	Name	Chemical Formula	Emission		Detection	Ambient conc.
			Source	Ref.	Ref.	
4.8-47	2-Indanone		diesel	138		
4.8-48	Indenone		diesel	138		
4.8-49	Dimethylindenone		diesel	138		
4.8-50	Pentamethyl-indenone		diesel	138		
4.8-51	Hydroxyindenone		diesel	138		
4.8-52	Methylhydroxy-indenone		diesel	138		
4.8-53	Dimethylhydroxy-indenone		diesel	138		

TABLE 4.8 AROMATIC KETONES AND QUINONES
Emission and Detection

| Species Number | Name | Chemical Formula | Emission | | Detection | Ambient |
			Source	Ref.	Ref.	conc.
4.8-54	1-Tetralone		diesel	138		
4.8-55	Methyltetralone		diesel	138		
4.8-56	Methoxytetralone		diesel	138		
4.8-57	Acetonaphthone		auto	311(a)		
			diesel	138		
4.8-58	2-Methyl-1,4-naphthoquinone		tobacco smoke	579(a)		
4.8-59	2,3,6-Trimethyl-1,4-naphthoquinone		tobacco smoke	579(a)		

TABLE 4.8 AROMATIC KETONES AND QUINONES
Emission and Detection

Species Number	Name	Chemical Formula	Emission		Detection	Ambient conc.
			Source	Ref.	Ref.	
4.8-60	2,6-Naphtho-quinone		diesel phthal anhyd. mfr.	138 64*	213(a),528(a)	
4.8-61	Methylnaphtho-quinone		diesel	138		
4.8-62	Fluorenone		diesel refuse comb.	309(a) 26(a)	466(a).	
4.8-63	Phenalene-9-one				98(a),363(a)	
4.8-64	9,10-Anthraquinone		auto diesel refuse comb.	311(a) 309(a) 26(a)	190*(a),213(a)	0.9-1.3pg/m^3
4.8-65	2-Methyl-9,10-anthraquinone		tobacco smoke	579(a)		

TABLE 4.8 AROMATIC KETONES AND QUINONES
Emission and Detection

Species Number	Name	Chemical Formula	Emission		Detection	Ambient conc.
			Source	Ref.	Ref.	
4.8-66	2,3-Dimethyl-9,10-anthraquinone		tobacco smoke	579(a)		
4.8-67	Benz[a]anthrone		auto	311(a)	179(a),188*(a)	0.23-15.52ng/m^3
4.8-68	Benz[de]anthracene-7-one		auto	311(a)	98(a),188(a)	
4.8-69	Triphenylene-1-one				188(a)	

TABLE 4.8 AROMATIC KETONES AND QUINONES
Emission and Detection

Species Number	Name	Chemical Formula	Emission		Detection Ref.	Ambient conc.
			Source	Ref.		
4.8-70	Benzo[a]pyrene-1, 6-quinone				190(a)	
4.8-71	Benzo[a]pyrene-3, 6-quinone				190(a)	
4.8-72	Benzo[a]pyrene-6, 12-quinone				190*(a),209(a)	38-88pg/m^3
4.8-73	Dibenzo[cd,jk]-pyrene-6,12-quinone				209(a),363(a)	
4.8-74	Dibenzo[b,def]-chrysene-7,14-quinone				190(a)	

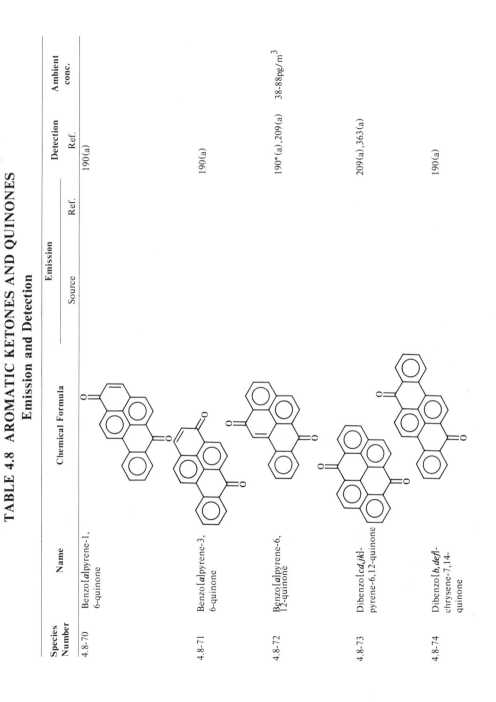

TABLE 4.8 AROMATIC KETONES AND QUINONES
Emission and Detection

Species Number	Name	Chemical Formula	Emission		Detection	Ambient conc.
			Source	Ref.	Ref.	
4.8-75	2-Coumarin		diesel refuse comb.	138 26(a)	528(a)	
4.8-76	Methylcoumarin		plant volatile refuse comb.	517e 26(a)		
4.8-77	Maraniol		plant volatile	531		
4.8-78	Dihydroxy-coumarin		refuse comb.	26(a)		

TABLE 4.8 AROMATIC KETONES AND QUINONES
Emission and Detection

Species Number	Name	Chemical Formula	Emission		Detection	Ambient conc.
			Source	Ref.	Ref.	
4.8-79	3-Coumarin		diesel plant volatile	138 507e		
4.8-80	Scopoletin		tobacco smoke	396(a)		
4.8-81	Xanthene-9-one				98(a),213(a)	

5

OXYGENATED ORGANIC COMPOUNDS

5.0. INTRODUCTION

The compounds presented and discussed in this chapter are of tremendous chemical diversity. They are emitted, in most cases in rather modest quantities, from a variety of natural and anthropogenic sources. Oxygenated organic compounds (except for the carbonyl compounds discussed in Chapter 4) are thought unlikely to be major participants in gas phase tropospheric processes. There are increasing indications that the organic component of atmospheric aerosols is primarily composed of these compounds, however. This subject is treated in somewhat more detail in the final chapter, given the perspective of the information assembled here.

Because information on organic oxygenates in the atmosphere is sparse, the text sections in this chapter are provided for overall chemical groups rather than for each of the tables. These groups are the organic acids (Tables $5.1-5.4$), esters (Tables $5.5-5.8$), alcohols (Tables $5.9-5.12$), peroxides (Table 5.13), and ethers (Tables $5.14-5.17$).

5.1-5.4. ORGANIC ACIDS

5.1.1-5.4.1. Identified Compounds

More than 80 organic acids participate in atmospheric processes. They are produced by vegetation, by combustion, by a variety of industrial sources, and, apparently, by atmospheric chemistry (see Section 3.4.3.). The aliphatic acids from C_1 to C_{32} are represented, as are the straight-chain diacids from C_2 to C_{11}. Many alkyl derivatives are also known. Most of the compounds have been detected as constituents of atmospheric aerosols.

TABLE 5.1 ALIPHATIC ACIDS
Emission and Detection

Species Number	Name	Chemical Formula	Emission		Detection Ref.	Ambient conc.
			Source	Ref.		
5.1-1	Formic acid	HCOOH	forest fire	354	90*,342*	4-72ppb
			lacquer mfr.	8		
			plant volatile	552e		
			plastics comb.	46,354		
			refuse comb.	26(a)		
			tobacco smoke	396(a)		
5.1-2	Acetic acid	CH_3COOH	animal waste	25,436*		
			auto body comb.	58*		
			diesel	1,176		
			forest fire	183,354		
			lacquer mfr.	8,427		
			plant volatile	552e		
			plastics comb.	46,354		
			refuse comb.	249		
			refuse comb.	26(a)		
			sewage tmt.	422*		
			starch mfr.	149		
			tobacco smoke	396(a)		
5.1-3	Ethanedioic acid	HOOCCOOH	tobacco smoke	396(a)		
5.1-4	Propionic acid	∕COOH	animal waste	25,436*		
			diesel	176		
			lacquer mfr.	427		
			plant volatile	552e		
			plastics comb.	354		
			starch mfr.	149		
			tobacco smoke	396(a)		
5.1-5	Propanedioic acid	HOOC∕⌒COOH	tobacco smoke	396(a)	480*(a)	~200ng/m^3

211

TABLE 5.1 ALIPHATIC ACIDS
Emission and Detection

Species Number	Name	Chemical Formula	Emission		Detection	Ambient
			Source	Ref.	Ref.	conc.
5.1-6	Butyric acid	COOH	animal waste	25,436*		
			diesel	176		
			food processing	156		
			industrial	540		
			microbes	552e		
			plant volatile	149		
			starch mfr.	396(a)		
			tobacco smoke			
5.1-7	Senecioic acid	COOH	plant volatile	552e		
5.1-8	Butanedioic acid	HOOC COOH	tobacco smoke	396(a)	480*(a)	200-500ng/m^3
5.1-9	2,3-Dimethylbutane-dioic acid	HOOC COOH			480*(a)	~200ng/m^3
5.1-10	Isobutyric acid	COOH	animal waste	25		
			microbes	302		
			tobacco smoke	396(a)		
5.1-11	Isobutanedioic acid	HOOC COOH			480*(a)	~200ng/m^3
5.1-12	Pentanoic acid	COOH	animal waste	25		
			diesel	176		
			food processing	540		
			microbes	396(a)		
			tobacco smoke	396(a)		
5.1-13	2-Methylpentanoic acid	COOH	tobacco smoke	396(a)		
5.1-14	Pentanedioic acid	HOOC COOH	tobacco smoke	396(a)	213*(a),532(a)	.12-1.35µg/m^3

TABLE 5.1 ALIPHATIC ACIDS
Emission and Detection

Species Number	Name	Chemical Formula	Emission Source	Emission Ref.	Detection Ref.	Ambient conc.
5.1-15	2-Methylpentanedioic acid	HOOC⌁COOH			480*(a)	~200ng/m^3
5.1-16	3-Methylpentanedioic acid	HOOC⌁COOH			480*(a)	~200ng/m^3
5.1-17	Isopentanoic acid	⌁COOH	animal waste microbes plant volatile	25 302 552e		
5.1-18	Isopentanedioic acid	HOOC⌁COOH	diesel food processing plant volatile tobacco smoke	176 552e 396(a)	480*(a)	~200ng/m^3
5.1-19	Hexanoic acid	HOOC⌁COOH	plant volatile	552e		
5.1-20	2-Methylhexanoic acid	⌁COOH				
5.1-21	Hexanedioic acid	HOOC⌁COOH	industrial tobacco smoke	60 396(a)	214*(a),532(a)	.12-1.35μg/m^3
5.1-22	2-Methylhexanedioic acid	HOOC⌁COOH	tobacco smoke		480*(a)	~200ng/m^3
5.1-23	3-Methylhexanedioic acid	HOOC⌁COOH			480*(a)	~200ng/m^3
5.1-24	Heptanoic acid	$CH_3(CH_2)_5COOH$	plant volatile tobacco smoke	552e 396(a)		
5.1-25	2-Methylheptanoic acid	⌁$(CH_2)_4COOH$	plant volatile	552e		
5.1-26	Heptanedioic acid	HOOC⌁COOH			214*(a),532(a)	30-150ng/m^3
5.1-27	Octanoic acid	$CH_3(CH_2)_6COOH$	plant volatile tobacco smoke	552e 396(a)		

TABLE 5.1 ALIPHATIC ACIDS
Emission and Detection

Species Number	Name	Chemical Formula	Emission Source	Emission Ref.	Detection Ref.	Ambient conc.
5.1-28	Octanedioic acid	$HOOC(CH_2)_6COOH$			$480^*(a), 570(a)$	$\sim 200\,ng/m^3$
5.1-29	Nonanoic acid	$CH_3(CH_2)_7COOH$	plant volatile tobacco smoke	$552e$ $396(a)$	$528(a), 570(a)$	
5.1-30	Nonanedioic acid	$HOOC(CH_2)_7COOH$	plant volatile	$552e$	$480^*(a), 570(a)$	$\sim 200\,ng/m^3$
5.1-31	2-Methylnonanoic acid	$\diagup\!\!-(CH_2)_6COOH$	plant volatile	$552e$		
5.1-32	Decanoic acid	$CH_3(CH_2)_8COOH$	plant volatile	$552e$	$528(a), 570(a)$	
5.1-33	Decanedioic acid	$HOOC(CH_2)_8COOH$	plant volatile	$552e$	$480^*(a), 570(a)$	$\sim 200\,ng/m^3$
5.1-34	Undecanoic acid	$CH_3(CH_2)_9COOH$	plant volatile	$552e$	$528(a), 570(a)$	
5.1-35	Undecanedioic acid	$HOOC(CH_2)_9COOH$			$570(a)$	
5.1-36	2-Methylundecanoic acid	$\diagup\!\!-(CH_2)_8COOH$	plant volatile	$552e$		
5.1-37	Dodecanoic acid	$CH_3(CH_2)_{10}COOH$	plant volatile tobacco smoke	$552e$ $396(a), 542(a)$	$98(a), 420^*(a)$	$0.5\text{-}10.3\,ng/m^3$
5.1-38	Tridecanoic acid	$CH_3(CH_2)_{11}COOH$	plant volatile tobacco smoke	$552e$ $542(a)$	$420^*(a), 528(a)$	$1.3\text{-}11.3\,ng/m^3$
5.1-39	Tetradecanoic acid	$CH_3(CH_2)_{12}COOH$	plant volatile tobacco smoke	$531, 552e$ $542(a)$	$269(a), 420^*(a)$	$0.3\text{-}19.5\,ng/m^3$
5.1-40	Pentadecanoic acid	$CH_3(CH_2)_{13}COOH$	tobacco smoke	$542(a)$	$363(a), 420^*(a)$	$0.2\text{-}17.6\,ng/m^3$
5.1-41	Hexadecanoic acid	$CH_3(CH_2)_{14}COOH$	plant volatile refuse comb. tobacco smoke veneer drying/m^3	$531, 552e$ $26(a)$ $396(a), 542(a)$	$98(a), 420^*(a)$	$1.2\text{-}49.7\,ng/m^3$

TABLE 5.1 ALIPHATIC ACIDS
Emission and Detection

Species Number	Name	Chemical Formula	Emission		Detection Ref.	Ambient conc.
			Source	Ref.		
5.1-42	Heptadecanoic acid	$CH_3(CH_2)_{15}COOH$	tobacco smoke	542(a)	363(a),420*(a)	$0.4\text{-}13.0 ng/m^3$
5.1-43	Octadecanoic acid	$CH_3(CH_2)_{16}COOH$	plant volatile refuse comb. tobacco smoke	552e 26(a),192(a) 396(a),542(a)	213(a),420*(a)	$0.2\text{-}30.7 ng/m^3$
5.1-44	Nonadecanoic acid	$CH_3(CH_2)_{17}COOH$	tobacco smoke	542(a)	363(a),528(a)	
5.1-45	Eicosanoic acid	$CH_3(CH_2)_{18}COOH$	tobacco smoke	396(a),542(a)	269(a),363(a)	
5.1-46	Heneicosanoic acid	$CH_3(CH_2)_{19}COOH$	tobacco smoke	542(a)	363(a),443(a)	
5.1-47	Docosanoic acid	$CH_3(CH_2)_{20}COOH$	tobacco smoke	542(a)	98(a),363(a)	
5.1-48	Tricosanoic acid	$CH_3(CH_2)_{21}COOH$	tobacco smoke	542(a)	570(a)	
5.1-49	Tetracosanoic acid	$CH_3(CH_2)_{22}COOH$	tobacco smoke	542(a)	443(a),570(a)	
5.1-50	Pentacosanoic acid	$CH_3(CH_2)_{23}COOH$	tobacco smoke	542(a)	443(a),570(a)	
5.1-51	Hexacosanoic acid	$CH_3(CH_2)_{24}COOH$	tobacco smoke	396(a),542(a)	443(a)	
5.1-52	Heptacosanoic acid	$CH_3(CH_2)_{25}COOH$	tobacco smoke	542(a)	570(a)	
5.1-53	Octacosanoic acid	$CH_3(CH_2)_{26}COOH$	tobacco smoke	542(a)		
5.1-54	Nonacosanoic acid	$CH_3(CH_2)_{27}COOH$	tobacco smoke	542(a)	570(a)	
5.1-55	Triacontanoic acid	$CH_3(CH_2)_{28}COOH$	tobacco smoke	542(a)		
5.1-56	Hentriacontanoic acid	$CH_3(CH_2)_{29}COOH$	tobacco smoke	542(a)		
5.1-57	Dotriacontanoic acid	$CH_3(CH_2)_{30}COOH$	tobacco smoke	542(a)		

TABLE 5.1 ALIPHATIC ACIDS
Reactions and Products

Species Number	Chemical reactions						
	Reactant	k	Ref.	Lifetime	Products	Ref.	Remarks
5.1-1							B
5.1-2							B

TABLE 5.2 OLEFINIC ACIDS
Emission and Detection

Species Number	Name	Chemical Formula	Emission		Detection	Ambient conc.
			Source	Ref.	Ref.	
5.2-1	Acrylic acid	$\diagup\!\!\!\!=\!\!\!\!\diagup$COOH	industrial	60		
5.2-2	Crotonic acid	$\diagup\!\!=\!\!\diagup$COOH	diesel lacquer mfr.	1 428		
5.2-3	Tridecenoic acid	$\diagup\!\!=\!\!(CH_2)_{10}COOH$			420*(a)	1.1-2.6ng/m^3
5.2-4	Pentadecenoic acid	$\diagup\!\!=\!\!(CH_2)_{12}COOH$			570(a)	
5.2-5	Hexadecenoic acid	$\diagup\!\!=\!\!(CH_2)_{13}COOH$			420*(a)	0.1-10.7ng/m^3
5.2-6	Heptadecenoic acid	$\diagup\!\!=\!\!(CH_2)_{11}COOH$			570(a)	
5.2-7	Octadecenoic acid	$\diagup\!\!=\!\!(CH_2)_{15}COOH$	tobacco smoke veneer drying	396(a),451(a) 68(a)	363(a),420*(a)	0.1-13.9ng/m^3
5.2-8	Eicosenoic acid	$\diagup\!\!=\!\!(CH_2)_{17}COOH$			420*(a)	1.4-2.0ng/m^3
5.2-9	Octadec-9,12-dienoic acid	COOH	tobacco smoke	396(a),451(a)	98(a)	
5.2-10	Octadec-9,12,15-trienoic acid	COOH	tobacco smoke	396(a),451(a)	98(a)	

217

TABLE 5.3 CYCLIC ACIDS
Emission and Detection

Species Number	Name	Chemical Formula	Emission		Detection	Ambient conc.
			Source	Ref.	Ref.	
5.3-1	Pinic acid				532(a)	
5.3-2	Pinonic acid				269(a),532(a)	
5.3-3	Nor-pinonic acid				532(a)	

TABLE 5.3 CYCLIC ACIDS

Reactions and Products

Species Number	Reactant	k	Ref.	Lifetime	Products	Ref.	Remarks
				Chemical reactions			
5.3-2							B

TABLE 5.4 AROMATIC ACIDS
Emission and Detection

Species Number	Name	Chemical Formula	Emission Source	Emission Ref.	Detection Ref.	Ambient conc.
5.4-1	Benzoic acid	COOH (benzene ring)	auto diesel phthal anhyd. mfr. plant volatile refuse comb. tobacco smoke	269(a) 137 64* 531,552e 26(a) 396(a)	307 213*(a),532(a)	90-380ng/m³
5.4-2	Methylbenzoic acid	COOH, CH₃ (benzene ring)			363(a)	
5.4-3	Ethylbenzoic acid	COOH, C₂H₅ (benzene ring)			570(a)	
5.4-4	Phthalic acid	COOH, COOH (benzene ring)	tobacco smoke vinyl chloride mfr.	396(a) 70	363(a),570(a)	
5.4-5	Methylphthalic acid	COOH, COOH, CH₃ (benzene ring)			363(a)	
5.4-6	Butylphthalic acid	COOH, COOH, C₄H₉ (benzene ring)			570(a)	
5.4-7	Isophthalic acid	COOH, COOH (benzene ring)			363(a)	
5.4-8	Terephthalic acid	COOH, COOH (benzene ring)	industrial	60	60 363(a)	
5.4-9	Phenylacetic acid	CH₂COOH (benzene ring)	auto plant volatile	269(a) 514e,531	214*(a),532(a)	410ng/m³

TABLE 5.4 AROMATIC ACIDS
Emission and Detection

| Species Number | Name | Chemical Formula | Emission | | Detection | Ambient conc. |
			Source	Ref.	Ref.	
5.4-10	Phenylpropionic acid				214t*(a), 532(a)	60-520ng/m³
5.4-11	Phenylbutyric acid				214t*(a), 532(a)	30ng/m³
5.4-12	Cinnamic acid		plant volatile	531		
5.4-13	Abietic acid		refuse comb. veneer drying	192(a) 68(a),197(a)		
5.4-14	Dehydroabietic acid		veneer drying	68(a)	570(a)	
5.4-15	Dihydroabietic acid		veneer drying	68(a)		
5.4-16	Pimaric acid		veneer drying	68(a)		

TABLE 5.4 AROMATIC ACIDS
Emission and Detection

| Species Number | Name | Chemical Formula | Emission | | Detection | Ambient |
			Source	Ref.	Ref.	conc.
5.4-17	β-Pimaric acid		veneer drying	68(a)		
5.4-18	Sandaracopimaric acid		veneer drying	68(a)		
5.4-19	Naphthalene-carboxylic acid				363(a)	
5.4-20	Phenanthrene-carboxylic acid				363(a)	
5.4-21	Anthracene-carboxylic acid				363(a)	
5.4-22	Pyrenecarboxylic acid				363(a)	

5.1.2-5.4.2. Ambient Concentrations

Formic acid is the only compound in this group whose atmospheric gas phase concentrations have been studied. The typical values are several tens of parts per billion (similar to those of formaldehyde and methanol on photochemically active days), which suggests that formic acid is a major product of atmospheric chemical chains.

Data are available for the concentrations of some 30 organic acids in atmospheric aerosols. It appears not to be unusual to have organic acid aerosol concentrations of a few $\mu g/m^3$ on sunny summer days in urban areas.

5.1.3-5.4.3. Chemistry

There have been no specific studies of the gas phase chemistry of the organic acids. There is every indication, however, that the compounds are primarily products of atmospheric chemistry rather than reactants. This is suggested by knowledge of normal chemical oxidation processes, by the detection of acids as end-products in smog-chamber simulations, and by the apparently ubiquitous presence of organic acids as constituents of the atmospheric aerosol. The relatively low vapor pressures of many of these compounds indicates that they would readily be deposited on aerosol surfaces. It is at least as likely, however, that the compounds reach the aerosols in less highly oxidized states and are transformed into acids by aqueous chemical processes. Perhaps more than any other evidence, the presence of the rich spectrum of organic acids in aerosols indicates the diverse and complex nature of the aqueous phase chemistry that must occur in the Earth's lower atmosphere.

5.5-5.8. ESTERS

5.5.1-5.8.1. Identified Compounds

More than 100 esters have been identified as emittants to the troposphere. Of these compounds, greater than 80% have no known source except volatilization from plants; many of the enticing smells of nature include esters as odor components. Methyl formate and methyl and ethyl acetate are the principal esters having anthropogenic sources; the fluxes appear to be quite small.

5.5.2-5.8.2. Ambient Concentrations

Four acetate esters have been detected in the atmosphere, as has methyl methacrylate (the latter as an aerosol component), but no atmospheric concentrations have been determined. It is possible, however, to infer some of the concentrations that must be present near ester sources by reference to olfactory detection thresholds. These indicate that esters are customarily detectable by humans when their concentrations are in the approximate range $1-100$ ppb (1220, 1257).

5.5.3-5.8.3. Chemistry

The rate constants for two esters with the hydroxyl radical are known from laboratory studies. The esters react somewhat more rapidly than do the corresponding alkanes, indicating that the CH bonds are less strong (588). The hydrogen abstraction will be followed in the atmosphere by peroxy radical formation. For ethyl formate,

The probability of 1:5 intramolecular H transfer in this system is large (439), and the transfer product will undergo β scission and hydrolysis processes:

Ester chemistry in the atmosphere can thus be expected to proceed along lines similar to those of the alkanes, with the ultimate products being small oxygenated organic molecules.

TABLE 5.5 ALIPHATIC ESTERS
Emission and Detection

Species Number	Name	Chemical Formula	Emission Source	Emission Ref.	Detection Ref.	Ambient conc.
5.5-1	Methyl formate	CH₃OOCH	animal waste	166		
			auto	217*		
			forest fire	354		
			plant volatile	552e		
			turbine	414		
5.5-2	Ethyl formate	CH₃CH₂OOCH	animal waste	166		
5.5-3	Citronellyl formate		plant volatile	506e		
5.5-4	Neryl formate		plant volatile	510e,517e		
5.5-5	Linalyl formate		plant volatile	514e		
5.5-6	Benzyl formate		plant volatile	552e		
5.5-7	Phenylethyl formate		plant volatile	552e		
5.5-8	Methyl acetate		animal waste	166,436*	591	
			forest fire	354		
			lacquer mfr.	428		
			petroleum mfr.	103,111		
			solvent	58*,439		
			tobacco smoke	396(a)		

TABLE 5.5 ALIPHATIC ESTERS
Emission and Detection

Species Number	Name	Chemical Formula	Emission — Source	Emission — Ref.	Detection Ref.	Ambient conc.
5.5-9	Ethyl acetate		animal waste	166,436*	358,591	
			glue vapor	11		
			microbes	302		
			petroleum mfr.	103,221		
			petroleum stor.	58*		
			plant volatile	509e,552e		
			tobacco smoke	396(a)		
			whiskey mfr.	32		
5.5-10	Propyl acetate		animal waste	327	358t	
			plant volatile	552e		
5.5-11	Acetoxyacetone		plant volatile	552e		
5.5-12	Isopropyl acetate		animal waste	166		
			plant volatile	552e		
			solvent	58*,88		
5.5-13	n-Butyl acetate		animal waste	327	591	
			chemical mfr.	462		
			plant volatile	552e		
			polymer mfr.	103		
			solvent	58*,78		
5.5-14	Isobutyl acetate		animal waste	166		
			microbes	302		
			solvent	88,134		
5.5-15	sec-Butyl acetate		solvent	588		
5.5-16	n-Amyl acetate		plant volatile	552e	358	

TABLE 5.5 ALIPHATIC ESTERS
Emission and Detection

Species Number	Name	Chemical Formula	Emission		Detection	Ambient conc.
			Source	Ref.	Ref.	
5.5-17	Isoamyl acetate		plant volatile whiskey mfr.	513e,552e 32	591	
5.5-18	n-Pentanol acetate		plant volatile	552e		
5.5-19	Hexyl acetate	$(CH_2)_5CH_3$	plant volatile	510e,552e		
5.5-20	Heptyl acetate	$(CH_2)_6CH_3$	turbine	414		
5.5-21	n-Octyl acetate	$(CH_2)_7CH_3$	plant volatile turbine	510e 414		
5.5-22	3-Octyl acetate		plant volatile	552e		
5.5-23	Nonyl acetate	$(CH_2)_8CH_3$	plant volatile	552e		
5.5-24	Decyl acetate	$(CH_2)_9CH_3$	plant volatile	514e,552e		
5.5-25	Undecyl acetate	$(CH_2)_{10}CH_3$	plant volatile	552e		
5.5-26	Hexadecyl acetate	$(CH_2)_{15}CH_3$	plant volatile	552e		
5.5-27	Octadecyl acetate	$(CH_2)_{17}CH_3$	plant volatile	552e		
5.5-28	Vinyl acetate		forest fire industrial	354 60	383	

227

TABLE 5.5 ALIPHATIC ESTERS
Emission and Detection

Species Number	Name	Chemical Formula	Emission		Detection	Ambient conc.
			Source	Ref.	Ref.	
5.5-29	1-Hydroxy-2-ethylene acetate		solvent	88,442		
5.5-30	*trans*-2-Hexenyl acetate		plant volatile	552e		
5.5-31	3-Hexenyl acetate		plant volatile	512e		
5.5-32	3-Octenyl acetate		plant volatile	552e		
5.5-33	Citronellyl acetate		plant volatile	506e		
5.5-34	Geranyl acetate		plant volatile	531		
5.5-35	Neryl acetate		plant volatile	510e,531		
5.5-36	Linalyl acetate		plant volatile	516e,531		
5.5-37	Methylcyclohexyl acetate		acetyl. mfr.	128		
5.5-38	*l*-Menthyl acetate		plant volatile	516e		
5.5-39	Bornyl acetate		plant volatile	506e		

228

TABLE 5.5 ALIPHATIC ESTERS
Emission and Detection

Species Number	Name	Chemical Formula	Emission		Detection	Ambient conc.
			Source	Ref.	Ref.	
5.5-40	α-Terpinyl acetate		plant volatile	510e		
5.5-41	β-Terpinyl acetate		plant volatile	510e		
5.5-42	γ-Terpinyl acetate		plant volatile	510e		
5.5-43	Phenyl acetate		solvent	134		
5.5-44	Benzyl acetate		plant volatile	500,501		
5.5-45	Methylbenzyl acetate		plant volatile	517e,552e		
5.5-46	2-Phenylethyl acetate		plant volatile	501,510e		
5.5-47	4-Methylphenyl acetate		plant volatile	507e		
5.5-48	4-Allyl-2-methoxy-phenyl acetate		plant volatile	510e		
5.5-49	Phenylpropyl acetate		plant volatile	510e		

229

TABLE 5.5 ALIPHATIC ESTERS
Emission and Detection

Species Number	Name	Chemical Formula	Emission		Detection	Ambient conc.
			Source	Ref.	Ref.	
5.5-50	Cinnamyl acetate		plant volatile	506e		
5.5-51	Dimethyl malonate		plant volatile	552e		
5.5-52	Ethyl propionate		plant volatile tobacco smoke	552e 396(a)		
5.5-53	Isopropyl propionate		animal waste	166		
5.5-54	Isoamyl propionate		plant volatile	514e		
5.5-55	Allyl propionate		turbine	414		
5.5-56	Neryl propionate		plant volatile	510e		
5.5-57	Linalyl propionate		plant volatile	514e		
5.5-58	α-Terpinyl propionate		plant volatile	510e		
5.5-59	β-Terpinyl propionate		plant volatile	510e		

230

Species Number	Name	Chemical Formula	Emission		Detection Ref.	Ambient conc.
			Source	Ref.		
5.5-60	γ-Terpinyl propionate	*(structure)*	plant volatile	510e		
5.5-61	Methyl butyrate	*(structure)*	plant volatile	552e		
5.5-62	Methyl-β-hydroxy butyrate	*(structure)*	plant volatile	552e		
5.5-63	Ethyl butyrate	*(structure)*	plant volatile / tobacco smoke	509e, 552e / 396(a)		
5.5-64	Butyl buterate	*(structure)*	plant volatile	552e		
5.5-65	2-Methylbutyl-2-Methyl buterate	*(structure)*	plant volatile	552e		
5.5-66	Hexyl buterate	*(structure)*	plant volatile	552e		
5.5-67	3-Octenyl buterate	*(structure)*	plant volatile	552e		
5.5-68	Citronellyl buterate	*(structure)*	plant volatile	506e		
5.5-69	Neryl butyrate	*(structure)*	plant volatile	510e		
5.5-70	Ethyl 3-methyl buterate	*(structure)*	tobacco smoke	396(a)		

TABLE 5.5 ALIPHATIC ESTERS
Emission and Detection

| Species Number | Name | Chemical Formula | Emission | | Detection | Ambient |
			Source	Ref.	Ref.	conc.
5.5-71	Bornyl 3-methyl buterate		plant volatile	510e		
5.5-72	Ethyl isobutyrate		plant volatile	552e		
5.5-73	Butyl isobutyrate		plant volatile	514e		
5.5-74	Octyl isobutyrate		plant volatile	514e		
5.5-75	Geranyl isobutyrate		plant volatile	512e		
5.5-76	Linalyl isobutyrate		plant volatile	514e		
5.5-77	Ethyl pentanoate		plant volatile	552e		
5.5-78	Ethyl 2-methyl pentanoate		tobacco smoke	396(a)		
5.5-79	Isobutyl pentanoate		plant volatile	552e		
5.5-80	Methyl-β-hydroxy hexanoate		plant volatile	552e		
5.5-81	Ethyl hexanoate		plant volatile tobacco smoke	552e 396(a)		

232

TABLE 5.5 ALIPHATIC ESTERS
Emission and Detection

Species Number	Name	Chemical Formula	Emission Source	Emission Ref.	Detection Ref.	Ambient conc.
5.5-82	Ethyl-β-hydroxy hexanoate		plant volatile	552e		
5.5-83	Methyl 4-methyl hexanoate		plant volatile	552e		
5.5-84	Methyl 5-methyl hexanoate		plant volatile	552e		
5.5-85	Methyl heptanoate		plant volatile	552e		
5.5-86	Ethyl heptanoate		plant volatile	514e		
5.5-87	Dioctal adipate		PVC mfr.	521*		
5.5-88	Ethyl dodecanoate	$CH_3(CH_2)_{10}$	plant volatile	511e		
5.5-89	Ethyl hexadecanoate	$CH_3(CH_2)_{14}$			570(a)	
5.5-90	Acetyl hexadecanoate	$CH_3(CH_2)_{14}$	plant volatile	552e		
5.5-91	Ethyl octadecanoate	$CH_3(CH_2)_{16}$			570(a)	
5.5-92	Ethyl nonadecanoate	$CH_3(CH_2)_{17}$			570(a)	
5.5-93	Ethyl eicosanoate	$CH_3(CH_2)_{18}$			570(a)	
5.5-94	Ethyl heneicosanoate	$CH_3(CH_2)_{19}$			570(a)	

233

TABLE 5.5 ALIPHATIC ESTERS
Reactions and Products

Species Number	Reactant	Chemical reactions					
		k	Ref.	Lifetime	Products	Ref.	Remarks
5.5-10	HO·	$(4.3 \pm 0.8) \times 10^{-12}$	588	5.7×10^6			
5.5-15	HO·	$(5.7 \pm 1.2) \times 10^{-12}$	588	2.6×10^6			

TABLE 5.6 OLEFINIC ESTERS
Emission and Detection

Species Number	Name	Chemical Formula	Emission		Detection Ref.	Ambient conc.
			Source	Ref.		
5.6-1	Ethyl acrylate		plant volatile	514e		
5.6-2	Neryl-*cis*-α, β-dimethyl acrylate		plant volatile	510e		
5.6-3	Methyl methacralate		industrial polymer mfr.	60 447*	399(a)	
5.6-4	Methyl crotonate		plant volatile	552e		
5.6-5	Ethyl crotonate		plant volatile	552e		
5.6-6	Isopropyl crotonate		plant volatile	552e		
5.6-7	Isobutyl crotonate		plant volatile	552e		
5.6-8	*n*-Butyl angelate		plant volatile	510e		
5.6-9	Isoamyl angelate		plant volatile	510e		
5.6-10	Allyl tiglate				399(a)	

TABLE 5.6 OLEFINIC ESTERS
Emission and Detection

| Species Number | Name | Chemical Formula | Emission | | Detection | Ambient |
			Source	Ref.	Ref.	conc.
5.6-11	Ethyl hex-2-enoate		plant volatile	552e		
5.6-12	Methyl-*cis*-4-octenoate		plant volatile	552e		
5.6-13	Ethyl octadecanoate				570(a)	

TABLE 5.7 CYCLIC ESTERS
Emission and Detection

Species Number	Name	Chemical Formula	Emission		Detection Ref.	Ambient conc.
			Source	Ref.		
5.7-1	Methyl thujate	COOCH₃	plant volatile	531		
5.7-2	Methyl furoate		microbes	302		

TABLE 5.8 AROMATIC ESTERS
Emission and Detection

Species Number	Name	Chemical Formula	Emission		Detection	Ambient
			Source	Ref.	Ref.	conc.
5.8-1	Methyl benzoate		microbes	302		
			plant volatile	499,552e		
			solvent	134		
5.8-2	Methyl-p-methoxy benzoate		plant volatile	516e		
5.8-3	Ethyl benzoate		plant volatile	552e		
5.8-4	Amyl benzoate		plant volatile	504e		
5.8-5	Neryl benzoate		plant volatile	510e		
5.8-6	Linalyl benzoate		plant volatile	516e		
5.8-7	Benzyl benzoate		plant volatile	506e,510e		
5.8-8	Phenylethyl benzoate		plant volatile	514e		

TABLE 5.8 AROMATIC ESTERS
Emission and Detection

Species Number	Name	Chemical Formula	Emission		Detection Ref.	Ambient conc.
			Source	Ref.		
5.8-9	Coniferyl benzoate		plant volatile	531		
5.8-10	Methyl cinnamate		plant volatile	500,531		
5.8-11	Ethyl cinnamate		plant volatile	509e		
5.8-12	Benzyl cinnamate		plant volatile	506e,510e		
5.8-13	Phenylpropyl cinnamate		plant volatile	510e		
5.8-14	Cinnamyl cinnamate		plant volatile	514e		
5.8-15	Ethyl naphthoate		auto	311(a)		

239

5.9-5.12. ALCOHOLS

5.9.1-5.12.1. Identified Compounds

Alcohols are emitted in a great variety of chemical forms by a wide number of natural and anthropogenic processes. These tables identify nearly 150 atmospheric alcohols, including the C_1 to C_{11} linear alkyl alcohols, olefinic alcohols with carbon numbers 1 to 6, plus 8, 10, and 15, and a large number of cyclic and aromatic compounds. The principal anthropogenic sources of alcohols are related to their extensive use as solvents, although combustion processes also appear to be significant emitters. Most of the atmospheric alcohols of high molecular weight are emitted by vegetation, producing many of the natural flower fragrances.

5.9.2-5.12.2. Ambient Concentrations

The tropospheric concentrations of several of the most chemically interesting alcohols have been determined on a few occasions. Methanol, which appears to be a product of atmospheric alkane chemistry (Section 3.1.) has been found at levels of as much as 100 ppb; this is sufficient to classify it as a major atmospheric trace species. Phenol, a common industrial chemical and a probable product of benzene-hydroxyl radical reactions, is present at a few ppb. Several difunctional alcoholic compounds are minor constituents of urban aerosols; they presumably are produced from cyclic olefin precursors (see Section 3.4.).

5.9.3-5.12.3. Chemistry

There is general agreement that the initial atmospheric reaction of the aliphatic alcohols is hydrogen abstraction by the hydroxyl radical (e.g., 1028). For methanol a plausible chain is thus

$$CH_3OH \xrightarrow{HO\cdot} CH_2OH\cdot \xrightarrow{O_2} O_2CH_2OH\cdot \xrightarrow{NO} OCH_2OH\cdot \xrightarrow{O_2} HCOOH .$$

For the general case,

$$CH_3(CH_2)_n OH \xrightarrow[\substack{(3)\ NO \\ (4)\ O_2}]{\substack{(1)\ HO\cdot \\ (2)\ O_2}} CH_3\overset{\overset{\text{O}}{\|}}{C}(CH_2)_{n-1}OH,$$

where the ketone occurs along the carbon backbone at the point of initial hydrogen abstraction. A similar process is envisioned for saturated cyclic alcohols.

In the cases of olefinic and aromatic alcohols and unsaturated cyclic alcohols, hydroxyl addition at the double bond is the expected process, leading to multifunctional oxygenated groups. These compounds will possess low vapor pressures and will thus be promptly deposited on aerosol and ground surfaces.

5.13. ORGANIC PEROXIDES

5.13.1. Identified Compounds

Only two organic peroxides are known to be emitted to the atmosphere, both as byproducts of polymer manufacture.

5.13.2. Ambient Concentrations

No data are available concerning the atmospheric concentrations of these compounds.

5.13.3. Chemistry

Peroxides are readily photodissociated in the troposphere, the cleavage occurring at the weak $O-O$ bond (1231). The rate of the analogous process for hydrogen peroxide suggests that lifetimes on the order of a day are likely.

An organic peroxide not detected thus far in the troposphere is methyl hydroperoxide, which is expected to be produced by the reaction

$$CH_3O_2 \cdot + HO_2 \cdot \rightarrow CH_3OOH + O_2.$$

In nonurban atmospheres where little NO is available to scavenge the peroxyl radicals, significant concentrations of this compound seem possible (1114,1125).

TABLE 5.9 ALIPHATIC ALCOHOLS
Emission and Detection

Species Number	Name	Chemical Formula	Emission		Detection	Ambient
			Source	Ref.	Ref.	conc.
5.9-1	Methanol	CH_3OH	animal waste	25,80	90*,376	8-100ppb
			auto	55,217*		
			charcoal mfr.	58*		
			fat subst. plant	52		
			forest fire	183,354		
			insects	471		
			microbes	302		
			petroleum stor.	58*		
			plant volatile	552e		
			plastics comb.	354		
			printing	543*		
			refuse comb.	439		
			refuse comb.	26(a)		
			solvent	78,88		
			tobacco smoke	396,446		
			turbine	414		
			volcano	234		
			wood pulping	19,267		

TABLE 5.9 ALIPHATIC ALCOHOLS
Emission and Detection

Species Number	Name	Chemical Formula	Emission Source	Emission Ref.	Detection Ref.	Ambient conc.
5.9-2	Ethanol	CH$_3$CH$_2$OH	animal waste	25,80	358t,591	
			auto	55,217*		
			fat subst. plant	52		
			forest fire	376,439		
			insects	471		
			microbes	302		
			petroleum stor.	58*		
			plant volatile	552e		
			plastics comb.	354		
			printing	543*		
			refuse comb.	439		
			refuse comb.	26(a)		
			solvent	88,439		
			tobacco smoke	396		
			turbine	414		
			volcano	234		
			whiskey mfr.	32		
			wood pulping	19,267		
5.9-3	1-Propanol	⌒OH	animal waste	25,80		
			diesel	176		
			fat subst. plant	52		
			microbes	302		
			plant volatile	552e		
			plastics comb.	354		
			printing	543*		
			sewage tmt.	422*		
			solvent	78,382*		
			whiskey mfr.	32		
			wood pulping	267,545		
			volcano	234		

TABLE 5.9 ALIPHATIC ALCOHOLS
Emission and Detection

Species Number	Name	Chemical Formula	Emission		Detection Ref.	Ambient conc.
			Source	Ref.		
5.9-4	Propylene glycol		industrial	60		
5.9-5	Lactic acid		tobacco smoke	396(a)		
5.9-6	2-Propanol		animal waste	25,80	358,591	
			auto	284		
			petroleum stor.	58*		
			plant volatile	552e		
			plastics comb.	354		
			printing	543*		
			sewage tmt.	422*		
			solvent	88,134		
			volcano	234		
			wood pulping	267,545		
5.9-7	Glycerol		industrial	60		
			printing ink mfr.	58*		
			tobacco smoke	396		
5.9-8	1-Butanol		animal waste	25,80	33*,566*	1.5-445ppb
			fat subst. plant	52		
			insects	471		
			microbes	302		
			plant volatile	33,552e		
			sewage tmt.	422*		
			solvent	88,439		
			turbine	414		
			wood pulping	19,267		

TABLE 5.9 ALIPHATIC ALCOHOLS
Emission and Detection

Species Number	Name	Chemical Formula	Emission		Detection	Ambient conc.
			Source	Ref.	Ref.	
5.9-9	Isobutanol		animal waste	25,80		
			microbes	302		
			petroleum mfr.	221		
			petroleum stor.	58*		
			plant volatile	552e		
			solvent	88,439		
			whiskey mfr.	32		
			wood pulping	19,267		
5.9-10	*tert*-Butyl alcohol		industrial	60		
			petroleum stor.	58*		
			turbine	414		
5.9-11	2-Methyl-1-butanol		plant volatile	552e		
5.9-12	3-Methyl-1-butanol		animal waste	80		
			industrial	108		
			microbes	302		
			plant volatile	552e		
			turbine	414		
			whiskey mfr.	32		
5.9-13	2-Butanol		auto	284		
			industrial	60		
5.9-14	Malic acid		tobacco smoke	396(a)		
5.9-15	1-Pentanol		fat subst. plant	52		
			plant volatile	552e		
			turbine	414		
			wood pulping	19		

TABLE 5.9 ALIPHATIC ALCOHOLS
Emission and Detection

Species Number	Name	Chemical Formula	Emission		Detection	Ambient conc.
			Source	Ref.	Ref.	
5.9-16	5-Hydroxypentanoic acid				214*(a),532(a)	.07-2.14 $\mu g/m^3$
5.9-17	1,5-Pentanediol				532(a)	
5.9-18	2-Methyl-1-pentanol		turbine	414		
5.9-19	3-Methyl-1-pentanol		turbine	414		
5.9-20	4-(p-Tolyl)-1-pentanol		wood pulping	267,545		
5.9-21	2-Pentanol		plant volatile	552e		
5.9-22	3-Pentanol		turbine	414		
5.9-23	1-Hexanol		fat subst. plant plant volatile	52 533,552e		
5.9-24	1,6-Hexanediol				532(a) ·	
5.9-25	6-Hydroxy-hexanoic acid		industrial river water odor	60 202	214*(a),532(a)	.42-3.4 $\mu g/m^3$
5.9-26	2-Ethylhexanol		turbine	414		

TABLE 5.9 ALIPHATIC ALCOHOLS
Emission and Detection

Species Number	Name	Chemical Formula	Emission Source	Emission Ref.	Detection Ref.	Ambient conc.
5.9-27	2-Ethoxy-hexanol		industrial	60		
5.9-28	2-Hexanol		plant volatile	552e		
5.9-29	Dimethyl-2,5-hexanediol		turbine	414		
5.9-30	1-Heptanol		plant volatile turbine	514e,552e 414		
5.9-31	2-Propyl-1-heptanol		turbine	414		
5.9-32	1,7-Heptanediol				532(a)	
5.9-33	7-Hydroxy-heptanoic acid				214*(a),532(a)	170-650ng/m^3
5.9-34	2-Heptanol		turbine	414		
5.9-35	2,6-Dimethyl-heptane-4-ol		river water odor	202		
5.9-36	1-Octanol		microbes plant volatile turbine	302 503e,552e 414	591	
5.9-37	3-Octanol		microbes plant volatile	302 552e		
5.9-38	Nonanol	$CH_3(CH_2)_7CH_2OH$	plant volatile	503e,552e		

TABLE 5.9 ALIPHATIC ALCOHOLS
Emission and Detection

Species Number	Name	Chemical Formula	Emission		Detection	Ambient conc.
			Source	Ref.	Ref.	
5.9-39	2-Methylnonanol	$CH_3(CH_2)_6CH(CH_3)$-CH_2OH	industrial	60		
5.9-40	Decanol	$CH_3(CH_2)_8CH_2OH$	plant volatile turbine	503e,552e 414		
5.9-41	Dodecanol	$CH_3(CH_2)_{10}CH_2OH$	plant volatile	503e		
5.9-42	Tetradecanol	$CH_3(CH_2)_{12}CH_2OH$	plant volatile	514e		
5.9-43	5-Pentadecanol		auto	309(a)		
5.9-44	Octadecanol	$CH_3(CH_2)_{16}CH_2OH$	plant volatile	552e		

TABLE 5.9 ALIPHATIC ALCOHOLS
Reactions and Products

Species Number	Reactant	k	Ref.	Chemical reactions Lifetime	Products	Ref.	Remarks
5.9-1	HO·	$(9.5\pm1.0)\times10^{-13}$	1028	2.3×10^6	HCHO	1100	B
	O	$(5.2\pm0.6)\times10^{-14}$	1048				
5.9-2	HO·	$(3.0\pm0.3)\times10^{-12}$	1028	7.3×10^5			
	O	$(1.5\pm0.6)\times10^{-13}$	1228				
5.9-3	HO·	$(3.8\pm0.3)\times10^{-12}$	1028	5.8×10^5			
5.9-6	HO·	$(7.2\pm2.2)\times10^{-12}$	1049	3.0×10^5			
	O	$(2.2\pm0.9)\times10^{-13}$	1228				
5.9-8	HO·	$(6.8\pm1.0)\times10^{-12}$	1028	3.2×10^5			
5.9-9							B
5.9-16							B

TABLE 5.10 OLEFINIC ALCOHOLS
Emission and Detection

Species Number	Name	Chemical Formula	Emission Source	Emission Ref.	Detection Ref.	Ambient conc.
5.10-1	Ethylene glycol		building resin industrial tobacco smoke	9* 60 396		
5.10-2	1-Ethoxyethene-2-ol		solvent	88,442		
5.10-3	1-Butoxyethene-2-ol		solvent	88		
5.10-4	1-Propene-3-ol		turbine	414		
5.10-5	2-Buten-1-ol		auto turbine	217* 414		
5.10-6	3-Methyl-2-buten-1-ol		turbine	414		
5.10-7	3-Buten-1,2-diol		turbine	414		
5.10-8	2-Methyl-3-buten-2-ol		plant volatile	552e		
5.10-9	3-Methyl-3-buten-2-ol		turbine	414		

TABLE 5.10 OLEFINIC ALCOHOLS
Emission and Detection

Species Number	Name	Chemical Formula	Emission		Detection Ref.	Ambient conc.
			Source	Ref.		
5.10-10	4-Penten-1-ol		turbine	414		
5.10-11	3-Methyl-1-pentyl-3-ol		diesel	1		
5.10-12	2-Hexen-1-ol		plant volatile turbine	510e,552e 414		
5.10-13	3-Hexen-1-ol		plant volatile	510e,533		
5.10-14	2,4-Hexadien-1-ol		turbine	414		
5.10-15	1-Hexyn-3-ol		turbine	414		
5.10-16	2-Octen-1-ol		microbes	302		
5.10-17	1-Octen-3-ol		microbes plant volatile	302 552e		
5.10-18	9-Decen-1-ol		plant volatile	503e		

251

TABLE 5.10 OLEFINIC ALCOHOLS
Emission and Detection

Species Number	Name	Chemical Formula	Emission Source	Emission Ref.	Detection Ref.	Ambient conc.
5.10-19	Citronellol		plant volatile	498,512e		
5.10-20	Geraniol		plant volatile	498,506e		
5.10-21	Nerol		plant volatile	505e,552e		
5.10-22	Linalool		plant volatile wood pulping	474,531 267,545		
5.10-23	Myrcenol		plant volatile	517e		
5.10-24	*trans*-2-Methyl-6-methylene-3,7-octadien-2-ol		plant volatile	552e		
5.10-25	Farnesol		plant volatile	514e		
5.10-26	Nerolidol		plant volatile	514e		

TABLE 5.11 CYCLIC ALCOHOLS
Emission and Detection

Species Number	Name	Chemical Formula	Emission Source	Emission Ref.	Detection Ref.	Ambient conc.
5.11-1	Cyclopentanol		diesel	1		
5.11-2	3-Methyl-1,2-cyclopentanediol		turbine	414		
5.11-3	Cyclohexanol		industrial solvent turbine	60 439 414	358	
5.11-4	*l*-Menthol		plant volatile	514e,516e		
5.11-5	Fenchyl alcohol		plant volatile tap water odor wood pulping	531 555 267,545		
5.11-6	*p*-Menthan-8-ol		plant volatile	508e		
5.11-7	*l*-Carveol		plant volatile	506e		
5.11-8	Isopulegol		plant volatile	514e		
5.11-9	α-Terpineol		plant volatile wood pulping	515e,552e 267,545		

253

TABLE 5.11 CYCLIC ALCOHOLS
Emission and Detection

Species Number	Name	Chemical Formula	Emission		Detection	Ambient conc.
			Source	Ref.	Ref.	
5.11-10	β-Terpineol		plant volatile	515e,517e		
5.11-11	γ-Terpineol		plant volatile	515e,517e		
5.11-12	Terpinene-4-ol		plant volatile wood pulping	552e 545		
5.11-13	Methylhydroxy cyclohexane		turbine	414		
5.11-14	α-Santalol		plant volatile	510e,531		
5.11-15	Civetol		animal secretion	510e		
5.11-16	Guaiol		plant volatile	510e,531		
5.11-17	Geosmin		plant volatile	555		
5.11-18	Cadinol		plant volatile	514e		

TABLE 5.11 CYCLIC ALCOHOLS
Emission and Detection

Species Number	Name	Chemical Formula	Emission		Detection	Ambient conc.
			Source	Ref.	Ref.	
5.11-19	Cedrol		plant volatile	510e		
5.11-20	Cedrenol		plant volatile	514e		
5.11-21	Caryophyllene alcohol		plant volatile	510e		

255

TABLE 5.12 AROMATIC ALCOHOLS
Emission and Detection

Species Number	Name	Chemical Formula	Emission Source	Emission Ref.	Detection Ref.	Ambient conc.
5.12-1	Phenol		auto	95,537	17,413*	2.8ppb
			auto	87(a)	214(a),458(a)	
			brewing	144		
			diesel	1,137		
			foundry	97*,279*		
			glass fibre mfr.	122*		
			lacquer mfr.	427		
			plastics comb.	46		
			refuse comb.	26(a)		
			solvent	121,439		
			tap water odor	555		
			tobacco smoke	4,396		
			wood pulping	267,545		
5.12-2	o-Cresol		auto	537	458(a)	
			auto	87(a)		
			diesel	138		
			tobacco smoke	396		
			wood pulping	267,545		
5.12-3	m-Cresol		auto	87(a)		
			diesel	138		
			glass fibre mfr.	122*		
			tap water odor	555		
			tobacco smoke	396		
			wood pulping	267,545		

TABLE 5.12 AROMATIC ALCOHOLS
Emission and Detection

Species Number	Name	Chemical Formula	Emission		Detection	Ambient
			Source	Ref.	Ref.	conc.
5.12-4	p-Cresol		auto	87(a)		
			brewing	144		
			diesel	137		
			glass fibre mfr.	122*		
			plant volatile	507e		
			tobacco smoke	396		
			wood pulping	267,545		
5.12-5	2,3-Xylenol		auto	537		
			auto	87(a)		
			brewing	144		
			diesel	137		
5.12-6	2,4-Xylenol		auto	87(a)		
			tobacco smoke	396(a)		
5.12-7	2,5-Xylenol		auto	87(a)		
5.12-8	2,6-Xylenol		auto	87(a)		
5.12-9	3,4-Xylenol		auto	87(a)		

TABLE 5.12 AROMATIC ALCOHOLS
Emission and Detection

Species Number	Name	Chemical Formula	Emission		Detection	Ambient conc.
			Source	Ref.	Ref.	
5.12-10	3,5-Xylenol		auto	87(a)		
			brewing	144		
			tobacco smoke	396(a)		
5.12-11	2,4,6-Trimethyl-phenol		auto	537		
			auto	87(a)		
			diesel	138		
			tobacco smoke	396(a)		
5.12-12	Tetramethyl-phenol		diesel	138		
5.12-13	o-Ethylphenol		auto	537		
			auto	87(a)		
5.12-14	m-Ethylphenol		auto	87(a)		
5.12-15	p-Ethylphenol		auto	87(a)		
5.12-16	Propylphenol		auto	537		
5.12-17	Carvacrol		plant volatile	510e,531		

TABLE 5.12 AROMATIC ALCOHOLS
Emission and Detection

Species Number	Name	Chemical Formula	Emission		Detection Ref.	Ambient conc.
			Source	Ref.		
5.12-18	Thymol		plant volatile	510e,531		
5.12-19	2,6-Di-*tert*-butyl-4-methylphenol		auto diesel	311(a) 309(a)	358	
5.12-20	Chavicol		diesel plant volatile	138 505e,531		
5.12-21	*o*-Phenylphenol		diesel plant volatile	309(a) 552e		
5.12-22	*m*-Phenylphenol		diesel	309(a)		
5.12-23	*p*-Phenylphenol		diesel	309(a)		
5.12-24	Pyrocatechol		plant volatile tobacco smoke	552e 396,405	213*(a),214(a)	<15-65ng/m^3
5.12-25	3-Methylcatechol		tobacco smoke	405		
5.12-26	4-Methylcatechol		tobacco smoke	405		

TABLE 5.12 AROMATIC ALCOHOLS
Emission and Detection

Species Number	Name	Chemical Formula	Emission Source	Emission Ref.	Detection Ref.	Ambient conc.
5.12-27	4-Ethylcatechol	(benzene ring with two OH, ethyl)	tobacco smoke	405		
5.12-28	4-n-Propyl-catechol	(benzene ring with two OH, propyl)	tobacco smoke	405		
5.12-29	Allylpyro-catechol	(benzene ring with two OH, allyl)	plant volatile	531		
5.12-30	Phenylpyro-catechol	(biphenyl ring with two OH)	auto	311(a)		
5.12-31	Resorcinol	(benzene ring with two OH, meta)	tobacco smoke	396(a)		
5.12-32	Hydroquinone	(benzene ring with two OH, para)	diesel, tobacco smoke	138, 396(a)	213*(a)	<15-125ng/m^3
5.12-33	Methoxyphenol	(benzene ring with OH and OCH$_3$)	brewing, diesel, forest fire, wood pulping	144, 138,227, 439, 267		
5.12-34	Nonylphenol (ethoxylated)	(benzene ring with OH, O, (CH$_2$)$_7$)	industrial	60		

TABLE 5.12 AROMATIC ALCOHOLS
Emission and Detection

Species Number	Name	Chemical Formula	Emission		Detection Ref.	Ambient conc.
			Source	Ref.		
5.12-35	Methylmethoxy-phenol		diesel	138		
5.12-36	4,6-Methoxy-pyrogallol		forest fire	439		
5.12-37	Benzyl alcohol		auto diesel plant volatile tobacco smoke	217*,537 138 533,552e 396	532(a)	
5.12-38	Methylbenzyl alcohol		auto	537		
5.12-39	Dimethylbenzyl alcohol		auto	537		
5.12-40	Cuminyl alcohol		plant volatile	510e		
5.12-41	Dihydrocuminyl alcohol		plant volatile	531		
5.12-42	p-Hydroxybenzyl alcohol		plant volatile	552e		

TABLE 5.12 AROMATIC ALCOHOLS
Emission and Detection

Species Number	Name	Chemical Formula	Emission Source	Emission Ref.	Detection Ref.	Ambient conc.
5.12-43	2-Hydroxybenzoic acid		refuse comb.	26(a)	363(a)	
5.12-44	4-Hydroxybenzoic acid				363(a)	
5.12-45	3,4-Dihydroxybenzoic acid				363(a)	
5.12-46	Methyl salicylate		plant volatile refuse comb.	501,518 26(a)		
5.12-47	Benzyl salicylate		plant volatile	506e		
5.12-48	2-Phenylethyl alcohol		plant volatile tobacco smoke	533,552e 396(a)		

TABLE 5.12 AROMATIC ALCOHOLS
Emission and Detection

Species Number	Name	Chemical Formula	Emission		Detection Ref.	Ambient conc.
			Source	Ref.		
5.12-49	Phenylmethyl-carbinol		river water odor	202		
5.12-50	Cinnamic alcohol		plant volatile	510e,533		
5.12-51	Bis(2-*o*-hydroxy-phenyl)propane		industrial	60		
5.12-52	Indanol		diesel	138		
5.12-53	1-Naphthol		tobacco smoke	396(a)		
5.12-54	2-Naphthol		tobacco smoke	396(a)	182(a)	
5.12-55	Geosmin		tap water odor	555		

263

TABLE 5.12 AROMATIC ALCOHOLS
Emission and Detection

Species Number	Name	Chemical Formula	Emission		Detection	Ambient conc.
			Source	Ref.	Ref.	
5.12-56	Cholesterol		tobacco smoke	396(a)		
5.12-57	Sigmasterol		tobacco smoke	396(a),542(a)		
5.12-58	β-Sitosterol		tobacco smoke veneer drying	396(a),542(a) 68(a)		
5.12-59	γ-Sitosterol		tobacco smoke	396(a)		

TABLE 5.12 AROMATIC ALCOHOLS
Reactions and Products

| Species Number | Reactant | Chemical reactions | | | | | Remarks |
		k	Ref.	Lifetime	Products	Ref.	
5.12-1							B
5.12-2	HO·	$(3.4 \pm 0.7) \times 10^{-11}$	1173	6.4×10^4			B
5.12-3							B
5.12-4							B

TABLE 5.13 ORGANIC PEROXIDES
Emission and Detection

| Species Number | Name | Chemical Formula | Emission | | Detection | Ambient |
			Source	Ref.	Ref.	conc.
5.13-1	Dodecanyl peroxide	$CH_3(CH_2)_{11}O_2H$	polymer mfr.	407*		
5.13-2	Cyclohexylperoxy-dicarbonate		polymer mfr.	407*		

5.14-5.17. ETHERS

5.14.1-5.17.1. Identified Compounds

Aliphatic ethers are widely utilized as industrial chemicals and solvents and the presence of most of the compounds on Table 5.14 can be attributed to these uses. Two of the simpler compounds of this class, diethyl ether and methyl propyl ether, have been detected in ambient air. The olefinic ethers and most of the cyclic ethers originate from combustion processes. A few of the aromatic ethers are also generated by combustion sources, but most are naturally produced by vegetation. They are extensively used in fragrance and flavoring applications (503-517).

5.14.2-5.17.2. Ambient Concentrations

No data are available concerning the atmospheric concentrations of these compounds.

5.14.3-5.17.3. Chemistry

Both oxygen atoms and hydroxyl radicals abstract hydrogen from aliphatic ethers (1048,1049). Aldehydes and ketones appear to be likely products. For dimethyl ether, for example,

$$CH_3OCH_3 \xrightarrow{HO\cdot} CH_3OCH_2\cdot \xrightarrow{O_2} CH_3OCH_2O_2\cdot \xrightarrow{NO} CH_3OCH_2O\cdot$$

$$CH_3OCH_2O\cdot \rightarrow HCHO + CH_3O\cdot \xrightarrow{O_2} HCHO$$

Olefinic and aromatic ethers will add the hydroxyl radical over the double bond, eventually forming multifunctional oxygenated compounds of low vapor pressure. Unsaturated cyclic ethers, however, are expected to mimic the ring cleavage chemistry of the cyclic olefins:

Because of the presence of an oxygen atom within the carbon rings and chains, the $C-H$ bonds are weaker in the ethers than in the corresponding hydrocarbon compounds (1049). As a result the ethers are much more reactive than the hydrocarbons and have estimated tropospheric lifetimes of about a day.

TABLE 5.14 ALIPHATIC ETHERS
Emission and Detection

Species Number	Name	Chemical Formula	Emission		Detection Ref.	Ambient conc.
			Source	Ref.		
5.14-1	Dimethyl ether	H_3COCH_3	propellant	559		
5.14-2	Methylpropyl ether				358	
5.14-3	Ethyl ether		industrial solvent	60 439	358,591	
5.14-4	Ethan-2-ol-ethyl ether		solvent	134		
5.14-5	Diethylene glycol		industrial solvent tobacco smoke	60 439 396		
5.14-6	Triethylene glycol		building resin industrial tobacco smoke	9* 60 396		
5.14-7	Diethoxyethane		plant volatile	552e		
5.14-8	Ethan-2-ol-butyl ether		industrial	60		

TABLE 5.14 ALIPHATIC ETHERS
Reactions and Products

Species Number	Reactant	Chemical reactions			Products	Ref.	Remarks
		k	Ref.	Lifetime			
5.14-1	HO·	$(3.5\pm0.4)\times10^{-12}$	1175	6.2×10^5			
	O	$(4.0\pm2.2)\times10^{-14}$	1174				
5.14-3	HO·	$(9.3\pm1.8)\times10^{-12}$	1049	2.4×10^5			

269

TABLE 5.15 OLEFINIC ETHERS
Emission and Detection

Species Number	Name	Chemical Formula	Emission		Detection Ref.	Ambient conc.
			Source	Ref.		
5.15-1	Vinyl methyl ether		landfill turbine	365 414		
5.15-2	Vinyl ethyl ether		turbine	414		
5.15-3	Vinyl butyl ether		turbine	414	358	
5.15-4	Vinyl isobutyl ether		turbine	414		
5.15-5	Vinyl hexyl ether		turbine	414		
5.15-6	Octadecyl vinyl ether		turbine	414		
5.15-7	Allyl ether		turbine	414		

TABLE 5.15 OLEFINIC ETHERS
Reactions and Products

Species Number	Reactant	Chemical reactions						Remarks
		k	Ref.	Lifetime	Products	Ref.	
5.15-1	HO·	$(3.4\pm0.3)\times10^{-11}$	1175	6.4×10^{4}			
	O	$(7.1\pm0.7)\times10^{-12}$	1162				
5.15-2	$O_2(^1\Delta)$				HCHO	1075	

TABLE 5.16 CYCLIC ETHERS
Emission and Detection

Species Number	Name	Chemical Formula	Emission Source	Emission Ref.	Detection Ref.	Ambient conc.
5.16-1	Ethylene oxide		auto diesel industrial	217* 261 60	525t 399(a)	
5.16-2	Propylene oxide		auto diesel petroleum mfr. turbine	284 261 111 414		
5.16-3	cis-2,3-Epoxybutane		turbine	414		
5.16-4	trans-2,3-Epoxybutane		turbine	414		
5.16-5	3-Isopropyloxetane		turbine	414		
5.16-6	Tetrahydrofuran		solvent	134		
5.16-7	2-Methyl-tetrahydrofuran		turbine	414		
5.16-8	2,2,4,4-Tetramethyl-tetrahydrofuran		auto turbine	217* 414		
5.16-9	Lilac alcohol-a		plant volatile	533		

272

TABLE 5.16 CYCLIC ETHERS
Emission and Detection

Species Number	Name	Chemical Formula	Emission		Detection Ref.	Ambient conc.
			Source	Ref.		
5.16-10	Lilac alcohol-b		plant volatile	533,552e		
5.16-11	Glucose		tobacco smoke	542(a)		
5.16-12	1,4-Dioxane		solvent	175		
5.16-13	2,6-Cineol		trees	198		
5.16-14	1,8-Cineol		plant volatile	500,501		
5.16-15	Furan		auto diesel fish oil mfr. forest fire tobacco smoke	284 1 110 354 396	358	
5.16-16	2-Methylfuran		diesel forest fire tobacco smoke wood pulping	1 354 396 19,267	358,582	
5.16-17	3-Methylfuran				208	

273

TABLE 5.16 CYCLIC ETHERS
Emission and Detection

Species Number	Name	Chemical Formula	Emission		Detection Ref.	Ambient conc.
			Source	Ref.		
5.16-18	Furfurylmethyl ether		refuse comb.	26(a)		
5.16-19	2,5-Dimethylfuran		auto diesel forest fire	284 138 354	358t	
5.16-20	Furoic acid		tobacco smoke	396(a)		
5.16-21	Methylfuroic acid		diesel	137		
5.16-22	Ethylfuran				591	
5.16-23	Diethylfuran				358t	
5.16-24	Benzofuran		auto polymer mfr. tobacco smoke	217* 251 339(a)	86	
5.16-25	2-Methylbenzofuran		tobacco smoke	339(a)	358	
5.16-26	Hydroxybenzofuran		diesel	138		

274

TABLE 5.16 CYCLIC ETHERS
Emission and Detection

Species Number	Name	Chemical Formula	Emission		Detection	Ambient conc.
			Source	Ref.	Ref.	
5.16-27	Ethylbenzofuran		tobacco smoke	339(a)		
5.16-28	Dimethylbenzofuran		diesel	138		
5.16-29	Dibenzofuran		tobacco smoke	130*(a),406(a)	86 209(a),466(a)	
5.16-30	Menthofuran		plant volatile	531		

TABLE 5.16 CYCLIC ETHERS
Reactions and Products

Species Number	Reactant	Chemical reactions				Products	Ref.	Remarks
		k	Ref.	Lifetime				
5.16-1	O	$(4.7\pm0.8)\times10^{-16}$	1230	8.5×10^{10}				
5.16-6	HO·	$(1.5\pm0.3)\times10^{-11}$	588	1.4×10^{5}				
5.16-19	$O_2(^1\Delta)$	2.7×10^{-11}	1032	1.9×10^{3}				

TABLE 5.17 AROMATIC ETHERS
Emission and Detection

Species Number	Name	Chemical Formula	Emission		Detection Ref.	Ambient conc.
			Source	Ref.		
5.17-1	Anisole		auto	217*	399(a),458t(a)	
5.17-2	p-Methylanisole		diesel plant volatile	138 507e	358	
5.17-3	Dimethylanisole		diesel	138		
5.17-4	Trimethylanisole		diesel	138		
5.17-5	Methylchavicol		plant volatile	509e,531		
5.17-6	Anethole		diesel plant volatile	138 326,515e		
5.17-7	Guaiacol		plant volatile tobacco smoke wood pulping	552e 396(a) 545(a)		
5.17-8	p-Methoxybenzyl alcohol		plant volatile	510e		
5.17-9	Zingiberene		plant volatile	510e		

277

TABLE 5.17 AROMATIC ETHERS
Emission and Detection

Species Number	Name	Chemical Formula	Emission		Detection	Ambient conc.
			Source	Ref.	Ref.	
5.17-10	Eugenol		plant volatile	326,500		
5.17-11	Isoeugenol		plant volatile	514e		
5.17-12	Methyleugenol		plant volatile	531,533		
5.17-13	1,2-Dimethoxy-benzene		diesel	138		
5.17-14	1,4-Dimethoxy-benzene		plant volatile	533		
5.17-15	Methylisoeugenol		plant volatile	514e,531		
5.17-16	Coniferyl alcohol		plant volatile	531		
5.17-17	i-Elemicin		plant volatile	531		
5.17-18	Elemicin		plant volatile	533		

TABLE 5.17 AROMATIC ETHERS
Emission and Detection

| Species Number | Name | Chemical Formula | Emission | | Detection | Ambient |
			Source	Ref.	Ref.	conc.
5.17-19	Methyl α-methoxy-β-phenylethyl ether		plant volatile	514e		
5.17-20	Methyl β-phenylethyl ether		plant volatile	531		
5.17-21	Benzyl ethyl ether		plant volatile	552e		
5.17-22	Zingerone		plant volatile	531		
5.17-23	Styrene oxide				358t	
5.17-24	Safrole		plant volatile	326,497e		
5.17-25	3,4-Methylenedioxy-1-propenylbenzene		plant volatile	515e		
5.17-26	Mericyl alcohol		plant volatile	531		
5.17-27	Myristicin		plant volatile	517e,531		

279

TABLE 5.17 AROMATIC ETHERS
Emission and Detection

Species Number	Name	Chemical Formula	Emission		Detection Ref.	Ambient conc.
			Source	Ref.		
5.17-28	Apiole		plant volatile	514e		
5.17-29	Methoxybiphenol		auto	311(a)		
5.17-30	Biphenyl ether		combustion phthal anhyd. mfr. turbine	431(a) 64* 414		
5.17-31	Methoxyphenanthrene		diesel	309(a)		
5.17-32	1,8,9-Perinaphtho-xanthene		tobacco smoke	396(a)		

TABLE 5.17 AROMATIC ETHERS
Reactions and Products

Species Number	Chemical reactions						Remarks
	Reactant	k	Ref.	Lifetime	Products	Ref.	
5.17-1	HO·	$(2.0\pm0.2)\times10^{-11}$	1173	1.1×10^5			
5.17-3							B

6

NITROGEN-CONTAINING ORGANIC COMPOUNDS

6.0. INTRODUCTION

The organic nitrogen compounds are both precursors and products in atmospheric chemical reactions. The precursors are the naturally emitted amines and nitriles, which can be expected to form mono- and difunctional oxygenated compounds as a result of atmospheric processes. A number of amines and nitriles are also emitted by a variety of anthropogenic sources.

Oxygenated organic nitrogen compounds are common products of atmospheric chemistry because of the chain termination reactions between NO or NO_2 and oxygenated radicals. As a result, reactions of the form

$$RO_2\cdot + NO_2 \rightarrow RO_2NO_2$$

are known to occur for several product compounds and are surmised to be responsible for a number of others. Although many of the products are not particularly long-lived under atmospheric conditions, they constitute "holding tanks" for reactive atmospheric free radicals and thus provide a natural modulation of the photochemical processes of the troposphere.

6.1. NITRILES

6.1.1. Identified Compounds

Fifteen nitriles comprise Table 6.1. Most are derivatives of the C_1 to C_3 aliphatic and olefinic hydrocarbons and the C_7 and C_8 aromatic hydrocarbons. The nitriles see widespread industrial use, with concomitant potential for atmospheric emission. Internal combustion engine effluent contains several aliphatic and olefinic nitriles.

Hydrogen cyanide is the most important of the atmospheric nitriles, having a variety of natural and anthropogenic sources. Few ambient studies of HCN have been performed, but it seems likely to be widely distributed throughout the troposphere.

6.1.2. Ambient Concentrations

The only nitrile whose ambient concentration has been determined is cyanogen (NCCN). The quoted numerical range, 10−25 ppb, seems unreasonably high, and the measurements should probably be regarded as preliminary.

6.1.3. Chemistry

Hydrogen cyanide reacts slowly with hydroxyl to form water and the cyano radical, which oxidizes to carbon monoxide and nitric oxide:

$$HCN \xrightarrow{HO\cdot} CN\cdot \xrightarrow{O_2} CO + NO \ .$$

The reaction of cyanogen with atomic oxygen has been found to be very slow; it appears unlikely that the compound has any significant chemical loss mechanism in the troposphere.

Nitrile chain and ring compounds will be much more reactive than are HCN and NCCN. Since the chemical attack will be at a ring or chain site rather than with the cyano group, probable chemical chains can be constructed by analogy with those of Chapter 3. The likely products are thus oxonitriles, cyanocarboxylic acids, and other multifunctional compounds.

6.2. AMINES

6.2.1. Identified Compounds

Amines are common atmospheric constituents. They are produced by microbial processes in decaying organic matter, and many of the identified sources (animal waste, rendering plants, sewage treatment) reflect some aspects of these processes. Amines are also widely used as solvents and as chemical intermediates; other entries in Table 6.2 reflect these anthropogenic sources.

The most common of the atmospheric amines are probably the methyl- and ethyl-derivatives, all of which have been found in emission. Straight- and branched-chain amines from C_3 to C_5 are also known. The most common industrial amine is aniline; other aromatic amines and several hydroxylamines also appear in Table 6.2.

TABLE 6.1 NITRILES
Emission and Detection

Species Number	Name	Chemical Formula	Emission		Detection	Ambient conc.
			Source	Ref.	Ref.	
6.1-1	Hydrogen cyanide	HCN	auto	205,415	242	
			foundry	279*		
			microbes	302		
			petroleum stor.	58*		
			plant volatile	531,583		
			plastics comb.	46,242		
			refuse comb.	13,416*		
			refuse comb.	26(a)		
			steel mfr.	105*		
			tobacco smoke	4,396		
			turbine	594		
6.1-2	Methyl cyanide	CH_3CN	industrial	58*,60		
			syn. rubber mfr.	58*		
			turbine	414		
6.1-3	Cyanogen	NCCN	auto	415	389*	10-25ppb
			refuse comb.	13		
			tobacco smoke	396		
6.1-4	Propanenitrile		turbine	414		
6.1-5	Glycolonitrile		turbine	414		
6.1-6	Acrylonitrile		auto	217*		
			industrial	58*,60		
			plastics mfr.	104		
			syn. rubber mfr.	58*		
6.1-7	2-Methyl-acrylonitrile		industrial	60		

TABLE 6.1 NITRILES
Emission and Detection

Species Number	Name	Chemical Formula	Emission		Detection	Ambient conc.
			Source	Ref.	Ref.	
6.1-8	Azodiisobutyl nitrile		polymer mfr.	407*		
6.1-9	Pentanenitrile		turbine	414		
6.1-10	Benzonitrile		refuse comb.	13	358 399(a), 458(a)	
6.1-11	Toluonitrile				458(a)	
6.1-12	Benzyl cyanide		plant volatile	531		
6.1-13	Fluorene carbonitrile		coal comb.	361(a)	362(a)	
6.1-14	Toluene-2,4-diisocyanate		industrial	60		
6.1-15	Toluene-2,6-diisocyanate		industrial	60		

TABLE 6.1 NITRILES
Reactions and Products

Species Number	Reactant	k	Ref.	Lifetime	Products	Ref.	Remarks
6.1-1	HO·	2.0×10^{-15}	1222	1.2×10^9	CO,NO	1222	
	O	$(1.1 \pm 0.2) \times 10^{-17}$	1052				
6.1-2	O	$(2.4 \pm 0.6) \times 10^{-16}$	1053	1.7×10^{11}			
6.1-3	O	$(4.0 +-0.5) \times 10^{-19}$	1051	1.0×10^{14}	CO,NO	1051	

6.2.2. Ambient Concentrations

No data are available concerning the atmospheric concentrations of these compounds.

6.2.3. Chemistry

The principal chemical loss mechanism for the amines is expected to be reaction with the hydroxyl radical. This assumption is consistant with recent measurements of the rate constants of HO· with several of the alkyl amines (1258). The reactions will presumably parallel those of the hydrocarbons, since the attack will be made away from the amino group and should produce a variety of oxygenated fragments.

6.3. NITRO COMPOUNDS

6.3.1. Identified Compounds

Nitro compounds in the atmosphere arise primarily from chemical reactions with precursor compounds rather than from direct emission. This is demonstrated by the fact that of the more than 30 compounds in Table 6.3, only nine have known sources. Six of these latter compounds are emitted from combustion processes; three arise from industrial use. The other compounds have been detected in gas or aerosol phases, and are products of atmospheric chemical chains (Section 6.3.3.).

6.3.2. Ambient Concentrations

Peroxyacetyl nitrate (PAN) is a widely measured constituent of photochemical smog, and is generally present in urban atmospheres on sunny, summer days at concentrations of several ppb. Peroxypropionyl nitrate and peroxybenzoyl nitrate are apparently somewhat less abundant; this reflects the relative concentrations of their hydrocarbon precursor radicals. Several nitro compounds have been detected in atmospheric aerosols; their combined masses may be as much as $1-2$ $\mu g/m^3$

6.3.3. Chemistry

Nitro compounds are formed by the combination of one of the oxides of nitrogen with an oxidized hydrocarbon radical, such as

$$RO_2· + NO_2 \rightarrow RO_2NO_2$$

where R is an alkyl, aryl, acyl, or hydrogen group (1132). PAN, at least, is subject to unimolecular decomposition. The compounds will

TABLE 6.2 AMINES
Emission and Detection

Species Number	Name	Chemical Formula	Emission Source	Emission Ref.	Detection Ref.	Ambient conc.
6.2-1	Hydrazine	H_2NNH_2	rocket	536		
6.2-2	Methylamine	CH_3NH_2	animal waste fish processing tobacco smoke	80,160 110,255* 396,534		
6.2-3	Dimethylamine	$(CH_3)_2NH$	animal waste fish processing tobacco smoke	160 255* 396,449		
6.2-4	Dimethylformamide	$(CH_3)_2NCHO$	industrial solvent	156 442		
6.2-5	Unsymm.dimethyl-hydrazine	$\begin{array}{c} H\ H \\ N{-}N \end{array}$	rocket	536		
6.2-6	N-Nitrosodimethyl-amine	$>NNO$	amine mfr. tobacco smoke	368,377 404,450	368,378*	16-116ng/m^3
6.2-7	Trimethylamine	$(CH_3)_3N$	animal waste fish processing microbes rendering plant sewage tmt. tobacco smoke	157,436* 58*,110 302 325,384 174* 396,534		
6.2-8	Ethylamine	$CH_3CH_2NH_2$	animal waste sewage tmt. tobacco smoke	80,160 422* 396,534	591	
6.2-9	Ethanolamine	$HOCH_2CH_2NH_2$	industrial natural gas proc.	60 423*		

288

TABLE 6.2 AMINES
Emission and Detection

Species Number	Name	Chemical Formula	Emission		Detection	Ambient conc.
			Source	Ref.	Ref.	
6.2-10	Diethylamine	$(CH_3CH_2)_2NH$	fish processing	254*,255*		
			solvent	134		
			tobacco smoke	534		
6.2-11	Diethanolamine	$(HOCH_2CH_2)_2NH$	industrial	60		
6.2-12	N-Nitrosodiethyl-amine	NNO			378	
6.2-13	Triethylamine	$(CH_3CH_2)_3N$	animal waste	80		
			sewage tmt.	422*		
			solvent	134		
6.2-14	Triethanolamine	$(HOCH_2CH_2)_3N$	industrial	60		
6.2-15	n-Propylamine	NH$_2$	animal waste	80,160		
			fish processing	254*,255*		
6.2-16	Isopropylamine	NH$_2$	animal waste	80,160		
6.2-17	Diisopropylamine	H N	sewage tmt.	422*		
6.2-18	n-Butylamine	NH$_2$	animal waste	160	358t	
			fertilizer mfr.	180		
			fish processing	254*		
			rendering plant	325		
			sewage tmt.	422*		
6.2-19	Putrescine	$H_2N(CH_2)_4NH_2$	food processing	62		
			rendering plant	325		

289

TABLE 6.2 AMINES
Emission and Detection

Species Number	Name	Chemical Formula	Emission Source	Emission Ref.	Detection Ref.	Ambient conc.
6.2-20	Dibutylamine	(structure)	sewage tmt.	422*		
6.2-21	Azo-*bis*-succinonitrile	(structure) NC–C≡N	plastics mfr.	46		
6.2-22	Diisobutylamine	(structure)	sewage tmt.	422*		
6.2-23	*sec*-Butylamine	(structure) NH$_2$	animal waste	160		
6.2-24	Pentylamine	(structure) NH$_2$	animal waste	80,160		
6.2-25	Cadaverine	$H_2N(CH_2)_5NH_2$	food processing rendering plant	62 325		
6.2-26	Glutamic acid	(structure) HOOC–...–COOH, NH$_2$	tobacco smoke	396(a)		
6.2-27	Glutamine	(structure) H$_2$N–...–COOH, NH$_2$	tobacco smoke	396(a)		
6.2-28	Hexylamine	(structure) NH$_2$	sewage tmt.	422*		
6.2-29	1,6-Hexanediamine	$H_2N(CH_2)_6NH_2$	industrial	60		
6.2-30	Formin	(structure)	explosives mfr.	60		

TABLE 6.2 AMINES
Emission and Detection

Species Number	Name	Chemical Formula	Emission Source	Emission Ref.	Detection Ref.	Ambient conc.
6.2-31	Aniline	(aniline structure, NH_2)	industrial plastics comb. tobacco smoke	60,250 354 534(a)	528(a)	
6.2-32	p-Toluidine	(p-toluidine structure, NH_2)	industrial plastics comb. tobacco smoke	250 354 534		
6.2-33	3,4-Xylidine	(3,4-xylidine structure, NH_2)			274	
6.2-34	N-sec-Butylaniline	(N-sec-butylaniline structure)	vulcanization	308		
6.2-35	Veritol	(veritol structure, OH / NH)	plant volatile	533		
6.2-36	Methylanthranilate	(methylanthranilate structure, H_2N)	plant volatile	531,552e		
6.2-37	Dimethylanthranilate	(dimethylanthranilate structure, CH_3, H_2N)	plant volatile	514e		
6.2-38	Ethylanthranilate	(ethylanthranilate structure, H_2N)	plant volatile	552e		
6.2-39	Propoxur	(propoxur structure)	pesticide	154*,573*		

TABLE 6.2 AMINES
Emission and Detection

Species Number	Name	Chemical Formula	Emission		Detection Ref.	Ambient conc.
			Source	Ref.		
6.2-40	OMS-33		pesticide	154*		
6.2-41	Diphenylamine				570(a)	
6.2-42	α-Naphthylamine				528(a)	
6.2-43	Carbaryl		pesticide pesticide	126 347(a)		

TABLE 6.2 AMINES
Reactions and Products

Species Number	Chemical reactions						Remarks
	Reactant	k	Ref.	Lifetime	Products	Ref.	
6.2-2	HO·	$(2.2 \pm 0.2) \times 10^{-11}$	1056	9.9×10^4			
6.2-3	HO· HNO_2	$(6.5 \pm 0.7) \times 10^{-11}$ 5.3×10^{-17}	1258 1054	3.8×10^4	$(CH_3)_2NNO$	1054	
6.2-6	$h\nu$				NO	1054	B
6.2-7	HO·	$(6.1 \pm 0.6) \times 10^{-11}$	1258	4.0×10^4			
6.2-8	HO·	$(2.8 \pm 0.3) \times 10^{-11}$	1258	8.7×10^4			

TABLE 6.3 NITRO COMPOUNDS
Emission and Detection

Species Number	Name	Chemical Formula	Emission		Detection Ref.	Ambient conc.
			Source	Ref.		
6.3-1	Nitromethane	CH_3NO_2	auto tobacco smoke turbine	217*,415 396 414		
6.3-2	Methylnitrate	CH_3ONO_2			173t	
6.3-3	Nitroethane	$CH_3CH_2NO_2$	auto	415		
6.3-4	Ethylnitrate	$CH_3CH_2ONO_2$			186,263	
6.3-5	Peroxyacetylnitrate	$CH_3C(O)O_2NO_2$			231*,329	0.1-66ppb
6.3-6	2-Nitropropane	(structure, NO_2)	solvent	88,134		
6.3-7	2-Methyl-2-nitropropane	(structure, NO_2)	turbine	414		
6.3-8	n-Propylnitrate	(structure, ONO_2)			307	
6.3-9	Peroxypropionyl-nitrate	(structure, O_2NO_2)			173,231*	1-5ppb
6.3-10	n-Butylnitrate	(structure, ONO_2)			307	
6.3-11	Neopentylnitrate	(structure, NO_2)	turbine	414		
6.3-12	5-Nitroxypentanal	(structure, OHC ... ONO_2)			214t*(a)	.3-1.01$\mu g/m^3$

TABLE 6.3 NITRO COMPOUNDS
Emission and Detection

Species Number	Name	Chemical Formula	Emission Source	Emission Ref.	Detection Ref.	Ambient conc.
6.3-13	5-Nitroxypentanoic acid				214*(a),532(a)	120ng/m^3
6.3-14	5-Peroxynitroso-pentanal				214*(a),532(a)	.07-1.01μg/m^3
6.3-15	5-Peroxynitro-pentanal				214t*(a)	140ng/m^3
6.3-16	5-Peroxynitro-pentanoic acid				532(a)	
6.3-17	6-Nitroxyhexanal				214*(a)	40-400ng/m^3
6.3-18	6-Nitroxyhexanoic acid				214*(a),532(a)	150ng/m^3
6.3-19	6-Peroxynitroso-hexanal				214*(a),532(a)	40-160ng/m^3
6.3-20	6-Peroxynitro-hexanal				214*(a)	40-240ng/m^3
6.3-21	6-Peroxynitro-hexanoic acid				214*(a),532(a)	30ng/m^3
6.3-22	7-Nitroxyheptanal				214t*(a)	40-270ng/m^3
6.3-23	7-Nitroxyhep-tanoic acid				532(a)	
6.3-24	7-Peroxynitroso-heptanal				532(a)	

TABLE 6.3 NITRO COMPOUNDS
Emission and Detection

Species Number	Name	Chemical Formula	Emission		Detection	Ambient conc.
			Source	Ref.	Ref.	
6.3-25	7-Peroxynitrohep-tanal	(formula)			214*(a)	40ng/m^3
6.3-26	7-Peroxynitrohep-tanoic acid	(formula)			214*(a),532(a)	170ng/m^3
6.3-27	2-Oxo-1-nitrooctane	(formula)	turbine	414		
6.3-28	Nitrobenzene	(formula)	industrial solvent	60 439		
6.3-29	Nitrophenol	(formula)			307,411	
6.3-30	4-Methyl-2-nitrophenol	(formula)	tobacco smoke	447*	411	
6.3-31	p-Nitroaniline	(formula)	industrial	250		
6.3-32	Peroxybenzoyl-nitrate	(formula)			549*	0.03-4.6ppb

TABLE 6.3 NITRO COMPOUNDS
Reactions and Products

Species Number	Reactant	Chemical reactions					Remarks
		k	Ref.	Lifetime	Products	Ref.	
6.3-1	HO·	$(9.2\pm1.0)\times10^{-13}$	1062	2.7×10^6			
6.3-2	O	$(3.2\pm0.1)\times10^{-15}$	1177	1.3×10^{10}			
6.3-4	O	$(1.2\pm0.1)\times10^{-14}$	1177	3.3×10^9			
6.3-5	U	2.6×10^{-4}	1132				B
	HO·	$\leq1.7\times10^{-13}$	588	3.9×10^3			
6.3-13							B
6.3-28	HO·					1136	B
6.3-29							B
6.3-32							B

also be susceptible to hydroxyl attack on alkyl and phenyl groups, but the measured reaction rate constants indicate that these loss paths are not particularly favored.

The difunctional compounds containing a nitro group are expected to have low vapor pressures, and thus to be susceptible to incorporation into the atmospheric aerosol. The detection of such compounds in aerosols appears to confirm that heterogeneous loss is the primary fate of many of the atmospheric nitro compounds.

6.4. HETEROCYCLIC NITROGEN COMPOUNDS

6.4.1. Identified Compounds

A surprisingly large number of heterocyclic nitrogen compounds are known to be emitted into the atmosphere, mostly by combustion processes. Many of these compounds have also been identified in ambient aerosols. The more common compounds of this group that are found in tobacco smoke are included here because other combustion of vegetation might be expected to produce them as well. An extensive review of all the tobacco smoke nitrogen compounds has recently appeared (534), however, and interested readers are referred to that work for more information. Table 6.4 demonstrates that the pyridines, pyrroles, indoles, carbazoles, and some of the larger polynuclear heterocyclic nitrogen compounds appear to be relatively common in the atmosphere.

6.4.2. Ambient Concentrations

The concentrations of the heterocyclic nitrogen compounds have not often been measured, but appear from the available evidence to be very small. Piperizine, carbazole, phenyl piperidine, and caffeine (presumably from coffee roasting) have been detected in the nanograms per cubic meter range; other compounds are present at picogram levels.

6.4.3. Chemistry

The atmospheric chemistry of the heterocyclic nitrogen compounds is unexplored. Their near-universal occurrence only in the aerosol phase suggests that gas phase chemistry can be ignored, however. In solution, the favored process would appear to be oxidation reactions similar to those of the polynuclear aromatic hydrocarbons (Section 3.7). As can be seen in Table 6.4, the ketones that would be typical products of such reactions have been detected in tobacco smoke.

TABLE 6.4 HETEROCYCLIC NITROGEN COMPOUNDS
Emission and Detection

Species Number	Name	Chemical Formula	Emission		Detection Ref.	Ambient conc.
			Source	Ref.		
6.4-1	Pyrrole		tobacco smoke	396	399(a), 458(a)	
6.4-2	Methylpyrrole				458(a)	
6.4-3	Methylpyrrolidone		solvent	134		
6.4-4	Imidazole		turbine	414		
6.4-5	2,4-Dimethyl-imidazoline		turbine	414		
6.4-6	Ethylimidazole				399(a)	
6.4-7	Dimethylpyrazole				399(a)	
6.4-8	Ethylpyrazole				399(a)	
6.4-9	Pyridine		coke oven tobacco smoke	387 396	528(a)	
6.4-10	2-Methylpyridine		tobacco smoke	396(a)		
6.4-11	3-Methylpyridine		tobacco smoke	396(a)	214(a)	
6.4-12	2,6-Dimethyl-pyridine		tobacco smoke	396(a)		

TABLE 6.4 HETEROCYCLIC NITROGEN COMPOUNDS
Emission and Detection

| Species Number | Name | Chemical Formula | Emission | | Detection | Ambient |
			Source	Ref.	Ref.	conc.
6.4-13	Ethylpyridine				458(a)	
6.4-14	2-Methyl-4-ethyl-pyridine		tobacco smoke	396(a)		
6.4-15	2-Methyl-5-ethyl-pyridine				202	
6.4-16	Pyridine-3-aldehyde		tobacco smoke	396(a)		
6.4-17	3-Pyridine carboxylic acid		tobacco smoke	396(a)		
6.4-18	α-Cyanopyridine				358	
6.4-19	Nicotinamide		tobacco smoke	396(a)		
6.4-20	2-(3-Pyridyl)-2-ethanone		tobacco smoke	396(a)		
6.4-21	3-(3-Pyridyl)-3-propanone		tobacco smoke	396(a)		
6.4-22	4-(3-Pyridyl)-4-butanone		tobacco smoke	396(a)		
6.4-23	Nicotine		tobacco smoke	356,421		
6.4-24	Nornicotine		tobacco smoke	396(a),449(a)		

TABLE 6.4 HETEROCYCLIC NITROGEN COMPOUNDS
Emission and Detection

Species Number	Name	Chemical Formula	Emission		Detection	Ambient conc.
			Source	Ref.	Ref.	
6.4-25	Nornicotyrine		tobacco smoke	396(a),534(a)		
6.4-26	Nicotyrine		tobacco smoke	396(a),534(a)		
6.4-27	Anabasine		tobacco smoke	396(a),534(a)		
6.4-28	Anatabine		tobacco smoke	396(a),534(a)		
6.4-29	Phenylpiperidine				214*(a)	3-90ng/m^3
6.4-30	Piperine		plant volatile	531		
6.4-31	Solanesol		tobacco smoke	396(a)		
6.4-32	Piperizine		diesel	1	214*(a)	10-60ng/m^3
6.4-33	2,5-Dimethyl-piperazine					
6.4-34	3,6-Dipropyl-1,2,4,-5-tetrazine		turbine	414		

TABLE 6.4 HETEROCYCLIC NITROGEN COMPOUNDS
Emission and Detection

Species Number	Name	Chemical Formula	Emission		Detection	Ambient conc.
			Source	Ref.	Ref.	
6.4-35	Indole		animal waste food processing plant volatile tobacco smoke	166,240 62 510e 534(a)		
6.4-36	Skatole		animal waste drug mfr. food processing rendering plant tobacco smoke	25,166 193 62 325 130*(a)		
6.4-37	Ethylindole		tobacco smoke	130*(a)		
6.4-38	Isoquinoline				363(a),488*(a)	140-180pg/m^3
6.4-39	Caffeine				546*(a),548(a)	3.4-7.0ng/m^3
6.4-40	Benzo[f]isoquinoline				488*(a)	34-110pg/m^3
6.4-41	11-H-Indeno[1,2-b]-quinoline		coal comb.	98(a)	98(a),488*(a)*	100pg/m^3
6.4-42	Indeno[1,2,3-i,j]-isoquinoline		coal comb.	98(a)		

302

TABLE 6.4 HETEROCYCLIC NITROGEN COMPOUNDS
Emission and Detection

Species Number	Name	Chemical Formula	Emission		Detection Ref.	Ambient conc.
			Source	Ref.		
6.4-43	Quinoline		tobacco smoke	396(a)	363(a),488*(a)	22-69pg/m³
6.4-44	Benzo[f]quinoline		coal comb.	98(a),361(a)	213(a),488*(a)	10-200pg/m³
6.4-45	Benzo[h]quinoline		coal comb.	98(a)	213(a),488*(a)	10-300pg/m³
6.4-46	Carbazole		aluminum mfr. tobacco smoke	551* 130*(a),534(a)	209(a),214*(a)	2-50ng/m³
6.4-47	Methylcarbazole		tobacco smoke	130*(a)		
6.4-48	Benzo[a]carbazole				182(a),363(a)	
6.4-49	Benzo[c]carbazole				363(a)	

TABLE 6.4 HETEROCYCLIC NITROGEN COMPOUNDS
Emission and Detection

Species Number	Name	Chemical Formula	Emission		Detection	Ambient conc.
			Source	Ref.	Ref.	
6.4-50	4-Azafluorene				488*(a)	5pg/m^3
6.4-51	Phenanthridine		coal comb.	98(a)	488*(a),528(a)	18-22pg/m^3
6.4-52	Benzo[l,m,n]phenanthridine		coal comb.	98(a)		
6.4-53	Acridine		coal comb refuse comb.	98(a),361(a) 26(a)	362(a),488*(a)	40-41pg/m^3
6.4-54	Methylacridine		refuse comb.	26(a)		
6.4-55	Benz[a]acridine		coal comb.	98(a)	98(a),343(a)	200pg/m^3
6.4-56	Benz[c]acridine		auto coal comb.	311(a) 98(a)	98(a)	600pg/m^3

TABLE 6.4 HETEROCYCLIC NITROGEN COMPOUNDS
Emission and Detection

Species Number	Name	Chemical Formula	Emission		Detection Ref.	Ambient conc.
			Source	Ref.		
6.4-57	Dibenz[a,h]-acridine		coal comb.	98(a)	98(a),363(a)	80pg/m^3
6.4-58	Dibenz[a,j]-acridine		coal comb.	98(a)	98(a),363(a)	40pg/m^3
6.4-59	1-Azafluoranthene				488*(a)	5pg/m^3
6.4-60	4-Azapyrene				488*(a)	21-22pg/m^3

7

SULFUR-CONTAINING ORGANIC COMPOUNDS

7.0. INTRODUCTION

The organic sulfur compounds appear to be only moderately common atmospheric species. They have not been extensively monitored and it is, in fact, uncertain whether any of them play significant roles in the natural atmospheric cycle of sulfur (1186). Most of the known compounds in this group contain sulfur in the -2 valence state (typical of the reduced species emitted by natural bacterial processes). They almost certainly undergo chemical transformations to more highly oxidized states and thus contribute to the natural background of sulfate aerosol and acid rain. Through carbonyl sulfide, this group of compounds also participates in stratospheric chemistry (1015). The sulfur-containing organic compounds thus enter, in very poorly understood ways, into a variety of interesting and important atmospheric chemical processes.

7.1. MERCAPTANS

7.1.1. Identified Compounds
A dozen different mercaptans are known to be emitted into the air, where they are readily detected by smell at low concentrations. Mercaptans are produced by natural microbial processes, and are found in emissions from animal waste, sewage treatment, and rendering.

Most of the anthropogenic mercaptan emissions come from two sources: wood pulping and natural gas additive emission. Wood is commonly processed by the Kraft technique that involves steam-cooking the fibers in a "liquor" containing sodium sulfite. The resulting gases are rich in reduced sulfur compounds. Mercaptans are added to natural gas in very small amounts because their low odor threshold makes them extremely useful as leak detectors. Mercaptan emissions from sources other than those mentioned above appear to be negligible.

7.1.2. Ambient Concentrations

Only methyl and propyl mercaptans have been detected in ambient air, and methyl mercaptan is the only compound of its class for which a concentration has been determined. The value (4 ppb) should probably be regarded as an upper limit to the concentration of this compound in the atmosphere.

7.1.3. Chemistry

Two distinct chemical processes appear likely for mercaptans in the troposphere. With oxygen atoms, the favored process is one of addition followed by cleavage to form HSO radicals (1178):

$$RSH \xrightarrow{O} R \cdot + HSO \cdot .$$

The HSO· radical will then be promptly oxidized to sulfur dioxide by (1014)

$$HSO \cdot \xrightarrow{O_2} SO \xrightarrow{O_2} SO_2 .$$

The mercaptan concentrations are probably not controlled by the oxygen chain, however, but by reaction with hydroxyl radicals, which abstract an alkyl hydrogen atom. Here the subsequent steps are less certain, but oxidation of the sulfur atom to SO_2 through a chain of the sort shown below seems likely:

$$CH_3SH \xrightarrow{HO \cdot} HSCH_2 \cdot \xrightarrow{O_2} HSCH_2O_2 \cdot \xrightarrow{NO} HSCH_2O \cdot$$

$$HSCH_2O \cdot \rightarrow HS \cdot \xrightarrow{O_2} SO \xrightarrow{O_2} SO_2 .$$

A review of atmospheric sulfur chemistry has indicated that the gas phase transformations of reduced sulfur compounds to sulfate aerosol can be described by a common chain consisting of five chemical steps followed by a heterogeneous loss process (1014). The schematic diagram of this chain is pictured in Fig. 7.1-1.

TABLE 7.1 MERCAPTANS
Emission and Detection

Species Number	Name	Chemical Formula	Emission		Detection		Ambient conc.
			Source	Ref.	Ref.		
7.1-1	Methyl mercaptan	CH_3SH	animal waste	25,157	165,566*		4ppb
			microbes	210,302			
			natural gas	262			
			petroleum mfr.	252*			
			plant volatile	584			
			rendering plant	325			
			sewage tmt.	174*,204			
			starch mfr.	523*			
			wood pulping	19,58*			
7.1-2	Ethyl mercaptan	CH_3CH_2SH	animal waste	25			
			microbes	302			
			natural gas	262,357			
			petroleum mfr.	252*			
			sewage tmt.	422*			
			wood pulping	238			
7.1-3	Propyl mercaptan	![SH structure]	animal waste	25	364(a)		
			fish processing	385*			
			natural gas	262,357			
			onion odor	326			
			sewage treatment	174*,422*			
			wood pulping	238			
7.1-4	Isopropyl mercaptan	![SH structure]	natural gas	262,357			
7.1-5	1-Mercapto-2-2-dimethylpropane	![SH structure]	sewage tmt.	422*			
7.1-6	n-Butyl mercaptan	![SH structure]	natural gas	262			
			skunk odor	326			

308

TABLE 7.1 MERCAPTANS
Emission and Detection

Species Number	Name	Chemical Formula	Emission		Detection Ref.	Ambient conc.
			Source	Ref.		
7.1-7	Isobutyl mercaptan	(structure, –SH)	natural gas	262,357		
7.1-8	*tert*-Butyl mercaptan	(structure, –SH)	natural gas sewage tmt.	262,357 422*		
7.1-9	*n*-Pentyl mercaptan	(structure, –SH)	sewage tmt.	174*,422*		
7.1-10	1-Mercapto-2-methyl-pentane	(structure, –SH)	turbine	414		
7.1-11	2-Methyl-2-mercapto-pentane	(structure, SH)	mercaptan mfr.	143		
7.1-12	*n*-Octyl mercaptan	$CH_3(CH_2)_7SH$	turbine	414		

TABLE 7.1 MERCAPTANS
Reactions and Products

Species Number	Reactant	Chemical reactions					Remarks
		k	Ref.	Lifetime	Products	Ref.	
7.1-1	HO·	$(3.4\pm0.3)\times10^{-11}$	1056	6.4×10^4			
	O	$(3.2\pm0.6)\times10^{-12}$	1178		SO_2	1059	
	O_3						
7.1-2	O	$(4.7\pm0.9)\times10^{-12}$	1178	8.5×10^6			

310

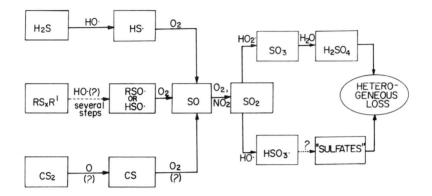

Fig. 7.1-1.   The principal homogeneous reaction paths of atmospheric sulfur compounds (from reference 1014).

7.2. ORGANIC SULFIDES

7.2.1. Identified Compounds

Although 11 organic sulfides have been identified in atmospheric emissions (Table 7-2), six are present only as odorant additives to natural gas. Of the other five, three are widely produced in nature. The processes are microbial, paralleling those for the lower mercaptans. Both oceanic and continental natural sources have been identified for dimethyl sulfide, which is probably the most common of the organic sulfur compounds in the atmosphere.

7.2.2. Ambient Concentrations

Dimethyl sulfide concentrations have recently been measured in the ambient atmosphere; the low concentrations, together with dimethyl sulfide's relatively short lifetime, suggest that the compound does not play a major role in the atmospheric sulfur cycle. More important roles are possible for CS_2 and COS which are rather widely emitted, in part by anthropogenic processes. Both are present in concentrations of several hundred ppt, and the uniformity of COS concentrations suggests a long atmospheric lifetime (1241) and possible participation in the chemistry of the stratospheric ozone layer (1015). The concentrations of CS_2, not yet as widely measured, are somewhat less homogeneous.

7.2.3. Chemistry

The atmospheric chemistry of the organic sulfides is presumably initiated by hydroxyl abstraction of a hydrogen atom. The subsequent processes have not been determined. It has been suggested, however (1014), that the sulfur is eventually separated from the carbon-containing molecular fragments in the form of SO and rapidly oxidized to SO_2 by O_2 or O_3.

A competing process for the hydroxyl reaction is the addition of atomic oxygen, followed by two cleavage reactions to form alkyl radicals (1178) and sulfur monoxide (1050). For dimethyl sulfide,

$$CH_3SCH_3 \xrightarrow{O} [CH_3S(O)CH_3]^* \rightarrow CH_3{\cdot}+[CH_3S(O)\cdot]^* \rightarrow CH_3{\cdot}+SO.$$

Ozone is also known to react with the sulfides to produce SO_2, presumably through an SO intermediate (1059).

Carbon disulfide reacts with atomic oxygen by addition (1179), followed by cleavage to form CO and SO. The reaction of CS_2 with HO· will be favored in the atmosphere, however. If the reaction proceeds by S abstraction, the resulting HOS· radical will probably add molecular oxygen and its subsequent chemistry will follow that of the sulfur dioxide chain.

$$CS_2 + HO\cdot \rightarrow CS + HOS\cdot$$

The CS molecule will presumably be oxidized to SO and hence to SO_2. An alternative mechanism (1264) involves addition

$$CS_2 + HO\cdot \longrightarrow \left[\begin{array}{c} S\diagdown{}_{\underset{OH}{|}}{C}\cdot{}^{\diagup S} \end{array} \right]^* \longrightarrow COS + SH\cdot,$$

in which case CS_2 serves as a chemical source for COS.

Carbonyl sulfide reacts slowly with tropospheric species. Its principal chemical chain is inaugurated by S abstraction by HO·:

$$COS + HO\cdot \rightarrow CO + HOS\cdot ,$$

and will then follow the path outlined above for the HOS· radical produced by CS_2.

TABLE 7.2 SULFIDES
Emission and Detection

Species Number	Name	Chemical Formula	Emission		Detection Ref.	Ambient conc.
			Source	Ref.		
7.2-1	Dimethyl sulfide	CH_3SCH_3	algae	464	140,491*	42-62ppt
			animal waste	25,436*	399(a)	
			microbes	210,302		
			natural gas	262,357		
			petroleum mfr.	252*		
			plant volatile	584		
			rendering plant	325		
			sewage tmt.	174*		
			starch mfr.	523*		
			trees	140,199		
			wood pulping	19,111*		
7.2-2	Ethylmethyl sulfide	$CH_3CH_2SCH_3$	wood pulping	238		
7.2-3	Diethyl sulfide	$(CH_3CH_2)_2S$	animal waste	327	199	
			natural gas	262		
			wood pulping	238		
7.2-4	Di-n-propyl sulfide		natural gas	262		
7.2-5	Diisopropyl sulfide		natural gas	262		
7.2-6	Diallyl sulfide		plant volatile	584		
			sewage tmt.	422*		
7.2-7	Dimethyl disulfide	CH_3SSCH_3	animal waste	436*,437	199,263	
			microbes	210,302		
			natural gas	262		
			plant volatile	584		
			wood pulping	19,58*		

TABLE 7.2 SULFIDES
Emission and Detection

| Species Number | Name | Chemical Formula | Emission | | Detection | Ambient |
			Source	Ref.	Ref.	conc.
7.2-8	Methylallyldi-sulfide	CH_3SS ⌇	plant volatile	584		
7.2-9	Diethyl disulfide	$CH_3CH_2SSCH_2CH_3$	natural gas	262	399(a)	
7.2-10	Ethylmethyl disulfide	$CH_3CH_2SSCH_3$	natural gas	262		
7.2-11	Diallyldisulfide	⌇SS⌇	plant volatile	584		
7.2-12	Dimethyl trisulfide	CH_3SSSCH_3	natural gas / plant volatile	262 / 584		
7.2-13	Methylpropyltri-sulfide	CH_3SSS ⌇	plant volatile	584		
7.2-14	Methylallyltri-sulfide	CH_3SSS ⌇	plant volatile	584		
7.2-15	Diethyl trisulfide	$CH_3CH_2SSSCH_2CH_3$	natural gas	262		

TABLE 7.2 SULFIDES
Emission and Detection

Species Number	Name	Chemical Formula	Emission		Detection	Ambient conc.
			Source	Ref.	Ref.	
7.2-16	Carbon disulfide	CS_2	animal waste	522	100*,258	70-370ppt
			fish processing	385*	399(a)	
			natural gas	423*,454*		
			petroleum mfr.	127*,252*		
			plastics comb.	354		
			rubber abrasion	331		
			rubber mfr.	119*		
			refuse comb.	439		
			refuse comb.	26(a)		
			starch mfr.	523*		
			synthetic fibre mfr.	58*,191		
			turbine	414		
			volcano	234		
7.2-17	Carbonyl sulfide	COS	animal waste	522	89*,100*	200-560ppt
			auto	418*	399(a)	
			fish processing	385*		
			forest fire	354		
			natural gas	423*,454*		
			petroleum mfr.	252*		
			plastics comb.	354		
			refuse comb.	26(a)		
			starch mfr.	523*		
			synthetic fibre mfr.	191,253*		
			tobacco smoke	396		
			volcano	234		

TABLE 7.2 SULFIDES
Reactions and Products

Species Number	Reactant	Chemical reactions					Remarks
		k	Ref.	Lifetime	Products	Ref.	
7.2-1	HO·	$(9.8 \pm 1.2) \times 10^{-12}$	1266	2.5×10^5			
	O	$(4.8 \pm 0.6) \times 10^{-11}$	1050				
	O_3				SO_2	1059	
7.2-16	HO·	$(1.85 \pm 0.34) \times 10^{-13}$	1264	1.0×10^7			
	O	$(4.0 \pm 0.3) \times 10^{-12}$	1179		CO, SO_2	1179	
	O_3				SO_2	1060	
7.2-17	HO·	$(5.66 \pm 1.21) \times 10^{-14}$	1264	4.0×10^7			
	O	1.3×10^{-14}	1058		CO, SO_2	1058	

7.3. HETEROCYCLIC SULFUR COMPOUNDS

7.3.1. Identified Compounds

This group consists of several five-membered ring structures and several polynuclear compounds. The former have a few identified sources, none of which appear to be substantial. No sources have been determined for the latter (except for benzothiazole, a byproduct of vulcanization). By analogy with other polynuclear compounds in the atmosphere (Sections 3.7. and 6.4.), however, it seems reasonable to assume that they are products of hydrocarbon combustion.

7.3.2. Ambient Concentrations

No data are available concerning the atmospheric concentrations of these compounds.

7.3.3. Chemistry

Experimental work provides some indication that the chemistry of the five-membered ring structures should be similar to that of the cyclic hydrocarbons. For thiophene, this would suggest ozone addition over a double bond, followed by formation of difunctional compounds. The observation of SO_2 chemiluminescence, presumably from the $SO+O_3$ reaction, in a thiophene $-O_3$ system (1060) indicates a chemical chain of the form

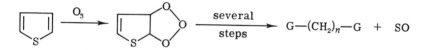

where G is CHO, CH_2OH, or COOH.

The reactions of the polynuclear compounds will occur in aerosol solution and should be analogous to those of the polynuclear aromatics. Oxygenated products, particularly ketones, are therefore to be expected.

TABLE 7.3 HETEROCYCLIC SULFUR COMPOUNDS
Emission and Detection

Species Number	Name	Chemical Formula	Emission		Detection Ref.	Ambient conc.
			Source	Ref.		
7.3-1	Thiophane		natural gas	262,357		
7.3-2	Thiophene		natural gas	454*		
			petroleum mfr.	252*		
			wood pulping	19,267		
7.3-3	Trimethylthiophene		diesel	176		
7.3-4	5-Propynyl-2-formylthiophene		microbes	302		
7.3-5	Benzothiophene				566*	40ppt
7.3-6	Dimethylbenzo-thiophene		diesel	137		
7.3-7	Dibenzo-thiophene				209(a), 408(a)	
7.3-8	Methyldibenzo-thiophene				408(a)	
7.3-9	Ethyldibenzo-thiophene				408(a)	

TABLE 7.3 HETEROCYCLIC SULFUR COMPOUNDS
Emission and Detection

Species Number	Name	Chemical Formula	Emission		Detection Ref.	Ambient conc.
			Source	Ref.		
7.3-10	Naphthobenzo-thiophene				408(a), 466(a)	
7.3-11	Methylnaphthobenzo-thiophene				408(a)	
7.3-12	Benzothiazole		vulcanization	308	86 488(a)	

TABLE 7.3 HETEROCYCLIC SULFUR COMPOUNDS
Reactions and Products

Species Number	Reactant	Chemical reactions					Remarks
		k	Ref.	Lifetime	Products	Ref.	
7.3-2	O_3				SO_2	1060	

7.4. THIO ACIDS, THIOCYANATES

7.4.1. Identified Compounds

Only four members of this minor class of atmospheric compounds are known and the identified sources do not suggest that they are likely to be ubiquitous or highly concentrated in the atmosphere.

7.4.2. Ambient Concentrations

No data are available concerning the atmospheric concentrations of these compounds.

7.4.3. Chemistry

No direct gas phase studies of any of these compounds have been reported. By analogy with other atmospheric sulfur and hydrocarbon compounds, however, the primary reactants are inferred to be HO· (for $HSCH_2COOH$), O (for HSCN), O (for $(SCN)_2$), and HO· (for $CH_2=CHCH_2NCS$). The lifetime of each of these compounds is probably of the order of a few days in the troposphere.

TABLE 7.4 THIO ACIDS, THIOCYANATES
Emission and Detection

| Species Number | Name | Chemical Formula | Emission | | Detection | Ambient |
			Source	Ref.	Ref.	conc.
7.4-1	Mercaptoacetic acid	$HSCH_2COOH$	food processing	62		
7.4-2	Hydrogen thiocyanide	HSCN	tobacco smoke	396(a)		
7.4-3	Thiocyanogen	$(SCN)_2$	tobacco smoke	396(a)		
7.4-4	Allylisothiocyanate	⟍⟍NCS	food processing	62		
			vegetation	531		

8

ORGANIC HALOGENATED COMPOUNDS

8.0. INTRODUCTION

The emission of halogenated organics into the air by natural processes is not common, the chlorinated derivatives of methane being the only such compounds known to have natural sources. As a result of the wide commercial and industrial utility of halogenated organics, however, many anthropogenically produced species are ubiquitous atmospheric constituents.

Organic halogenated compounds as a group play a very minor role in tropospheric chemistry. This is in part because the concentrations of most of the compounds are very low and in part because those that are emitted in large quantities are fully halogenated compounds that are chemically inert in the troposphere. This latter property, however, renders them extremely important to the chemistry of the stratosphere.

Three groups of halogenated organics are designated: halogenated alkanes and alkenes, halogenated aromatics, and chlorinated pesticides.

8.1. HALOGENATED ALKANES AND ALKENES

8.1.1. Identified Compounds

This grouping comprises more than 40 compounds, nearly all of them anthropogenic. The widespread use of many of these materials as propellants, refrigerants, and solvents is responsible for their ubiquitous occurrence. Methyl chloride is produced by burning vegetation, and both methyl chloride and methyl iodide are apparently emitted from the oceans. Several of the halocarbons have been detected in active volcano emissions.

TABLE 8.1 HALOGENATED ALIPHATIC AND OLEFINIC COMPOUNDS
Emission and Detection

Species Number	Name	Chemical Formula	Emission		Detection Ref.	Ambient conc.
			Source	Ref.		
8.1-1	Carbon tetrafluoride	CF_4	aluminum mfr.	256*	72*	0.1-1ppb
8.1-2	Methyl chloride	CH_3Cl	chemical mfr.	559,560	85*,321*	0.8-2.2ppb
			forest fire	354,495*		
			plant volatile	547		
			polymer comb.	211,304		
			propellant	562		
			refuse comb.	26(a)		
			solvent	439		
			tobacco smoke	298,396		
			turbine	414		
			volcano	234		
8.1-3	Methylene chloride	CH_2Cl_2	chemical mfr.	462*,559	358,392	<5ppt
			foaming agent	562		
			landfill	365		
			solvent	50,60		
8.1-4	Chloroform	$CHCl_3$	chemical mfr.	559	85*,135*	4-250ppt
			landfill	365	213(a)	
			plant volatile	547		
			refuse comb.	26(a)		
			sewage tmt.	204		
			solvent	60,559		
			water treatment	457,487		

TABLE 8.1 HALOGENATED ALIPHATIC AND OLEFINIC COMPOUNDS
Emission and Detection

Species Number	Name	Chemical Formula	Emission		Detection Ref.	Ambient conc.
			Source	Ref.		
8.1-5	Carbon tetrachloride	CCl_4	chemical mfr. fumigating agent petroleum stor. plant volatile refuse comb. solvent	559 560 58* 547 26(a) 60,439	89*,224* 213(a)	0.4-260ppt
8.1-6	Methyl bromide	CH_3Br	fumigating agent turbine	560,595 414	373	<5-30ppt
8.1-7	Bromoform	$CHBr_3$			358,462	
8.1-8	Methyl iodide	CH_3I			135*,322	1.2-40ppt
8.1-9	Fluorochloromethane	CH_2ClF			358	
8.1-10	Fluorodichloro-methane	$CHCl_2F$	air cond. plastics comb. volcano	323 354 234	358	
8.1-11	Chlorodifluoro-methane	$CHClF_2$	propellant refrigeration volcano	559 323,560 234		
8.1-12	Fluorotrichloro-methane	$CFCl_3$	air cond. foaming agent plastics comb. propellant refrigeration volcano	323 559,560 354 559,560 559,560 234	85*,135*	50-800ppt

TABLE 8.1 HALOGENATED ALIPHATIC AND OLEFINIC COMPOUNDS
Emission and Detection

Species Number	Name	Chemical Formula	Emission		Detection	Ambient conc.
			Source	Ref.	Ref.	
8.1-13	Dichlorodifluoro-methane	CF_2Cl_2	air cond. foaming agent propellant refrigeration	323 559,560 559,560 559,560	85*,135*	80-1000ppt
8.1-14	Chlorotrifluoro-methane	CF_3Cl	refrigeration	562		
8.1-15	Bromotrifluoro-methane	CF_3Br	refrigeration	563		
8.1-16	Chloropentafluoro-ethane	$CClF_2CF_3$	propellant refrigeration	559,562 563		
8.1-17	Perfluoroethane	CF_3CF_3	aluminum mfr.	256*		
8.1-18	Ethyl chloride	CH_3CH_2Cl	chemical mfr. forest fire plastics comb. refuse comb. solvent	559 354 354 26(a) 60	213,358	<5ppt
8.1-19	1,1-Difluoro-ethane	CH_3CHF_2	propellant	562		
8.1-20	Methyl chloroform	CH_3CCl_3	chemical mfr. landfill solvent	559 365 60,369	89,135* 213(a)	30-370ppt
8.1-21	1-Chloro-1,1-difluoroethane	CH_3CClF_2	propellant	562		
8.1-22	1,1,2-Trichloro-ethane	$CHCl_2CH_2Cl$	industrial	462	462	

326

TABLE 8.1 HALOGENATED ALIPHATIC AND OLEFINIC COMPOUNDS
Emission and Detection

Species Number	Name	Chemical Formula	Emission		Detection	Ambient conc.
			Source	Ref.	Ref.	
8.1-23	1,2,2-Trichloro-1,1, 2-Trifluorethane	$CClF_2CCl_2F$	foaming agent	559	135*,373*	16-30ppt
			refrigeration	561		
			solvent	264,559		
			turbine	414		
			volcano	234		
8.1-24	1,2-Difluorotetra- chloroethane	CCl_2FCCl_2F	dry cleaning	561		
			solvent	561		
8.1-25	1,2-Dichlorotetra- fluoroethane	CF_2ClCF_2Cl	foaming agent	559,561		
			propellant	559,562		
			refrigeration	559,561		
8.1-26	Perfluoroethylene	$CF_2=CF_2$	volcano	234		
8.1-27	Vinyl chloride	$CH_2=CHCl$	chemical mfr.	559,560	135,275	<5ppt
			landfill	365		
			plastics mfr.	58*,59*		
			polymer comb.	211,304		
			solvent	439		
			tobacco smoke	298		
8.1-28	Ethylene dichloride	CH_2ClCH_2Cl	chemical mfr.	462,564	333,429	<5ppt
			petrochem. mfr.	111,369		
			plastics comb.	354		
			refuse comb.	26(a)		
			solvent	60,439		

TABLE 8.1 HALOGENATED ALIPHATIC AND OLEFINIC COMPOUNDS
Emission and Detection

Species Number	Name	Chemical Formula	Emission		Detection	Ambient conc.
			Source	Ref.	Ref.	
8.1-29	Trichlorethylene	$CHCl=CCl_2$	chemical mfr.	559	135*,163*	14-350ppt
			dry cleaning	426		
			landfill	365		
			solvent	7,324		
			turbine	414		
			volcano	234		
8.1-30	Perchloroethylene	$CCl_2=CCl_2$	chemical mfr.	559,560	135*,233	10-1250ppt
			dry cleaning	426,560		
			landfill	365		
			solvent	50,60		
			tobacco smoke	421		
8.1-31	Ethylene dibromide	$CHBr=CHBr$	solvent	60		
8.1-32	Chlorotrifluoroethylene	$CFCl=CF_2$	volcano	234		
8.1-33	1,3-Dichloropropane	(structure)			566*	2ppb
8.1-34	2-Fluoro-2-methyl-propane	(structure)			358	
8.1-35	1,2-Dichloropropene	(structure)	solvent	60		
			turbine	414		
8.1-36	1,3-Dichloropropene	(structure)	solvent	60		
8.1-37	1,2,3,3-Tetrachloro-propene	(structure)			358	
8.1-38	Hexafluoropropene	(structure)	volcano	234		
8.1-39	Propyl chloride	(structure)	refuse comb.	26(a)		
8.1-40	n-Butyl fluoride	(structure)	turbine	414		

TABLE 8.1 HALOGENATED ALIPHATIC AND OLEFINIC COMPOUNDS
Emission and Detection

Species Number	Name	Chemical Formula	Emission		Detection Ref.	Ambient conc.
			Source	Ref.		
8.1-41	2-Chlorobutane	(structure)	turbine	414t	263	
8.1-42	1-Chloro-3-methyl-butane	(structure)	turbine	414		
8.1-43	2-Chloro-3-methyl-butane	(structure)	turbine	414		
8.1-44	Chlorobutene	(structure)	plastics comb.	354		
8.1-45	2-Chloro-1,3-butadiene	(structure)	glue vapor / plastics comb.	11 / 354		
8.1-46	1-Chlorohexane	(structure)	turbine	414		
8.1-47	1-Fluoroheptane	$CH_3(CH_2)_5CH_2F$	turbine	414		
8.1-48	Amyl-2,2-dichloro-propionate	(structure)	turbine	414		
8.1-49	Perfluorocyclobutane	(structure)	propellant	559,562		
8.1-50	Cyclohexyl fluoride	(structure)	turbine	414		
8.1-51	Phosgene	$COCl_2$	chemical mfr. / plastics comb. / refuse comb.	167,177 / 46,184 / 31	394*	21.6-31.7ppt
8.1-52	Carbonyl fluoride	COF_2	polymer mfr.	124		
8.1-53	Chloromethyl ether	CH_3OCH_2Cl	ion xch. resin mfr.	38		
8.1-54	Bis(chloromethyl) ether	$ClCH_2OCH_2Cl$	ion xch. resin mfr.	38		
8.1-55	Bis(2-chloroethyl)ether	$(ClCH_2CH_2)O$	river water odor	202		

TABLE 8.1 HALOGENATED ALIPHATIC AND OLEFINIC COMPOUNDS
Emission and Detection

Species Number	Name	Chemical Formula	Emission		Detection	Ambient conc.
			Source	Ref.	Ref.	
8.1-56	Epichlorhydrin		building resin industrial	9 60		

TABLE 8.1 HALOGENATED ALKANES AND ALKENES
Reactions and Products

Species Number	Reactant	Chemical reactions					Remarks
		k	Ref.	Lifetime	Products	Ref.	
8.1-1	$O(^1D)$ $HO\cdot$	$(3.0\pm0.4)\times10^{-11}$ $<4\times10^{-16}$	1201 1073	6.7×10^{10}			
8.1-2	$HO\cdot$ $Cl\cdot$ O	4.8×10^{-14} $(5.2\pm1.1)\times10^{-13}$ $(1.3\pm0.3)\times10^{-17}$	1170 1203 1072	4.6×10^{7}	$HCOCl$	1097	
8.1-3	$HO\cdot$	1.3×10^{-13}	1170	1.6×10^{7}	CO_2,HCl	1097	
8.1-4	$HO\cdot$	1.1×10^{-13}	1170	2.0×10^{7}	$COCl_2$	1097	
8.1-5	$O(^1D)$ $HO\cdot$	$(8.5\pm1.1)\times10^{-10}$ $<1.0\times10^{-16}$	1201 1202	2.4×10^{9}	$COCl_2$	1097	
8.1-6	$HO\cdot$ O	4.0×10^{-14} $(1.3\pm0.3)\times10^{-17}$	1170 1072	5.4×10^{7}			
8.1-9	$HO\cdot$	4.1×10^{-14}	1170	5.3×10^{7}			
8.1-10	$HO\cdot$ $O(^1D)$	3.0×10^{-14} $(4.8\pm0.5)\times10^{-10}$	1170 1201	7.3×10^{7}			
8.1-11	$HO\cdot$ $O(^1D)$	4.6×10^{-15} $(2.4\pm0.3)\times10^{-10}$	1170 1201	4.8×10^{8}			

TABLE 8.1 HALOGENATED ALKANES AND ALKENES
Reactions and Products

Species Number	Chemical reactions							Remarks
	Reactant	k	Ref.	Lifetime	Ref.	Products	Ref.	
8.1-12	$O(^1D)$	2.3×10^{-10}		8.7×10^9	1170			
	HO·	$<4.8\times10^{-18}$			1170			
8.1-13	$O(^1D)$	2.0×10^{-10}		1.0×10^{10}	1170			
	HO·	$<6.5\times10^{-18}$			1170			
8.1-14	$O(^1D)$	$(2.5\pm0.3)\times10^{-10}$		8.0×10^9	1201			
8.1-18	HO·	$(3.9\pm0.7)\times10^{-13}$		5.5×10^6	1150			
8.1-19	HO·	$(3.1\pm0.7)\times10^{-14}$		6.5×10^7	1150			
	O_3	$(1.3\pm0.1)\times10^{-27}$			1204			
8.1-20	HO·	1.9×10^{-14}		1.2×10^8	1170			
8.1-21	HO·	$(2.8\pm0.4)\times10^{-15}$		7.1×10^8	1150			
8.1-23	$O(^1D)$	$(5.3\pm0.6)\times10^{-10}$		3.8×10^9	1201			
	HO·	$<3\times10^{-16}$			1006			
8.1-25	$O(^1D)$	$(3.7\pm0.4)\times10^{-10}$		5.4×10^9	1201			
	HO·	$<5\times10^{-16}$			1150			
8.1-26	O	1.1×10^{-12}		3.6×10^7	1206	CF_2O	1070	
	O_3	$(2.1\pm0.1)\times10^{-21}$			1074	CF_2O	1074	

TABLE 8.1 HALOGENATED ALKANES AND ALKENES
Reactions and Products

Species Number	Reactant	Chemical reactions			Products	Ref.	Remarks
		k	Ref.	Lifetime			
8.1-27	O HO·	$(6.0\pm0.6)\times10^{-13}$	1205	6.7×10^{7}	HCHO,HCl HCHO	1098 1099	
8.1-28	HO·	$(2.2\pm0.5)\times10^{-13}$	1150	9.9×10^{6}			
8.1-29	HO· O	2.3×10^{-12} 6.7×10^{-14}	1170 1245	9.5×10^{5}	COCl$_2$ CHCl$_2$COCl	1101 1098	
8.1-30	HO· O	1.7×10^{-13} 6.7×10^{-14}	1170 1245	1.3×10^{7}	CCl$_3$COCl	1098	
8.1-32	HO· O	$(6.5\pm1.3)\times10^{-12}$ 3.3×10^{-14}	1207 1245	3.4×10^{5}	CFClO	1070	
8.1-38	O$_3$	2.2×10^{-17}	1143	4.6×10^{4}			
8.1-51	O(^{1}D) O	1.7×10^{-10} $<3.3\times10^{-14}$	1170 1070	1.2×10^{10}			B
8.1-52	O(^{1}D) O	4.5×10^{-11} $<3.3\times10^{-14}$	1170 1070	4.4×10^{10}			B

8.1.2. Ambient Concentrations

Methyl chloride is the most abundant of the atmospheric halocarbons, with concentrations of $1-2$ ppb occurring at different locations around the world. The other halocarbons are much less common, with the highest concentrations being commonly observed only near large population centers. Because of their potential influence on the stratospheric ozone shield, the fully halogenated methanes and ethanes have been perhaps more intensively monitored than any other atmospheric organic compounds. The references in Table 8.1 provide extensive details.

8.1.3. Chemistry

Halogenated alkanes that have one or more hydrogen atoms are susceptible to hydrogen abstraction by the hydroxyl radical. For methyl chloride, the most abundant of the halocarbons, the favored reaction chain is thought to be (1125)

$$CH_3Cl \xrightarrow{HO \cdot} CH_2Cl \cdot \xrightarrow{O_2} CH_2ClO_2 \cdot \xrightarrow{NO} CH_2ClO \cdot \xrightarrow{O_2} CO + HCl \ .$$

The relative importances of this chemical source of HCl and the physical emission of HCl from sea-salt aerosol (470) have not yet been determined.

Halogenated alkenes add hydroxyl across the double bond and proceed to a variety of oxygenated products, principally phosgene and the chloroacetyl chlorides. The details of the reaction chains are not yet determined.

Fully halogenated alkanes are among the most unreactive of all the atmospheric compounds, thus their involvement in tropospheric chemistry is minimal. As they diffuse to the stratosphere they are decomposed by the increasingly intense ultraviolet radiation and participate in the complex processes of stratospheric chemistry. This subject has received extensive treatment $(1126-1129)$ in the literature, to which the interested reader is referred for details.

8.2. HALOGENATED AROMATIC COMPOUNDS

8.2.1. Identified Compounds

Nearly a dozen halogenated aromatic compounds are known to be present in the troposphere; several are abundant enough to have been detected in ambient air samples. All of these compounds are anthropogenic; with the possible exception of chlorobenzene none is widely emitted.

TABLE 8.2 HALOGENATED AROMATIC COMPOUNDS
Emission and Detection

Species Number	Name	Chemical Formula	Emission		Detection Ref.	Ambient conc.
			Source	Ref.		
8.2-1	Chlorobenzene		chemical mfr. landfill plastics comb. solvent	462* 365 181,354 60,439	358	
8.2-2	o-Dichlorobenzene				358	
8.2-3	m-Dichlorobenzene		tobacco smoke	421	358,591	
8.2-4	p-Dichlorobenzene		industrial	60	86,334	
8.2-5	Dichlorobenzoic acid		refuse comb.	26(a)		
8.2-6	p-Chloroaniline		industrial	250		
8.2-7	1,2,4-Trichlorobenzene				358	
8.2-8	Tetrachlorophenol				363(a)	
8.2-9	Pentachlorophenol		fertilizer mfr.	272*	363(a),443(a)	

TABLE 8.2 HALOGENATED AROMATIC COMPOUNDS
Emission and Detection

Species Number	Name	Chemical Formula	Emission		Detection	Ambient conc.
			Source	Ref.	Ref.	
8.2-10	Methylchloro-indan one				399(a),458(a)	
8.2-11	Bromoxylene		vitamin mfr.	274		
8.2-12	p-Bromoaniline		industrial	250		
8.2-13	4,6-Dibromoxylene		vitamin mfr.	274		
8.2-14	p-Iodoaniline		industrial	250		

336

TABLE 8.2 HALOGENATED AROMATIC COMPOUNDS
Reactions and Products

Species Number	Reactant	k	Ref.	Lifetime	Products	Ref.	Remarks
8.2-1	O	$(3.3 \pm 0.8) \times 10^{-13}$	1071	1.2×10^8			

8.2.2. Ambient Concentrations

No data are available concerning the atmospheric concentrations of these compounds.

8.2.3. Chemistry

The only gas phase chemical study of any of these compounds is of oxygen atoms and chlorobenzene (1071); the reaction was found to be moderately rapid. It is likely, however, that the principal atmospheric reaction of these compounds will be with the hydroxyl radical. Such reactions have been shown for similar compounds to proceed by addition to the aromatic ring (1161); one thus anticipates the formation of halogenated phenols that will be largely lost to surfaces (i.e., aerosols and the ground).

8.3. CHLORINATED PESTICIDES

8.3.1. Identified Compounds

More than two dozen compounds are included in this group, which can be expected to grow as more sensitive and specific detection methods are applied to atmospheric samples. The compounds have much greater molecular weights (typically >200 amu) than do most other species present in the atmosphere; the structures consist of one or more aromatic rings with substituted chlorine atoms and sometimes with additional epoxide or ester groupings.

Because these compounds are all anthropogenically produced and because of their toxic influences on food chains, their study has become a subspecialty in the atmospheric sciences. Volatilization, both from soil (567) and water (1246), is a general property of these aerosol-applied pesticides, as is their transport on small aerosol particles; the compounds are present in ambient air as far as 2000 km from their point of application (91). As a result, although the pesticides are generally applied in aerosol form, they have been frequently found in both gas and aerosol phases.

The rapid development of new compounds of this type ensures that Table 8.3 is representative rather than comprehensive.

8.3.2. Ambient Concentrations

The concentrations for many of the chlorinated pesticides have been measured in a variety of field locations. For most compounds, ambient levels of a few tens of ng/m^3 are typical. Concentrations are sometimes higher in agricultural regions where these chemicals are applied in large quantity: 2−4D and DDT have both been detected in $\mu g/m^3$ amounts in such locations. In general, the concentrations decrease with distance from the point of application.

8.3.3. Chemistry

The chemistry of the chlorinated pesticides in the air environment is complex and incompletely understood. The compounds absorb photons at tropospheric wavelengths, and subsequent isomerization (1244) and oxidation (568) processes have been identified. The degree of participation of photochemical products in interactive atmospheric chemistry is less certain. Evidence has been presented, however, for a first-order photochemical decomposition of trifluralin with a 70-sec half-life (1243) and for photocleavage of carbon-chlorine bonds in *o* -chlorobiphenyl (569). In this latter case, hydrogen chloride and phosgene were among the identified products. It thus appears that the products of atmospheric chlorinated pesticide chemistry may be involved to some degree in more general chemical cycles.

TABLE 8.3 CHLORINATED PESTICIDES
Emission and Detection

Species Number	Name	Chemical Formula	Emission		Detection	Ambient conc.
			Source	Ref.	Ref.	
8.3-1	Toxaphene	(structure)	pesticide	126(a)	21*(a),230*(a)	$<.02\text{-}7.0 ng/m^3$
8.3-2	Poly chloropinene	(structure)	pesticide	126(a)		
8.3-3	Benzene hexachloride	(structure)	pesticide	239,245	577*(a) 230*(a)	$1.8\text{-}9.9 ng/m^3$
8.3-4	2,4-D	(structure)	pesticide	194*	230*(a)	$4.0 ng/m^3$
8.3-5	2,4-D isopropyl ester	(structure)	pesticide	63,303*	303*(a) 303*(a)	$10\text{-}130 ng/m^3$
8.3-6	2,4-D butyl ester	(structure)	pesticide	63,303*	63*,303* 194*(a),303*(a)	$.06\text{-}455 ng/m^3$
8.3-7	2,4-D isobutyl ester	(structure)	pesticide.	63,303*	303*	$10\text{-}40 ng/m^3$
8.3-8	2,4-D butoxyethanol ester	(structure)	pesticide	63,303*	303*(a) 303*(a)	$10\text{-}780 ng/m^3$
8.3-9	2,4-D isooctyl ester	(structure)	pesticide	63,303*	303*(a) 303*(a)	$10\text{-}80 ng/m^3$

TABLE 8.3 CHLORINATED PESTICIDES
Emission and Detection

Species Number	Name	Chemical Formula	Emission		Detection Ref.	Ambient conc.
			Source	Ref.		
8.3-10	2,4-D propylene glycol butyl ether ester	(structure; OH, OH, O, O, Cl, Cl)	pesticide	63,303*	303* 303*(a)	10-160ng/m^3
8.3-11	Dacthal	(structure; Cl, Cl, Cl, Cl, O, O)	pesticide	567		
8.3-12	Chlorpropham	(structure; Cl, N, H, O)	pesticide	575		
8.3-13	Trifluralin	(structure; CF$_3$, NO$_2$, O$_2$N, N)	pesticide	567		
8.3-14	Tetrachlorobiphenyl	(structure; (Cl)$_4$)	refuse comb.	26(a)		
8.3-15	Pentachlorobiphenyl	(structure; (Cl)$_5$)	refuse comb.	26(a)		
8.3-16	Hexachlorobiphenyl	(structure; (Cl)$_6$)	refuse comb.	26(a)		
8.3-17	Polychlorobiphenyl	(structure; (Cl)$_n$)	fluorescent lamps / paint plasticizer / pesticide / refuse comb.	229* / 120 / 20,91 / 237*,557*	20*,91* / 54(a),569*(a)	.05-50ng/m^3

341

TABLE 8.3 CHLORINATED PESTICIDES
Emission and Detection

Species Number	Name	Chemical Formula	Emission		Detection Ref.	Ambient conc.
			Source	Ref.		
8.3-18	Polychlorotriphenyl		refuse comb.	237*		
8.3-19	p,p'-DDT		pesticide	69*,265*	20*, 216(a),230*(a)	.009-500ng/m³
8.3-20	o,p'-DDT		pesticide	20,568	20*, 54*(a),230*(a)	.009-500ng/m³
8.3-21	p,p'-TDE		pesticide / tobacco smoke	265*,572 / 130*	577	
8.3-22	o,p'-TDE		pesticide / tobacco smoke	130 / 130*		
8.3-23	p,p'-DDE		pesticide	572,596	216(a),230*(a)	2.4-131ng/m³
8.3-24	o,p'-DDE		pesticide	230,596	230*(a)	1.9-9.6ng/m³
8.3-25	p,p'-TDEE		pesticide / tobacco smoke	130 / 130*		
8.3-26	Heptachlor		pesticide	574	230*(a)	2.3-19.2ng/m³

TABLE 8.3 CHLORINATED PESTICIDES
Emission and Detection

Species Number	Name	Chemical Formula	Emission		Detection	Ambient conc.
			Source	Ref.	Ref.	
8.3-27	Heptachlor epoxide		pesticide	577	577	
8.3-28	Chlordane		pesticide	20,567	20*	<5-250pg/m^3
8.3-29	Aldrin		pesticide	230,572	230*(a)	8.0ng/m^3
8.3-30	Endrin		pesticide	230	230*(a)	58.5ng/m^3
8.3-31	Dieldrin		pesticide	212,265*	577 230*(a)	29.7ng/m^3

TABLE 8.3 CHLORINATED PESTICIDES
Reactions and Products

| Species Number | Chemical reactions | | | | | Remarks |
	Reactant	k	Ref.	Lifetime	Products	Ref.	
8.3-14	hν				HCl, COCl$_2$	569	

9

ORGANOMETALLIC COMPOUNDS

9.0. INTRODUCTION

Organometallic compounds are not known to be abundant in the atmosphere, but very few data are available to permit an accurate assessment to be made. The alkyl derivatives of several of the elements may be of interest in atmospheric chemical cycles; additional information is sorely needed. Most of the other atmospheric organometallic compounds are anthropogenic in origin. Their atmospheric abundances seem likely to be very small.

9.1. ORGANOMETALLICS:
ALKANES, CARBONYLS, SILOXANES

9.1.1. Identified Compounds

This chemical group is very diverse. It includes alkyl compounds of three elements (arsenic, mercury, and selenium) that are volatilized by natural processes. Natural atmospheric cycles for these elements are thus in operation, but the data are so sparse the cycles cannot presently be quantified. Three carbonyl compounds are known to be present in cigarette smoke and presumably also in natural vegetative combustion; information on these compounds is also insufficient to assess their atmospheric importance.

The alkyl lead compounds are wholly anthropogenic, being present as gasoline additives in motor vehicle exhaust. Their average concentrations are expected to decrease as the use of lead additives becomes progressively diminished. Several silanes have been tentatively detected in the atmosphere; they are presumably derived from industrial manufacturing processes and doubtless are only rarely present

TABLE 9.1 ORGANOMETALLICS: ALKANES, CARBONYLS, AND SILOXANES
Emission and Detection

Species Number	Name	Chemical Formula	Emission		Detection Ref.	Ambient conc.
			Source	Ref.		
9.1-1	Methylsilane	CH_3SiH_3	turbine	414		
9.1-2	Tetramethylsilane	$(CH_3)_4Si$			358t	
9.1-3	Hexamethyl-cyclotrisiloxane				358	
9.1-4	Octamethylcyclo-tetrasiloxane				358t	
9.1-5	Tetradecamethyl-hexasiloxane	$(CH_3)_3Si-O-[Si(CH_3)_2-O]_1-Si(CH_3)_2$			358t	
9.1-6	Isopropylmethylphos-phoro-fluoridate		nerve gas mfr.	441		
9.1-7	Iron carbonyl	$Fe(CO)_5$	tobacco smoke	493(a)		
9.1-8	Cobalt carbonyl	$Co(CO)_4$	tobacco smoke	493(a)		
9.1-9	Nickel carbonyl	$Ni(CO)_4$	tobacco smoke	493(a)		
9.1-10	Dimethylarsine	$HAs(CH_3)_2$	microbes	34,302	112*	0.1-0.4ng/m^3
9.1-11	Trimethylarsine	$As(CH_3)_3$	microbes	34,302	112*,367	0.1-0.4ng/m^3
9.1-12	Dimethylselenide	$(CH_3)_2Se$	bacteria	393		

TABLE 9.1 ORGANOMETALLICS: ALKANES, CARBONYLS, AND SILOXANES
Emission and Detection

Species Number	Name	Chemical Formula	Emission		Detection Ref.	Ambient conc.
			Source	Ref.		
9.1-13	Dimethyldiselenide	$CH_3SeSeCH_3$	bacteria plant volatile	393 236,587		
9.1-14	Methylmercury	CH_3HgH	auto exhaust	397	388	
9.1-15	Methylmercuric-chloride	CH_3HgCl	sewage tmt.	529t*	529t	
9.1-16	Dimethylmercury	CH_3HgCH_3	sewage tmt.	529t	215*,529	3.4-38pg/m^3
9.1-17	Tetramethyllead	$Pb(CH_3)_4$	gasoline vapor	346,370		
9.1-18	Ethyltrimethyllead	$Pb(CH_3)_3C_2H_5$	gasoline vapor	346,370		
9.1-19	Dimethyldiethyl-lead	$Pb(CH_3)_2(C_2H_5)_2$	gasoline vapor	346,370		
9.1-20	Methyltriethyl-lead	$Pb(C_2H_5)_3CH_3$	gasoline vapor	346,370		
9.1-21	Tetraethyllead	$Pb(C_2H_5)_4$	gasoline vapor	346,370		

347

as atmospheric constituents.

9.1.2. Ambient Concentrations

Quantitative measurements of ambient concentrations have been made only for some of the alkyl arsenic and mercury compounds. The concentrations are so low that it is necessary to invoke the use of the unit "ppq", parts per quadrillion, to describe them. A few measurements near known or suspected sources give much higher concentrations.

9.1.3. Chemistry

Most of the compounds in Table 9.1 are not photochemically active in the troposphere, since their absorption spectra have wavelength upper limits below the 290 nm ozone cutoff [e.g., 240 nm for the selenium compounds (1265), 230 nm for dimethyl mercury (1155), etc.]. In the absence of photochemical processes, the principal atmospheric reactions of the organometallic alkanes will be abstraction of a hydrogen atom by the hydroxyl radical. The reaction chains would be expected to proceed as do those of the alkyl sulfides (Section 7.2). The resulting compounds will be oxides or other oxygenated compounds of arsenic, mercury, lead, selenium, and silicon.

The metal carbonyls are apparently rare in the troposphere, and will in any case be unstable if present. Sunlight will be absorbed by the $-C=O$ group, with subsequent production of carbon monoxide and metal oxides. The silanes are also unstable in the atmosphere and will be promptly removed by rapid oxidation reactions.

9.2. ORGANOPHOSPHORUS PESTICIDES

9.2.1. Identified Compounds

The organophosphorus pesticides comprise a group of compounds characterized by a phosphorus atom with alkoxy ligands, often with other organic and organosulfur groups attached as well. Nearly a dozen of these compounds have been detected in ambient air. All are the result of application as pesticides to crops and soil.

9.2.2. Ambient Concentrations

The concentrations of these chemicals in ambient air are typically several tens or hundreds of ng/m^3. Measurements near sources demonstrate that the concentrations show a marked decrease with distance from the point of application.

TABLE 9.2 ORGANOPHOSPHORUS PESTICIDES
Emission and Detection

Species Number	Name	Chemical Formula	Emission Source	Emission Ref.	Detection Ref.	Ambient conc.
9.2-1	Butyphos	$(C_4H_9)_3PO$	pesticide / pesticide	228* / 126(a)		
9.2-2	DDVP	$(CH_3O)_2POCHCCl_2$	pesticide	578		
9.2-3	Trichlorfon	$(CH_3O)_2P(O)$ — CHCCl$_3$ / OH	pesticide	228*		
9.2-4	DEF	$(C_4H_9OS)_3PO$	pesticide		230*(a)	16ng/m^3
9.2-5	Methylmercaptophos	$(CH_3OS)_2P(O)S$ — (structure)	pesticide / pesticide	228* / 126(a)		
9.2-6	Dimethoate	$(CH_3O)_2P(S)S$ — (structure)	pesticide / pesticide	228* / 126(a)		
9.2-7	Malathion	$(CH_3O)_2P(S)S$ — (structure)	pesticide	228*	230*(a)	2.0ng/m^3
9.2-8	Methylparathion	$(CH_3O)_2P(S)O$ — (structure with NO$_2$)	pesticide		230*(a)	5.4-129ng/m^3
9.2-9	Fenitrothion	$(CH_3O)_2P(S)O$ — (structure with NO$_2$)	pesticide	276		

TABLE 9.2 ORGANOPHOSPHORUS PESTICIDES
Emission and Detection

Species Number	Name	Chemical Formula	Emission		Detection Ref.	Ambient conc.
			Source	Ref.		
9.2-10	Parathion	$(C_2H_5O)_2P(S)O$—◯—NO_2	pesticide	228*	230*(a)	465ng/m^3
9.2-11	Triphenyl phosphate	$(C_6H_5O\!-\!)_3$ PO	auto refuse comb.	311(a) 26(a)		
9.2-12	Diazinon	$(CH_3O)_2P(S)O$—[pyrimidine]	pesticide	578		

9.2.3. Chemistry

Studies of the atmospheric chemistry of the organophosphorus pesticides are few. There is no current evidence to suggest the direct involvement of these compounds in any of the principal atmospheric chemical cycles.

10

SYNTHESIS OF DATA ON
ATMOSPHERIC COMPOUNDS

10.0. INTRODUCTION

The primary purpose of this book has been to assemble a substantial amount of widely distributed material on atmospheric compounds and their chemistry into a format suitable for reference and instruction. As with most taxonomic efforts, however, this assemblage has allowed certain patterns to emerge and permitted certain conclusions to be drawn from the grouped data. In this concluding chapter, some of these topics of broader scope will be presented and discussed.

10.1. COMPOUND OCCURRENCE SUMMARY

More than 1600 different compounds are listed in Tables 2.1 to 9.2. The size of this data collection is such that attempts to summarize its characteristics are useful. Accordingly, Table 10.1-1 presents statistics for the major chemical groupings that have been used. The categories indicate whether a compound is known to be emitted or detected only as an aerosol component, only as a gas, or in both forms. (The entries are thus quite dependent on the analytical effort devoted to each of the compounds, as well as to its chemical and physical properties.) The hydrocarbon grouping contains the largest number of compounds, reflecting principally the tremendous diversity of hydrocarbon species in the combustion products of petroleum derivatives. Oxygenated ($\equiv$noncarbonyl) organics are nearly as numerous, however; many of these compounds are natural vegetation emittants. Carbonyls have both anthropogenic and natural sources, and are quite numerous. Nitrogen-containing organic compounds and the inorganic compounds are common, although many species in each group seem unlikely to be

TABLE 10.1-1 COMPOUND OCCURRENCE SUMMARY[a]

Sections	Classification	Emitted				Detected			
		A[b]	A+G[c]	G[d]	T[e]	A	A+G	G	T
2.1-5	Inorganics	51	9	45	105	34	8	23	65
3.1-7	Hydrocarbons	174	17	234	425	175	72	69	316
4.1-8	Carbonyls	21	18	159	198	40	2	19	61
5.1-17	Oxyg. organics	63	41	285	389	57	5	23	85
6.1-4	N-Cont. organics	34	7	66	107	48	1	17	66
7.1-4	S-Cont. organics	2	2	31	35	7	4	3	14
8.1-3	Halogen. organics	7	5	71	83	12	10	29	51
9.1-2	Organometallics	3	3	18	24	4	0	9	13
	Totals	355	102	909	1366	377	102	192	671

[a] Tabular entries indicate the number of individual compounds in the various subgroups.

[b] Aerosol [c] Both aerosol and gas [d] Gas [e] Total

important participants in atmospheric chemical processes. More than 80 halogenated organic compounds are known to be present in the atmosphere; this grouping is heavily influenced by anthropogenic emissions. Sulfur-containing organic compounds and organometallic compounds are relatively uncommon.

A feature of Table 10.1-1 that emerges after close study is that the ratio of gas phase compounds to aerosol phase compounds is systematically different for the emitted fraction than for the detected fraction. To study this aspect more systematically, data of the form of Table 10.1-1 for each individual data table have been plotted histogramatically in Figs. 10.1-1 to 10.1-6. Fig. 10.1-1 presents data on the inorganic compounds. The tendency for inorganic nitrogen compounds (Table 2.2) to be emitted in the gas phase but often detected in the aerosol phase (usually as nitrates) is evident. The sulfur-containing and halogen-containing inorganic compounds are emitted in both gas and aerosol phases, but are more likely to be detected in the aerosol phase. The compounds of Table 2.5 (largely oxide and carbonate salts) are heavily weighted toward the aerosol phase both in emission and detection.

Hydrocarbon compound data are shown in Fig. 10.1-2. For Tables 3.1 to 3.5, the transition from gas to aerosol is particularly striking. Other characteristics of interest can also be noted readily. For example, many terpenes (Table 3.3) are known to be emitted, but few have been detected in ambient air. This fact is consistent with the relatively short chemical lifetimes of all members of this class of compounds. Also, if terpenes are combined with other alkenes (Table 3.2) and if benzene (Table 3.5) and naphthalene (Table 3.6) derivatives are combined, the total numbers of compounds in the classes alkanes, alkenes, one- and two-ring aromatics, and polynuclear aromatics are similar.

Carbonyl compound data are displayed in Fig. 10.1-3. Experimental field effort on certain of these subgroups has been sparse, and some aspects of the data doubtless represent incomplete sampling. Nonetheless, the gas to aerosol transition is apparent for the aliphatic and olefinic aldehydes (Tables 4.1 and 4.2) and for the aromatic ketones (Table 4.8).

The noncarbonyl oxygenates (Fig. 10.1-4) have been the subjects of rather little field study, although their emission as vegetation volatiles is widely recognized. The carboxylic acids found in the aerosol phase (Tables 5.1-5.4) show up prominently, as do the aliphatic (Table 5.9) and aromatic (Table 5.12) alcohols. Certain of the esters (Tables 5.5-5.8) and ethers (Tables 5.14-5.17) should be capable of detection if directed searches are made.

Fig. 10.1-5 displays data for nitrogen-containing and sulfur-containing compounds. The nitriles (Table 6.1) are not uncommon

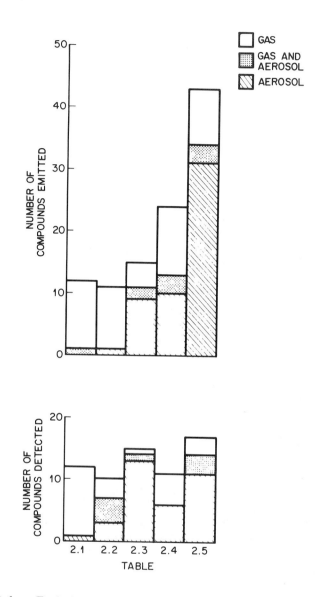

Fig. 10.1-1.   Emission and detection data for inorganic compounds, grouped by phase.

emittants in urban areas, yet almost none have been detected in ambient air. Since their chemical removal is likely not to be overly rapid, a search for selected compounds may prove rewarding, as is the case with the amines (Table 6.2). Table 6.3 shows dramatically the

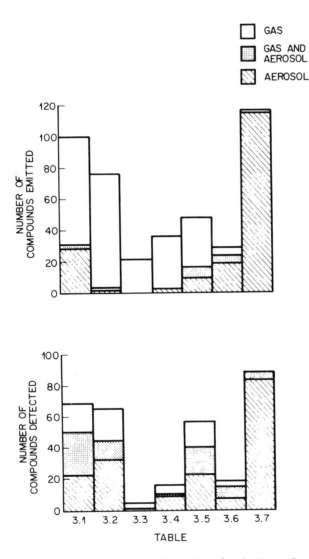

Fig. 10.1-2.   Emission and detection data for hydrocarbons, grouped
              by phase.

large numbers of nitro compounds found in ambient aerosols, despite
the lack of evidence for the emission of more than a few. This doubtless reflects the chemical combination of organic radicals with the ubiquitous oxides of nitrogen (see Section 6.3). The heterocyclic nitrogen
compounds (Table 6.4) are chiefly polynuclear structures and appear
not to be chemically important in the troposphere.

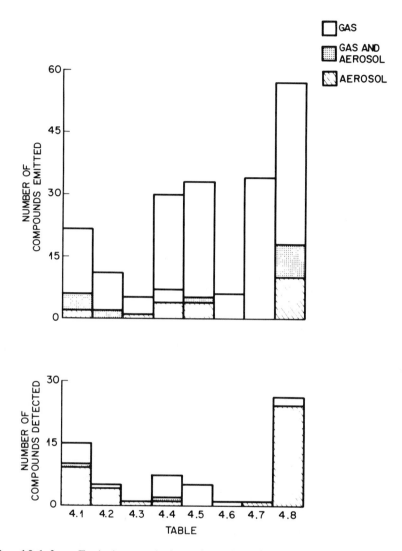

Fig. 10.1-3. Emission and detection data for carbonyl compounds, grouped by phase.

Mercaptans (Table 7.1) and sulfides (Table 7.2) come from a rather wide variety of sources and and in a number of chemical forms. Ambient measurements have been rare. In view of the substantial uncertainties still present in the atmospheric sulfur cycle (see Section 10.4.), increased measurement attention to these groups would seem justified. The more complex organic sulfur compounds of Tables 7.3

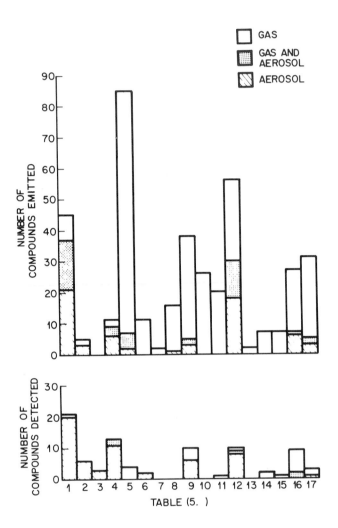

Fig. 10.1-4.   Emission and detection data for oxygenated organic compounds, grouped by phase.

and 7.4 are not numerous and are of interest here chiefly for the gas-to-aerosol conversion seen in the Table 7.3 data.

The halogenated organic compounds are the most completely studied of the chemical groups, because of their largely anthropogenic origin and because of their potentially detrimental effects on stratospheric ozone. The data display of Fig. 10.1-6 does not provide new insights into the chemistry of these compounds, nor of the few organometallic

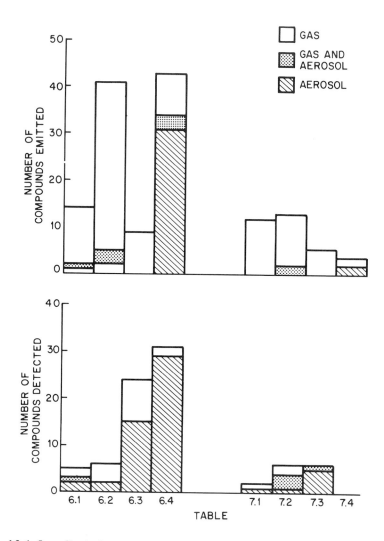

Fig. 10.1-5. Emission and detection data for nitrogen-containing and sulfur-containing organic compounds, grouped by phase.

compounds of Chapter 9.

In order to derive a quantitative estimate of the gas to aerosol conversion taking place in the atmosphere, it is useful to define a *gas phase fraction* F_g for both the emitted and detected compounds. Within any one table, these fractions are given by

$$F_{ge} = N_{ge}/(N_{ge} + N_{ae}) \quad ,$$

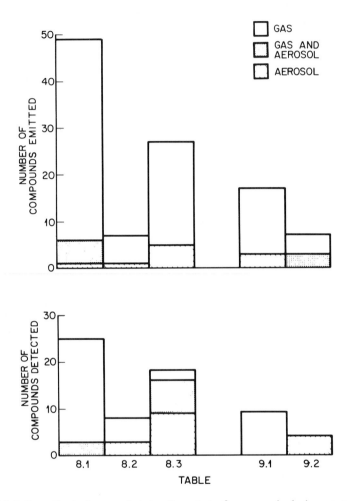

Fig. 10.1-6.   Emission and detection data for organic halogenated and
organometallic compounds, grouped by phase.

$$F_{gd} = N_{gd} / (N_{gd} + N_{ad}) \ ,$$

where N indicates the number of compounds and the substripts a, g, e,
d indicate aerosol, gas, found in emission, and found in detection. A
group conversion ratio C_g is then defined as

$$C_g = F_{gd} / F_{ge} \ .$$

A value of C_g less than unity indicates a lower fraction of compounds
are detected in the gas phase than are emitted in that phase. On Fig.

10.1-7, histograms of F_{ge}, F_{gd}, and C_g are displayed; the median values are 0.92, 0.40, and 0.46. These results indicate the substantial chemical processing of compounds in the atmosphere, and the existence of the majority of the products (almost always more highly oxygenated and with lower vapor pressures than their precursors) in the atmospheric aerosol.

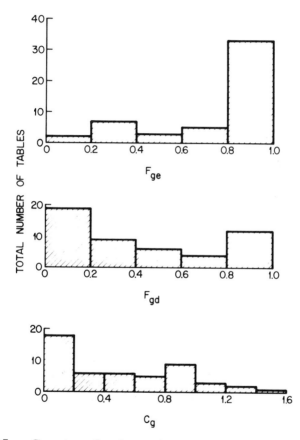

Fig. 10.1-7.   Gas phase fraction and group conversion ratio histograms for atmospheric compounds.

10.2. SOURCES OF ATMOSPHERIC COMPOUNDS

The sources of atmospheric compounds are as diverse as are the compounds themselves. A number of sources, both anthropogenic and natural, produce only a few different compounds, some only one. At the opposite end of the spectrum are sources that produce hundreds. To assess this diversity (in addition to providing useful reference material), the source data contained in the tables of Chapters 2-9 have been cross-indexed by source name instead of by chemical compound. This information is presented in Appendix A, where it is available to those concerned with emissions from a specific source.

As with most tabulations, it is of interest to select and examine those entries that stand out from the others. Table 10.2-1 lists the fifteen sources producing the largest numbers of different compounds. Leading the list is vegetation, whose emission of atmospheric compounds is amazingly diverse. Despite the position of the source in the table, it seems safe to say that the spectrum of natural atmospheric emissions is even broader than is indicated. Detailed analyses of the chemical constituents of vegetative fragrances (e.g. 500,533) typically show that scores of compounds are involved. Few such analyses have

TABLE 10.2-1. SOURCES THAT EMIT NUMEROUS ATMOSPHERIC COMPOUNDS

Ranking	Source	Number of Compounds
1.	Vegetation[a]	367
2.	Tobacco smoke	349
3.	Auto	285
4.	Turbine	234
5.	Diesel	212
6.	Refuse combustion	81
7.	Solvent	73
8.	Microbes	67
9.	Gasoline vapor	62
10.	Wood pulping	62
11.	Coal combustion	61
12.	Forest fire	58
13.	Petroleum mfr.	57
14.	Animal waste	56
15.	Volcano	56

[a] "Vegetation" includes the Appendix A listings for "plant volatile", "trees", and "vegetation".

been performed, however, and even fewer estimates of the *amounts* of such emissions are available. The importance of the natural organic emissions on atmospheric chemistry cannot thus be accurately assessed at present, although it seems likely to be substantial.

Tobacco smoke ranks second in this compilation, and would rank first if the many polynuclear compounds detected on smoke aerosols (534) but apparently not generally present in the ambient atmosphere were included. The ranking reflects the fact that the combustion process is a prolific source of different compounds and also that concerns about the biologically harmful effects of cigarette smoking have inspired considerable analytical activity.

Internal combustion engines occupy the next three places in Table 10.2-1. Three factors are responsible for this situation. The first is that the fossil fuels used by automobiles, turbines, and diesels are themselves chemically diverse. The second is that this diversity is enhanced during combustion by high temperature fragmentation and rearrangement of the fuel molecules. Finally, the intensive scientific studies of vehicle combusion have provided many detailed chemical analyses of the exhaust gases.

The remainder of the table contains sources representative of several different classifications. Combustion processes of various kinds (refuse combustion, wood pulping, coal combustion, forest fire emissions) occupy several places. Microbial processes, eighth on the list, occur in many different systems. Although separately indexed, animal waste emissions are largely due to microbial processes as well. The vaporization of gasoline during storage and handling produces many atmospheric compounds, as does solvent evaporation. Petroleum manufacture, which combines aspects of both petroleum handling and petroleum combustion, is also represented. Concluding the list is a diverse group of emittants from volcanic plumes and eruptions.

10.3. OXIDATION IN THE TROPOSPHERE

In the basic tables presented earlier, known gas phase reactants were listed for each species, together with the stable products of the reactions (if known). In Appendix B these data are assembled in a different way for cross-referencing purposes; the chemical products are the primary listing, with the species numbers of the precursors that form them being given for each product. Perhaps the most striking characteristic of Appendix B is that nearly every product results from an *oxidation* process. Most of the relevant processes for the inorganic compounds have been described in Chapter 2. In this section, the atmospheric oxidation of organic compounds is discussed in more detail.

The initial atmospheric reactions of hydrocarbons were discussed in Chapter 3. They are of three types: hydrogen abstraction followed by molecular oxygen addition, hydroxyl addition followed by molecular oxygen addition, and ozone addition followed by cleavage to carbonyl compounds and oxygenated radicals. For the simplest case, hydrogen abstraction from the end (alpha) carbon, gas phase oxidation eventually produces an α -aldehyde (Section 3.1). This compound has two principal fates. One is photodissociation to free radical fragments (Section 4.1). The second is incorporation into the atmospheric aerosol. The experimental evidence suggests that the solid aerosol core, if present, is surrounded in most or all cases by a liquid shell (1237). Further oxidation of incorporated compounds is thus expected to occur by liquid phase processes.

The liquid atmospheric aerosol component is acidic (1238). It appears reasonable to thus anticipate that the oxidation chain would proceed from alcohol to aldehyde to carboxylic acid. This process is indicated schematically in Fig. 10.3-1a, where the hydrocarbon to aldehyde transition (step 1) is assumed to occur in the gas phase. The aldehyde is then oxidized in either the gas phase or (more likely) in aerosol solution to the acid (step 3). An alcohol proceeds through both steps 2 and 3.

More vigorous oxidation processes are demonstrated by the difunctional oxygenated compounds, thought to result primarily from cleavage of gas phase cyclic precursors (Section 3.4). In such a cleavage, the oxidation would take place at the two terminal carbon atoms (the α and ω carbons), and would likely produce a spectrum of mixed alcohols, aldehydes, and acids. The vapor pressures of these compounds are low (1063), and their prompt incorporation into aerosols is probable. In the case of an alpha−omega dialdehyde, the aldehydic groups would undergo sequential oxidation to form the dicarboxylic acid (steps 4 and 5, Fig. 10.3-1a). With an alpha−omega dialcohol, one of the terminal carbons would be expected to undergo oxidation to form the α -alcohol- ω -acid (steps 6 and 7, Fig. 10.3-1b), followed by a similar process on the second end carbon to form the diacid (steps 8 and 5). The difunctional forms actually found in aerosols will be related to the kinetic parameters of processes $3-8$, and to the chemical nature of the incorporated compounds.

It is to be expected that monofunctional alcohols and aldehydes will be found principally in the gas phase in the atmosphere (particularly for low carbon number compounds, whose vapor pressures are high), and that difunctional compounds will occur primarily in aerosols. The data can thus be examined with these hypotheses in mind. In Figs.

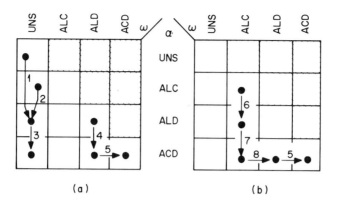

Fig. 10.3-1.   Atmospheric oxidation paths for difunctional organic compounds.

10.3-2 to 10.3-4, information for 11 hydrocarbon precursors and mono- and difunctional oxygenated products is presented. Monofunctional oxidation is indicated by entries in the leftmost column and difunctional oxidation by entries in the rightmost three columns. The entries for the unsubstituted hydrocarbons refer for simplicity to alkane or aromatic compounds, but the cyclic olefins [thought to be likely difunctional precursors (1063)] could also be entered for C_5, C_6, and C_7. The C_1 and C_2 compounds are so ubiquitous that little information relevant to this discussion can be derived from their occurrence. No ambient detection for any difunctional C_3 or C_4 compounds has been reported, thus suggesting that progressive oxidation of these compounds is less likely than is cleavage of the carbon chain. For the remainder of the groups, however, the anticipated pattern of gas phase detection of unsubstituted and monofunctional compounds and aerosol phase detection of difunctional compounds appears to be generally reproduced by the data.

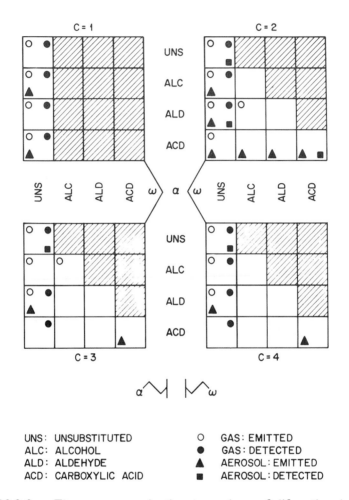

UNS: UNSUBSTITUTED
ALC: ALCOHOL
ALD: ALDEHYDE
ACD: CARBOXYLIC ACID

○ GAS: EMITTED
● GAS: DETECTED
▲ AEROSOL: EMITTED
■ AEROSOL: DETECTED

Fig. 10.3-2.   The occurrence in the atmosphere of difunctional derivatives of the C_1-C_4 alkanes.

Further evidence for atmospheric oxidation processes involving
these compounds is provided by the Los Angeles Basin aerosol observations of Appel, et al. (1273). They collected two-hour aerosol samples
over a two-day period of substantial photochemical activity and
analyzed them by thermal mass spectrometric techniques. The results
for aldehyde nitrate and acid nitrate compounds are reproduced in Fig.
10.3-5. Although it is uncertain whether all the structure in the data is
significant, the sequential occurrence of the principal aldehyde and acid
peaks is striking. The authors point out that this may be the result of
oxidation of the initially produced aldehyde, an inference completely

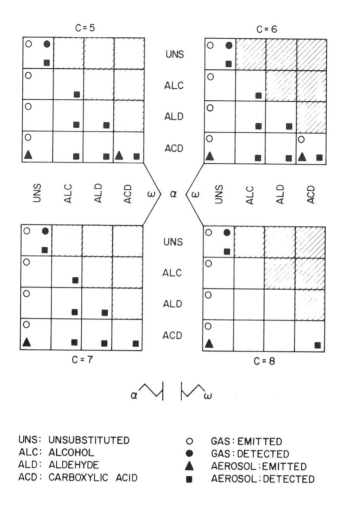

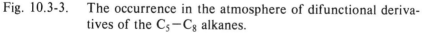

UNS:	UNSUBSTITUTED	○	GAS: EMITTED
ALC:	ALCOHOL	●	GAS: DETECTED
ALD:	ALDEHYDE	▲	AEROSOL: EMITTED
ACD:	CARBOXYLIC ACID	■	AEROSOL: DETECTED

Fig. 10.3-3.   The occurrence in the atmosphere of difunctional derivatives of the C_5-C_8 alkanes.

consistent with the data presented above.

The discussion in this section, together with Grosjean's comple-
mentary review of aerosol formation (1063), suggests the complexity
and importance of atmospheric liquid phase chemistry. Few of the
specific details have been studied, however, and they would be
expected to be topics of active research interest within the coming
years.

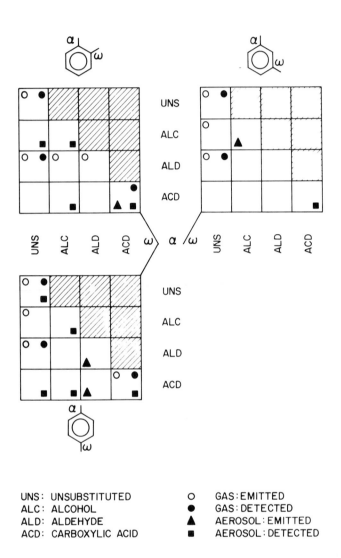

UNS: UNSUBSTITUTED   ○   GAS:EMITTED
ALC: ALCOHOL   ●   GAS:DETECTED
ALD: ALDEHYDE   ▲   AEROSOL:EMITTED
ACD: CARBOXYLIC ACID   ■   AEROSOL:DETECTED

Fig. 10.3-4.   The occurrence in the atmosphere of difunctional deriva-
                tives of the xylenes.

10.4. ACID RAIN

It is widely recognized that the acidity of precipitation has under-
gone an increase in recent years(1272). From pH values of 5.6 (established by the equilibrium of atmospheric carbon dioxide and pure
water), precipitation acidity in many regions has now reached values
near 4.0. The effects of such a change on natural ecosystems are

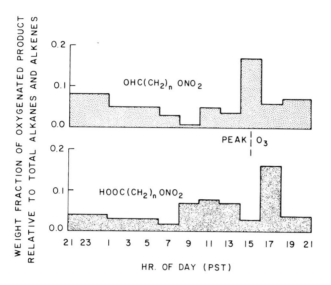

Fig. 10.3-5. The diurnal variation of aldehydic nitrate and acidic nitrate aerosol compounds in West Covina, CA, July 23-24, 1973 (from reference 1273).

thought likely to be substantial.

The proton donors in the atmosphere that are the major sources of acidity in precipitation are the inorganic acids HNO_3, H_2SO_4, and possibly HCl. Lesser effects result from other inorganic compounds and from organic acids (1272). A number of these compounds have now been detected in precipitation (1268,1272), and it seems likely that many others are present as well. Information on the occurrence of and sources for more than one hundred acidic compounds is contained in Tables 2.2-5 and 5.1-4. Such data may well be of assistance in analyses of the long-term effects of and prospects for precipitation acidity.

10.5. TROPOSPHERIC CYCLES

The concentrations of tropospheric compounds are established by the balance struck between sources and sinks (removal mechanisms) for each of them. The sources are such processes as emission, chemical formation, and transport [e.g., the injection of ozone into the troposphere from the stratosphere (1037)]. The sinks are such processes as chemical reaction, adsorption on surfaces, absorption into the oceans, and transport into the stratosphere. If all of these processes are well understood for a particular compound, a *tropospheric compound cycle* can be drawn up to describe the flow of that compound through the troposphere. Alternatively, all compounds containing a particular element can be combined to derive a *tropospheric element cycle*

Although it is outside the scope of this book to study these tropospheric cycles, the information contained herein is obviously pertinent to such efforts. For tropospheric compound cycles, the listing for an individual compound contains information on its emission sources, its chemical production and removal, and its ambient concentration. For tropospheric element cycles the information in the tables can be combined appropriately. Table 10.5-1 lists the tables in which information pertaining to element cycles of possible interest (and with the probable involvement of chemical transformation aspects) may be found. The elements and compounds for which cycles have been presented to date are listed in Table 10.5-2. Many of the processes involved are extremely difficult to specify, however, and as of this writing it appears that none of the cycles of elements or compounds in the troposphere are very well known. The attempts to define the cycles have nevertheless been useful in illuminating those processes in which additional investigations are needed. As new information becomes available, tropospheric element and compound cycles will become better understood and will serve as useful tools in assessing a variety of difficult atmospheric problems.

TABLE 10.5-1.   INFORMATION FOR TROPOSPHERIC ELEMENT CYCLES

Element	Tables
Hydrogen	All
Carbon	2.5,3.1-9.2
Nitrogen	2.2,6.1-4
Oxygen	All
Halogens(F,Cl,Br,I)	2.4,8.1-3
Sulfur	2.3,7.1-4
Arsenic	2.5,9.1
Selenium	2.5,9.1
Lead	2.5,9.1
Mercury	2.5,9.1

TABLE 10.5-2.   TROPOSPHERIC CYCLE ANALYSES

Species	Reference
Element	
O	1181,1208
N	1182,1183
C	1184,1185
S	1186,1187
Se	1210
Compound	
H_2	291,1188
O_3	410,1189
N_2O	1190,1191
NH_3	1183,1192
CO	1193,1194
CO_2	1195,1196
CH_4	1106,1197
$CFCl_3$	1096,1209
CF_2Cl_2	560,1096
CCl_4	1198
CH_3CCl_3	1199,1200

10.6. LIFETIMES OF COMPOUNDS IN THE TROPOSPHERE

The chemical lifetimes of atmospheric species are determined by the compounds with which they react and the rates at which those reactions proceed. The lifetime set by any one reaction is defined (for bimolecular reactions, the usual type) as

$$\tau_A = \frac{1}{k_{AB}[C_B]}$$

where τ_A is the lifetime of species A, k_{AB} the rate constant for the reaction of A and B, and $[C_B]$ the average concentration of species B.

Most atmospheric compounds of interest react with more than one other species and several τ_A values can thus be calculated. For ethene, for example, the results are as given in Table 10.6-1.

TABLE 10.6-1 ETHENE LIFETIMES AS DETERMINED BY DIFFERENT ATMOSPHERIC REACTANTS

B	$k_{AB}{}^a$	$C_B{}^b$	$\tau_A(\text{sec})$
HO·	8.7×10^{-12}	4.1×10^5	2.5×10^5
O_3	1.9×10^{-18}	1.0×10^{12}	5.3×10^5
NO_3·	9.3×10^{-16}	3.0×10^8	3.6×10^6
O	7.8×10^{-13}	2.5×10^4	5.1×10^7
HO_2·	1.7×10^{-17}	6.5×10^8	9.1×10^7
CH_3O·	6.2×10^{-17}	1.3×10^6	1.3×10^{10}
$O_2(^1\Delta)$	$\leqslant1.7\times10^{-17}$	2.0×10^7	$\geqslant2.9\times10^9$

^{a} Table 3.2., cm^3molecule^{-1}sec^{-1} ^{b} Table 1.1., molecule cm^{-3}.

The tables throughout this book show only the shortest of the τ_A values (that is, the limit to the lifetime set by the removal reaction with the fastest average rate). It is of interest, however, to plot all the τ_A values for all species to study the chemical species that control atmospheric chemistry. This display is presented in Fig. 10.6-1. The hydroxyl radical, known to be an effective atmospheric "vacuum cleaner" (1011), is seen to produce large numbers of τ_A values of a few hours or days (i.e., $10^4 - 10^6$ sec). For those compounds that react with ozone, the ozone reaction is often limiting, however. Other reactants producing some very short lifetimes are solar photons ($h\nu$) , and the hydroperoxyl (HO_2·) and nitrate (NO_3·) radicals. The very low concentrations of both ground state (^3P) and excited state (^1D) oxygen atoms and of the excited ($^1\Delta$) oxygen molecule render them ineffective atmospheric scavengers, as seen in Fig. 10.6-1. Conversely, the HO_2·

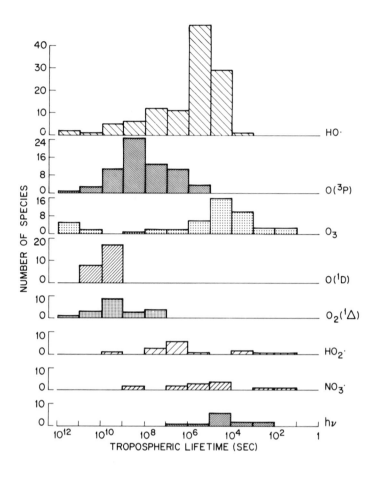

Fig. 10.6-1.   Histograms of the lifetimes of atmospheric compounds as limited by reaction with the species shown at the right of each display.

and NO₃· radical reactions, few of which have been measured, appear to have strong potential as lifetime-limiting processes for those atmospheric species with which they react.

In Fig. 10.6-2, the shortest τ_A values for all of the compounds in this book are plotted histogrammatically. It will be seen that with few exceptions, the lifetimes shorter than 10^6 sec are set by reaction with HO· or O₃, or by photodissociation. The presence of O(^{3}P) and O₂($^1\Delta$) as lifetime limiters in the $10^6 - 10^{12}$ sec range is largely artificial, since it results almost entirely in cases where rate constants with HO· and other reactive species have not yet been determined.

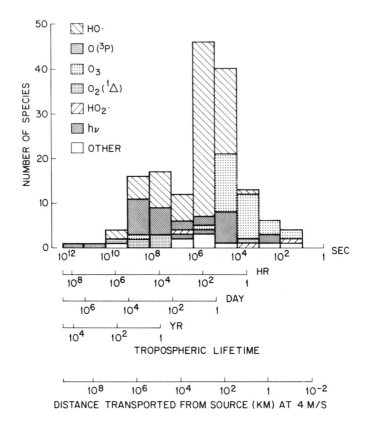

Fig. 10.6-2.   Histogram of the shortest reaction-limited lifetime for atmospheric compounds. The species responsible for the
limit is indicated by hatching. See text for discussion of
distance scale.

The lowest abscissa on Fig. 10.6-2 indicates the average distance a
molecule would be transported from its source prior to reaction if it
traveled at the typical urban ground level wind velocity of 4 m/sec
(1240). It is immediately evident that nearly all molecules travel
several kilometers from their sources prior to reaction, and that most
are transported several hundred kilometers. The result emphasizes the
nonlocal nature of atmospheric chemical precursors and products.

The most uncertain component by far of an atmospheric lifetime
calculation is the value chosen for the average concentration of the
reactant. Such species are, in general, short-lived in the atmosphere
and thus difficult to measure. The concentrations chosen for Table 1.1
are therefore rather uncertain in several cases. Furthermore, lifetimes

that are of the order of a few hours should ideally be calculated with diurnal peak reactant concentrations, those of a few days with diurnally averaged reactant concentrations, and those of a few months with seasonally or annually averaged reactant concentrations. New information or modified computational techniques could thus shift the τ_A positions in Fig. 10.6-1 somewhat, and alter some of the conclusions made in this section. There seems little doubt, however, that HO· and O_3 are the major atmospheric scavengers, and that chemically-limited lifetimes for a wide variety of species are in the range $10^4 - 10^6$ sec.

10.7. SUMMARY

The chemistry of the atmosphere is so incredibly diverse that most general summary comments have little meaning, or are so broad as to be of little value. Once grouping and ordering of the trace atmospheric compounds is performed, however, certain patterns and trends in the data become apparent and justify comment. These features have been discussed in this chapter and, to some extent, throughout the book. In this final section some of the most significant of the results are briefly reviewed.

The inorganic compounds (Chapter 2) and the hydrocarbons (Chapter 3) are, for the most part, the "feedstocks" of atmospheric chemistry. Their interactions, often driven by sunlight, produce a very wide spectrum of oxygenated products. These products constitute significant portions of the compounds of Chapters 4 and 5. The atmospheric chemistry of the organic compounds of nitrogen (Chapter 6), sulfur (Chapter 7), and the halogens (Chapter 8) seems largely secondary in the chemical system of the lower atmosphere. It should be noted, however, that these compounds have been measured in the atmosphere with much less frequency than have the compounds of Chapters 2 and 3 and their atmospheric chemical chains are less well understood. In certain chemical regimes and near certain sources, their chemistry is doubtless of significance.

It is clear that the chemistry of the natural troposphere has been little explored, yet the large numbers of naturally emitted compounds suggest that that chemistry is rich indeed. Further, the major influence estimated for natural emissions on many atmospheric cycles (1125) suggests the importance of natural atmospheric processes. Much work remains to be done in defining natural atmospheric sources and source strengths, and in exploring the subsequent chemistry of the natural atmospheric compounds.

The transition of atmospheric compounds from gas to aerosol phase is difficult to quantify but rather easy to demonstrate qualitatively. Figs. 10.1-1 to 10.1-7 show that this conversion occurs over all chemical classes, with a spectrum of elements, vapor pressures, and functional groups being represented.

Within and on the atmospheric aerosol is found a chemical system that is very complex and about which little is known. It seems unlikely, however, that individual components of that system, such as the transition of sulfur dioxide to sulfate, can be studied independently of the others. Efficient oxidation processes seem common in aerosols (see Figs. 10.3-1 to 10.3-4), and the organic compounds are potentially capable of participating in many aspects of aerosol chemistry. Innovative theoretical and experimental studies in this topic are badly needed.

The lifetimes of the gas phase compounds in the troposphere span a very wide range. For the bulk of the compounds, however, lifetimes of a few days are set by reaction with the hydroxyl radical. HO· is without doubt the atmospheric "vacuum cleaner" and its concentrations and diurnal variations need to be much better determined over a wide spectrum of geographical areas, altitudes, and seasons than they are at this writing.

This book has presented an impressive array of information on atmospheric compounds and their chemistry, reflecting the efforts of a large and dedicated group of scientists. It is apparent that there exists a rather high degree of understanding of many atmospheric chemical processes. Nonetheless, much remains to be done. Ambitious measurement programs, laboratory efforts, and computational studies are needed to assess a number of potential concerns. For few subjects, however, will the results be as interesting or as potentially vital to mankind as will be those from the study of the chemical compounds in the atmosphere.

REFERENCES

The references are an integral part of this book, since they provide the detailed information on detection, identification, site location, source type and proximity, and reaction kinetics and processes. Full reference information, including complete titles and inclusive pages, is given. In order to permit future expansion of the tables in the most convenient manner, the numbering is divided into two sections. References with one-, two-, or three-digit numbers are used for information on emissions of compounds from sources or to the detection of compounds in the atmosphere. References with four-digit numbers provide information on chemical rates or processes of significance in atmospheric chemistry.

Three abbreviations are used in the reference lists to aid the reader in obtaining access to specific documents. These are as follows:

APTIC   Air Pollution Technical Information Center, United States Environmental Protection Agency, Research Triangle Park, North Carolina 27711.

NTIS   National Technical Information Service, U.S. Department of Commerce, 5285 Port Royal Road, Springfield, Virginia 22151

SAHB   Smoking and Health Bulletin, Technical Information Center, National Clearinghouse for Smoking and Health, Center for Disease Control, 1600 Clifton Road, N.E., Bldg. 14, Atlanta, Georgia 30333

1. Aaronson, A.E. and R.A. Matula., Diesel odor and the formation of aromatic hydrocarbons during the heterogeneous combustion of pure cetane in a single-cylinder diesel engine, *Proc. 13th Int. Comb. Symp.,* Combustion Inst. Pittsburgh, Pa., p. 471-481, 1971.

2. Abdoh, Y., N. Aghdaie, M.R. Darvich and M.H. Khorgami, Detection of some polynuclear aromatic hydrocarbons and determination of benzo(a) pyrene in Teheran atmosphere, *Atmos. Environ.,* **6,** 949-952, 1972.

3. Abeles, F.B. and H.E. Heggested, Ethylene: an urban air pollutant, *J. Air Poll. Contr. Assoc.,* **23,** 517-521, 1973.

4. Abelson, P.H., A damaging source of air pollution, *Science,* **158,** 1527,1967.

5. Altshuller, A.P. and S.P. McPherson, Spectrophotometric analysis of aldehydes in the Los Angeles atmosphere, *J. Air Poll. Contr. Assoc.,* **13,** 109-111, 1963.

6. Altshuller, A.P., W.A. Lonneman, F.D. Sutterfield and S.L. Kopczynski, Hydrocarbon composition of the atmosphere in the Los Angeles basin-1967, *Environ. Sci. Technol.,* **5,** 1009-1016, 1971.

7. Anonymous, Ford plant filters noxious vapors with activated carbon system, *Filtration Eng.,* **1,** 4-5, 1969.

8. Anonymous, Environmental protection and cultivation of the environment in lower Saxony, *Staedtehygiene (Uelzen/Hamburg),* **22,** 266-267, 1971. (APTIC No. 37190).

9. Antonyuk, O.K., Hygienic assessment of glycols present in polymer type building materials, *Gigena i Sanit.,* **9,** 106-107, 1974. (APTIC No. 71466).

10. Augustine, F.E., *Airborne sampling of particles emitted to the atmosphere from Kraft paper mill processes and their characterization by electron microscopy,* Oregon State University (Corvallis, Oregon), Ph.D Dissertation, 1974,

11. Babina, M.D., Determination of volatile substances emitted into the air by shoe industry plants, *Nov. Obl. Prom.-Sanit. Khim. (S.I. Muraveva, ed.),* 227-232, 1969. (APTIC No. 33508).

12. Balabaeva, L. and G. Petrova, Contamination of atmospheric air with fluoride compounds and their exeration in urine of persons, *Khig. Zdraueopazyane (Bulgaria),* **15,** 162-168, 1972. (APTIC No. 47982).

13. Ball, G. and E.A. Boettner, Combustion products of plastics and their contribution to incineration problems, *Am. Chem. Soc., Div. Water, Air, Waste Chem., Gen. Papers,* **10,** 236-239, 1970.

14. Bandy, A.R., Studies of the importance of biogenic hydrocarbon emissions to the photochemical oxidant formation in Tidewater, Va., *Scientific Seminar on Automotive Pollutants,* EPA 600/9-75-003, Environmental Protection Agency, Washington, D.C., 1975.

15. Barber, E.D. and J.P. Lodge, Jr., Paper chromatographic identification of carbonyl compounds as their 2,4- dinitrophenylhydrazones in automobile exhaust, *Anal. Chem.,* **35,** 348-350, 1963.

16. Bentley, M.D., I.B. Douglass, J.A. Lacadie and D.R. Whittier, The photolysis of dimethyl sulfide in air, *J. Air Poll. Contr. Assoc.,* **22,** 359-363, 1972.

17. Berglund, B., U. Berglund and T. Lindvall, Measurement of rapid changes of odor concentration by a signal detection approach, *J. Air Poll. Cont. Assoc.,* **24,** 162-164, 1974.

18. Bethea, R.M. and R.S. Narayan, Identification of beef cattle feedlot odors, *Trans. Am. Soc. Agr. Engrs.,* **15,** 1135-1137, 1972.

19. Bethge, P.O. and L. Ehrenborg, Identification of volatile compounds in Kraft mill emissions, *Svensk Papperstidning,* **70,** 347-350, 1967.

20. Bidleman, T.F. and C.E. Olney, Chlorinated hydrocarbons in the Sargasso Sea atmosphere and surface water, *Science,* **183,** 516-518, 1974.

21. Bidleman, T.F. and C.E. Olney, Long range transport of toxaphene insecticide in the atmosphere of the western North Atlantic, *Nature,* **257,** 475-477, 1975.

22. Bigg, E.K., A. Ono and J.A. Williams, Chemical tests for individual submicron aerosol particles, *Atmos. Environ.,* **8,** 1-13, 1974.

23.Breeding, R.J., J.P. Lodge, Jr., J.B. Pate, D.C. Sheesley, H.B. Klonis, B. Fogle, J.A. Anderson, T.R. Englert, P.L. Haagenson, R.B. McBeth, A.L. Morris, R. Pogue and A.F. Wartburg, Background trace gas concentrations in the Central United States, *J. Geophys. Res., 78*, 7057-7064, 1973.

24.Brinkmann, W.L.F. and U. de M Santos, The emission of biogenic hydrogen sulfide from Amazonian floodplain lakes, *Tellus, 26*, 261-267, 1974.

25.Burnett, W.E., Air pollution from animal wastes: determination of maloders by gas chromatographic and organoleptic techniques, *Environ. Sci. Technol., 3*, 744-749, 1969.

26.Busso, R.H., *Identification and determination of pollutants emitted by the principal types of urban waste incinerator plants (final report)*, Centre d Etudes Et Recherches des Charbonnages de France, Laboratoire du CERCHAR, Creil, France, Contract 69-01-758, A.R. 144, 1971.

27.Caban, R. and T.W. Chapman, Losses of mercury from chlorine plants: a review of a pollution problem, *Am. Inst. Chem. Engrs. J., 18*, 892-903, 1972.

28.Cadle, R.D., A.F. Wartburg, W.H. Pollock, R.W. Gandrud and J.P. Shedlovsky, Trace constituents emitted to the atmosphere by Hawaiian volcanos, *Chemosphere, 2*, 231-234, 1973.

29.Cadle, R.D., H.H. Wickman, C.B. Hall and K.M. Eberle, The reaction of atomic oxygen with formaldehyde, crotonaldehyde, and dimethyl sulfide, *Chemosphere, 3*, 115-118, 1974.

30.Cadle, R.D., Volcanic emissions of halides and sulfur compounds to the troposphere and stratosphere, *J. Geophys. Res., 80*, 1650-1652, 1975.

31.Carotti, A.A. and E.R. Kaiser, Concentrations of 20 gaseous chemical species in the flue gas of a municipal incinerator, Paper presented at 64th Annual Meeting, Air Poll. Contr. Assoc. (Atlantic City, NJ), 1971.

32.Carter, R.V. and B. Linsky, Gaseous emissions from whisky fermentation units, *Atmos. Environ., 8*, 57-62, 1974.

33.Cavanagh, L.A., C.F. Schadt and E. Robinson, Atmospheric hydrocarbon and carbon monoxide measurements at Point Barrow, Alaska, *Environ. Sci. Technol., 3*, 251-257, 1969.

34.Challenger, F., Biological methylation, *Chem. Rev., 36*, 315-361, 1936.

35.Challis, E.J., The approach of industry to the assessment of environmental hazards, *Proc. Roy. Soc. B, 185*, 183-197, 1974.

36.Charlson, R.J., A.P. Waggoner, N.C. Ahlquist, D.S. Covert and R. Husar, Sulfates in lower tropospheric aerosols, Paper 76-20.2, 69th Annual Meeting, Air Poll. Contr. Assoc. (Portland, OR), 1976.

37.Clemons, C.A., A.I. Coleman and B.E. Saltzman, Concentration and ultrasensitive chromatographic determination of sulfur hexafluoride for application to meteorological tracing, *Environ. Sci. Technol., 2*, 551-556, 1968.

38.Collier, L., Determination of bis-chloromethyl ether at the ppb. level in air samples by high resolution mass spectroscopy, *Environ. Sci. Technol., 6*, 930-932, 1972.

39.Collins-Williams,C., H.K. Kuo, E.A. Varga, S. Davidson, D. Collins-Williams and M. Fitch, Atmospheric pollen counts in Toronto, Canada, 1971, *Ann. Allergy, 31(2)*, 65-68, 1973.

40.Collins-Williams, C., H.K. Kuo, D.N. Garey, S. Davidson, D. Collins-Williams, M. Fitch and J.B. Fischer, Atmospheric mold counts in Toronto, Canada, 1971, *Ann. Allergy, 31(2)*, 69-71, 1973.

41.Cooper, R.L. and A.J. Lindsey, Atmospheric pollution by polycyclic hydrocarbons, *Chem. Ind. (London)*, 1177-1178, 1953.

42.Cox, R.A. and F.J. Sandalls, The photo-oxidation of hydrogen sulphide and dimethyl sulphide in air, *Atmos. Environ., 8*, 1269-1281, 1974.

43.Davis, D.D., G. Smith and G. Klauber, Trace gas analysis of power plant plumes via air-craft measurement: O_3, NO_x, SO_2 chemistry, *Science*, **186**, 733-736, 1974.

44.Davison, A.W., A.W. Rand and W.E. Betts, Measurement of atmospheric fluoride concentrations in urban areas, *Environ. Pollut.*, **5**, 23-33, 1973.

45.Dept. of Nature Conservation, *Air protection problems in the forestry industry*, Comm. on Air Protection in the Forestry Industry, (Stockholm), 1969. (APTIC No. 65119).

46.Dept. of Trade and Industry, Great Britain, Programmer Analysis Unit, Appendix II: The contribution of plastics to air pollution in the United Kingdom, *An Economic and Technical Appraisal of Air Pollution in the United Kingdom*, (H.M. Stationery Office, London), p. 233-243, 1972.

47.Dimitriades, B., B.H. Eccleston and R.W. Hurn, An evaluation of the fuel factor through direct measurement of photochemical reactivity of emissions, *J. Air Poll. Contr. Assoc.*, **20**, p. 150-160, 1970.

48.Dimitriades, B. and T.C. Wesson, Reactivities of exhaust aldehydes, *J. Air Poll. Contr. Assoc.*, **22**, 33-38, 1972.

49.Dishart, K.T., Exhaust HC composition: its relation to gasoline composition, Paper presented at 35th Midyear Meeting, Amer. Petroleum Inst. (Houston, Texas), 1970.

50.Dow Chemical U.S.A., Dow chlorinated solvents and the clean air act, Inorganic Chemicals Dept, (Midland, Mich.), 1972.

51.Dubois, L., T. Teichman and J.L. Monkman, The sulfuric acid content of soot, *Sci. Total Environ.*, **2**, 97-100, 1973.

52.Dubrovskaya, F.I., *Gigiena i Sanit.* **31(1)**, 97-98, 1966.

53.Dudley, H.C. and J.M. Dalla Valle, A study of the odors generated in the manufacture of Kraft paper, *Paper Trade J.*, **108**, 30-33, 1939.

54.Ekstedt, J. and S. Oden, *Chlorinated hydrocarbons in the lower atmosphere in Sweden*, Forskningsavdelningen for Miljoward Inst. for Markvetenskap, Lantbrukshogskolan (Uppsala, Sweden), (APTIC No. 68784).

55.Ellis, C.F., R.F. Kendall and B.H. Eccleston, Identification of some oxygenates in automobile exhausts by combined gas liquid chromatography and infrared techniques, *Anal. Chem.*, **37**, p. 511-516, 1965.

56.Engdahl, R.B., Stationary combustion sources, in *Air Pollution,III*, (A.C. Stern, ed.) 2nd ed., New York: Academic Press, p. 3-54, 1968.

57.Environmental Protection Agency, *Air quality criteria for hydrocarbons*, Report No. AP-64, Washington, D.C., 1970.

58.Environmental Protection Agency, *Air Pollution Emission Factors*, (original plus supplements), Report No. AP-42, Research Triangle Park, (N.C.), 1973-75.

59.Environmental Protection Agency, *Vinyl Chloride Monomer Investigation*, Tenneco Chemical, Inc., Flemington, New Jersey, Office of Enforcement and General Counsel, EPA, New Jersey, 1974.

60.Environmental Protection Agency, *Guideline on use of reactivity criteria in control of organic emissions for reduction of atmospheric oxidants*, Washington,D.C., August 13, 1975.

61.Eshleman, A., S.M. Siegel and B.Z. Siegel, Is mercury from Hawaiian volcanoes a natural source of pollution?, *Nature*, **233**, 471-472, 1971.

62.Faith, W.L., Food and feed industries, in *Air Pollution,III*, (A.C. Stern, ed.) 2nd ed., New York: Academic Press, p. 269-288, 1968.

63.Farwell, S.O., F.W. Bowes and D.F. Adams, Determination of chlorophenoxy herbicides in air by gas chromatography/mass spectrometry: selective ion monitoring, *Anal. Chem.*, **48**, 420-426, 1976.

64.Fawcett, R.L., Air pollution potential of phthalic anhydride manufacture, *J. Air Poll. Contr. Assoc.*, **20**, 461-465, 1970.

65.Finkelsteyn, D.H., A.P. Yelenevich and V.N. Dymchenko, Chemical composition of the dispersed phase of smokes generated in the process of manganese steel meeting, *Gigiena i Sanit,* **11** (1), 25, 1953. (APTIC No. 40757).

66.Foote, R.S., Mercury vapor concentrations inside buildings, *Science,* **177,** 513-514, 1972.

67.Fracchia, M.F., F.J. Schuette and P.K. Mueller, A method for sampling and determination of organic carbonyl compounds in automobile exhaust, *Environ. Sci. Technol.,* **1,** 915-922, 1967.

68.Fraser, H.S. and E.P. Swan, *Chemical analysis of veneer-dryer condensates,* Western Forest Products Lab. (Canadian Forestry Service, Vancouver, B.C.). (NTIS Document PB-212661).

69.Freed, V.H., R. Haque and D. Schmedding, Vaporization and environmental contamination by DDT, *Chemosphere,* **1,** 61-66, 1972.

70.Fujii, T., Offensive odors. 13. Constituents of, and countermeasures for, the odoriferous fumes from vinyl chloride 'leather' factories, *J. Japan Soc. Air Poll.,* **2** (1), 48-59, 1967.

71.Fujii, T., N. Tajima, K. Yoshimura, K. Inoue, K. Taguchi, F. Kure, Y. Nishikawa and K. Oka, Studies on the air pollution by exhaust gases of aircraft(1), *J. Japan Soc. Air Poll.,* **8,** 515, 1973. (APTIC No. 58368).

72.Gassmann, M., Freon 14 in purest krypton and in the atmosphere, *Naturwissenschaften,* **61,** 127, 1974.

73.Gautier, A., Fluorine is an element always present in emissions from the Earth's core, *Compt. Rend.,* **157,** 820-825, 1913.

74.Gerhold, H.D. and G.H. Plank, Monoterpene variations in vapor from white pines and hybrids, *Phytochemistry,* **9,** 1393-1398, 1970.

75.Gerstle, R.W., *Atmospheric emissions from asphalt roofing processes,* PEDCo-Environmental, Inc. (Cincinnati, Ohio), 1974. (NTIS Document, PB-238445).

76.Gerstle, R.W. and T.W. Devitt, Chlorine and hydrogen chloride emissions and their control, Paper 71-25, 64th Annual Meeting, Air Poll. Contr. Assoc. (Atlantic City, NJ), 1971.

77.Giam, C.S., H.S. Chan and G.S. Neff, Rapid and inexpensive method for detection of polychlorinated biphenyls and phthalates in air, *Anal. Chem.,* **47,** 2319-2320, 1975.

78.Gilbert, T.E., Rate of evaporation of liquids into air, *J. Paint Technol.,* **43,** 93-97, 1971.

79.Gilbert, J.A.S. and A.J. Lindsey, Polycyclic hydrocarbons in tobacco smoke: pipe smoking experiments, *Brit. J. Cancer (London),* **10,** 646-648, 1956.

80.Goedseels, V., Evaluation of the odorous emissions in relation to the infrastructure of intense cattle raising, *Ingenieursblad (Holland),* **42,** 557-564, 1973. (APTIC No. 57342).

81.Gordon, R.J. and R.J. Bryan, Ammonium nitrate in airborne particles in Los Angeles, *Environ. Sci. Technol.,* **7,** 645-647, 1973.

82.Gorodinskiy, S.M., M.I. Vakar, G.A. Gaziyev, Y.Y. Sotnikov and A.N. Mazin, Hygienic and chemical investigation of the hydrocarbons in the gas mixture in isolating pressure chambers, *Gigiena i Sanit.,* **7,** 106-109, 1974. (APTIC No. 66648).

83.Graedel, T.E., B. Kleiner and C.C. Patterson, Measurements of extreme concentrations of tropospheric hydrogen sulfide, *J. Geophys. Res.,* **79,** 4467-4473, 1974.

84.Gray, E.W., L.G. McKnight and J.M. Sawina, Identity and interactions of ions from relay break arcs, *J. Appl. Phys.,* **45,** 661-666, 1974.

85.Grimsrud, E.P. and R.A. Rasmussen, Survey and analysis of halocarbons in the atmosphere by gas chromatography-mass spectroscopy, *Atmos. Environ.,* **9,** 1014-1017, 1975.

86.Grob, K. and G. Grob, Gas-liquid chromatographic-mass spectrometric investigation of C_6-C_{20} organic compounds in an urban atmosphere, *J. Chromatogr.,* **62,** 1-13, 1971.

87.Gross, G.P., *Gasoline composition and vehicle exhaust gas polynuclear aromatic content,* CRC-APRAC Project No. CAPE-6-68, Coordinating Research Council (New York, New York), 1973.

88.Hansen, C.M., Solvents for coatings, *Chemtech,* 547-553, 1972.

89.Hanst, P.L., L.L. Spiller, D.M. Watts, J.W. Spence and M.F. Miller, Infrared measurement of fluorocarbons, carbon tetrachloride, carbonyl sulfide, and other atmospheric trace gases, *J. Air Poll. Contr. Assoc.,* **25,** 1220-1226, 1975.

90.Hanst, P.L., W.E. Wilson, R.K. Patterson, B.W. Gay, Jr., L.W. Chaney and C.S. Burton, *A spectroscopic study of California smog,* EPA 650/4-75-006, Environmental Protection Agency (Washington, D.C.), 1975. (NTIS Report PB 241022).

91.Harvey, G.R. and W.G. Steinhauer, Atmospheric transport of polychlorobiphenyls to the North Atlantic, *Atmos. Environ.,* **8,** 777-782, 1974.

92.Hayashi, M. and S. Hayashi, Environmental pollution caused by processing polyvinyl chloride resin, *Proc. Osaka Pub. Health Inst. (Ed. Ind. Health),* 7, 46-55, 1969. (APTIC No. 30687).

93.Heller, A.N., S.T. Cuffe and D.R. Goodwin, Inorganic chemical industry, in *Air Pollution, III,* (A.C. Stern, ed.) 2nd ed., New York: Academic Press, p. 191-242, 1968.

94.Hendrickson, E.R. and C.I. Harding, Air pollution problems associated with Kraft pulping, *Proc. Int. Clean Air Congress (London),* p. 95-97, 1966.

95.Hinkamp, J.B., M.E. Griffing and D.W. Zutaut, Aromatic aldehydes and phenols in the exhaust from leaded and unleaded fuels, *Am. Chem. Soc., Div. Petrol. Chem. (preprints),* **16** ,(2), E5-E11, 1971.

96.Hirschler, D.A., L.F. Gilbert, F.W. Lamb and L.M. Niebylski, Particulate lead compounds in automobile exhaust gas, *Ind. Eng. Chem.,* **49,** 1131-1142, 1957.

97.Hoefig, R., Air pollution problems caused by the use of plastic binders, *Giessereitechnik,* **15,** 250-254, 1969. (APTIC No. 40595).

98.Hoffman, D. and E.L. Wynder, Chemical analysis and carcinogenic bioassays of organic particulate pollutants, *Air Pollution, II,* (A.C. Stern, ed.) 2nd ed., New York: Academic Press, p. 187-242, 1968.

99.Hollingdale-Smith, P.A., *Gaseous atmospheric pollutants: a literature survey,* Chemical Defense Estab. (Porton Dowes, Salisbury, Witts), Tech. Note 144, Nov., 1972. (NTIS Document AD 905797).

100.Sandalls, F. J., and S. A. Penkett, Measurements of carbonyl sulphide and carbon disulphide in the atmosphere, *Atmos. Environ.,* **11** ,197-199, ,1977.

101.Hori, M., N. Tanigawa and Y. Kobayashi, Investigation of lead compounds in exhaust gases from various kinds of sources, *J. Japan Soc Air Poll.,* **8,** 344, 1973. (APTIC No. 56970).

102.Hoshika, Y., T. Ishiguro and Y. Shigeta, Analysis of odor components from feather rendering plants, *J. Japan Soc. Air Poll.,* 5 (1), 5(1), 99, 1970.

103.Hoshika, Y., T. Ishiguro, Y. Katori, S. Futaki and Y. Shigeta, An example of investigation methods for odor pollution, *J. Japan Soc. Air Poll.,* **6,** 227, 1971. (APTIC No. 36910).

104.Hoshika, Y. and S. Kadowaki, Identification of evolved component from ABS copolymer by g.c., *J. Japan Soc. Air Poll.,* **8,** 271, 1973. (APTIC No. 58564).

105.Hosono, Y., G. Kawaski, M. Yamazaki, S. Inada, M. Ishisaka, S. Maeda, S. Taka, R. Ishisaka, K. Senda, I. Yamazaki and I. Shinmura, On the hydrogen cyanide in exhaust gas of electric furnaces, *J. Japan Soc. Air Poll.,* **6,** 246, 1971. (APTIC No. 37322).

106.Huntingdon, A.T., The collection and analysis of volcanic gases from Mount Etna, *Phil. Trans. R. Soc. (Lond.),* **A274,** 119-128, 1973.

107.Husar, J.D., R.B. Husar and P.K. Stubits, Determination of submicrogram amounts of atmospheric particulate sulfur, *Anal. Chem.,* **47,** 3062-3064, 1975.

108.Ishiguro, T., K. Hishida and T. Yajima, Present state of public nuisance caused by offensive odors in Tokyo, *J. Water Waste (Japan),* 13(8), 972-978, 1971.

109.Jacobson, A.R., Viable particles in the air, *Air Pollution,I,* (A.C. Stern, ed.) 2nd ed., New York: Academic Press, p. 95-119, 1968.

110.Japan Environ. Health Center, *Reports on investigations of bad odors from animal offal processing and poultry farms in the Soka and Koshigaya areas, Saitama Prefect, and future countermeasures,* 1971. (APTIC No. 65669).

111.Japan Environ. San. Center, *Report of survey of the specified poisonous substances and the prevention of offensive odor,* Report 4 (Tokyo), 67 p. ,1969. (APTIC No. 32475).

112.Johnson, D.L. and R.S. Braman, Alkyl and inorganic arsenic in air samples, *Chemosphere,* **6,** 333-338, 1975.

113.Just, J., S. Maziarka and H. Wyschinska, Benzpyrene and other aromatic hydrocarbons in the dust of Polish towns, *Wiss. Z. Humboldt Univ. (Berlin Math. Naturw. Raike),* **19,** 513-515, 1970. (APTIC No. 29569).

114.Kasparov, A.A. and V.G. Kiriy, *Gigiena i Sanit,* **37,** 57, 1972. Cited by Saltzman and Cuddeback (206).

115.Katari, V., G. Isaacs and T.W. Devitt, *Trace pollutant emissions from the processing of metallic ores,* EPA 650/2-74-115, Environmental Protection Agency (Washington, D.C.), 1974. (NTIS Document PB 238-655).

116.Kipling, M.D. and R. Fothergill, *Br. J. Ind. Med.,* **21,** 74-77, 1964. Cited by Hollingdale-Smith (99).

117.Kirsch, H. and K. Schwinkowski, Emission measurement of styrol and possibilities for its limitation in the waste gas, *Z. Hyg.,* **18,** 193-194, 1972. (APTIC No. 39340).

118.Kitaoka, Y. and K. Murata, Experiments on the thermal degradation of ethylene low polymer, *J. Japan Fuel Soc. (Tokyo),* **50,** 791-799, 1971. (APTIC No. 36274).

119.Kitagawa, Y., A. Furukawa, A. Sueda, H. Ryoken and M. Ito, Investigation of odorous air pollution from factory of vulcanization accelerator, *J. Japan Soc. Air Poll.,* **8,** 379, 1973. (APTIC No. 58375).

120.Klee, O., PCB in the wake of DDT, *Kosmos (Stuttgart),* **2,** 65-66, 1972. (APTIC No. 46111).

121.Kobayashi, S., H. Matsushita, H. Osaka and K. Morimura, Measurements of odorous pollutants dryer emissions and study on its deodorization, *J. Japan Soc. Air Poll.,* **8,** 374, 1973. (APTIC No. 58544).

122.Kobayashi, Y., S. Tsukada, H. Hirobe, M. Takahashi and N. Ito, Report on the odor investigation in factory areas, *Mie Prefect. Pub. Nuisance Center Ann. Rpt.,* **2,** 102-107, 1974. (APTIC No. 69042).

123.Kopczynski, S.L., W.A. Lonneman, F.D. Sutterfield and P.E. Darley, Photochemistry of atmospheric samples in Los Angeles, *Environ. Sci. Technol,* **6,** 342-347, 1972.

124.Kremnera, S.N., et al., *Toksikol. Novykh. Press. Khim. Veshchestv.,* **5,** 123-135, 1963. Cited by Hollingdale-Smith (99).

125.Krey, P.W. and R.J. Lagomarsino, Stratospheric concentrations of SF_6 and CCl_3F , *Environmental Quarterly (Health and Safety Laboratory, Atomic Energy Commission New York, New York),* HASL-294, 1975.

126.Kuchak, Y.A., Electro-aerosol pesticides and the hygienic aspects of their application, *Gigiena i Sanit.,* **1,** 10-13, 1974. (APTIC No. 62867).

127.Kutuzova, L.N., A.F. Kononenko and G.P. Sokulskii, Discharges to the atmosphere from benzole plant, *Coke Chem. (USSR),* **8,** 39-42, 1970. (APTIC No. 3508).

128.Lang, O. and T. zur Muehlen, Air pollution by organic acids and esters and their analytical determination, *Zbl. Arbeitsmed.,* **2** , 39-45, Feb. 1971.

129.Farmer, C.B., O.F. Raper and R.H. Norton, Spectroscopic detection and vertical distribution of HCl in the troposphere and stratosphere, *Geophys. Res. Lett.,* **3,** 13-16, 1976.

130.Lee, M.L., M. Novotny and K.D. Bartle, Gas chromatography/mass spectrometric and nuclear magnetic resonance spectrometric studies of carcinogenic polynuclear aromatic hydrocarbons in tobacco and marijuana smoke condensates, *Anal. Chem.,* **48,** 405-416, 1976.

131.Leonardos, G., D. Kendall and N. Barnard, Odor threshold determinations of 53 odorant chemicals, *J. Air Poll. Contr. Assoc.,* **19,** 91-95, 1969.

132.Levaggi, D.A., Private communication, 1976.

133.Levaggi, D.A. and M. Feldstein, The determination of formaldehyde, acrolein, and low molecular weight aldehydes in industrial emissions on a single collection sample, *J. Air Poll. Control Assoc.,* **20,** 312-314, 1970.

134.Levy, A., S.E. Miller and F. Schofield, The photochemical smog reactivity of solvents, *Proc. 2nd. Int. Clean Air Congress* , (H. Englund, ed.) New York: Academic Press, p. 305-316, 1971.

135.Lillian, D., H.B. Singh, A. Appleby, L. Lobban, R. Arnts, R. Gumpert, R. Hague, J. Toomey, J. Kazazis, M. Antell, D. Hansen and B. Scott, Atmospheric fates of halogenated compounds, *Environ. Sci. Technol.,* **9,** 1042-1048, 1975.

136.Linnell, R.H. and W.E. Scott, Diesel exhaust analysis, *Arch. Environ. Health,* **5,** 616-625, 1962.

137.Little, A.D., Inc., *Chemical identification of the odor components in diesel engine exhaust,* Report C-71407/71475, HEW Contract No. CPA 22-69-63 (Cambridge, Mass), 1970.

138.Little, A.D., Inc., *Chemical identification of the odor components in diesel engine exhaust,* Report ADL 62561-5, Cambridge, Mass., 1971.

139.Lonneman, W.A., T.A. Bellar and A.P. Altshuller, Aromatic hydrocarbons in the atmosphere of the Los Angeles basin, *Environ. Sci. Technol.,* **2,** 1017-1020, 1968.

140.Lovelock, J.E., R.J. Maggs and R.A. Rasmussen, Atmospheric dimethyl sulphide and the natural sulphur cycle, *Nature,* **237,** 452-453, 1972.

141.Luebs, R.E., A.E. Laag and K.R. Davis, Ammonia and related gases emanating from a large dairy area, *Calif. Agr.,* **27,** (2), 10-12, 1973.

142.Luzin, Yu.P. and V.V. TsarKov, Reduction of atmospheric discharges during the etching of steel, *Metallurg. (USSR),* **16,** (12), 32, 1972. (APTIC No. 38011).

143.Maarse, H. and M.C. Ten Noever de Brauw, Another catty odour compound causing air pollution, *Chem. Ind. (London),* **1,** 36-37, 1974.

144.Macfarlane, C., J.B. Lee and M.B. Evans, The qualitative composition of peat smoke, *J. Inst. Brewing,* **79,** 202-209, 1973.

145.Masek, V., New findings concerning the properties of fly dust from coking plants. Part IV. Hard coal tar distillation plants, *Zbl. Arbeitsmed.,* **22,** (11), 332-337, 1972. (APTIC No. 47099).

146.Matheson Gas Products, *Ethylene FRP (Fruit Ripening Purity) in safe, easy-to-use Matheson lecture bottles,* Pamphlet 25M/11/72, East Rutherford, N.J., 1972.

147.Mathu, M., S.K. Majumder and H.A.B. Parpia, *J. Agric. Food Chem.,* **21,** 184, 1973. Cited by Saltzman and Cuddeback (206).

148.Matsuda, Y., On airborne true fungi, *J. Japan Air Cleaning Assoc. (Tokyo),* **9,** (7), 42-58, 1972. (APTIC No. 50583).

149.Matsumoto, H., Analysis of odor composition in starch lees, *J. Japan Soc. Air Poll.,* **9,** 461, 1974. (APTIC No. 70309).

150.Matsuo, K., A study of air pollution at Wakayama City. Part 3. *J. Meteor. Res. (Tokyo),* **25,** 481-486, 1973. (APTIC No. 61204).

151.Matsuyama, T., Air pollution with dust, *J. Res. Assoc. Powder Technol. (Japan),* **11,** 40-42, 1974. (APTIC No. 61569).

152.Matteucci, M., Peculiarities of the eruption of Vesuvius, *C.R. Acad. Sci. (Paris),* **129,** 65-67, 1899.

153.McEwen, D.J., Automobile exhaust hydrocarbon analysis by gas chromatography, *Anal. Chem.,* **38,** 1047-1053, 1966.

154.Miller, C.W. and T.M. Shafik, *Concentrations of OMS-33 in air following repeated indoor applications,* Dept. of Health, Educ. and Welfare (Atlanta, Ga.) and Environmental Protection Agency (Research Triangle Park, N.C.), 1973. (APTIC No. 67476).

155.Mill, R.A., J.M. Robertson and B. Walker, A technique for airborne aerobiological sampling, *J. Environ. Health,* **35,** 51-53, 1972.

156.Ministry of Labor, Health and Welfare of North Rhine-Westphalia (West Germany), Concept of clean air maintenance in North Rhine-Westphalia, Reine luft fuer morgen. Utopia odor Wirklichkeit?, Ein Konzept fuer das Nordrhein-Westfalen bis 1980, 13-14, 1972. (APTIC No. 59773).

157.Mizutani, H., A. Kamiya, M. Kitasi, T. Aoyama and E. Ito, Odor pollution research in Nagoya City, *J. Japan Soc. Air Poll.,* **8,** 381, 1973. (APTIC No. 58583).

158.Morie, G.P., Determination of hydrogen sulfide in cigarette smoke with a sulfide ion electrode, *Tobacco Sci.,* **15,** (29), 34, 1971.

159.Morris, W.E. and K.T. Dishart, Influence of vehicle emission control systems on the relationship between gasoline and vehicle exhaust hydrocarbon composition, in *Effect of Automotive Emission Requirements on Gasoline Characteristics,* ASTM STP 487, Amer. Soc. for Testing and Materials, Philadelphia, Pa., p. 63-93, 1971.

160.Mosier, A.R., C.E. Andre and F.G. Viets, Jr., Identification of aliphatic amines volatilized from cattle feedyard, *Environ. Sci. Technol.,* **7,** 642-644, 1973.

161.Muhlrad, D., *The fight against air pollution caused by electro-metallurgic furnaces,* Eidgenoessische Kommission fuer Lufthygiene, Zurich, Probleme der Luftverunreinigung duroh die Industrie, 30-37, 1968. (APTIC No. 39309).

162.Murata, M. Bad odor in Yokkaichi City, *Mie Prefect. (Japan) Pub. Nuisance Center Ann. Rep.,* **1,** 71-76, 1973. (APTIC No. 52659).

163.Murray, A.J. and J.P. Riley, Occurrence of some chlorinated aliphatic hydrocarbons in the environment, *Nature,* **242,** 37-38, 1973.

164.Naka, K., Y. Hasegawa and H. Hirobe, Studies on hydrocarbons present in the Yokkaichi City area, *Mie Prefect. (Japan) Pub. Nuisance Center Ann. Rep.,* **1,** 77-93, 1973. (APTIC No. 52669).

165.Nakayama, H. *Report on the investigation on the offensive odor in Yokkaichi City,* Yokkaichi, Japan, 122 p., 1969. (APTIC No. 29349).

166.Narayan, R.S., Identification and control of cattle feedlot odors, *Texas Tech. Univ. (Lubbock, Tex.),* M.S. Thesis, 1971.

167.National Air Pollution Control Administration; Publication AP-54, oo c Cited by Hollingdale-Smith (99).

168.Naumann, R.J., Smoking and air pollution standards, *Science,* **182,** 334, 336, 1973.

169.Nazyrov, G.N. and K.Y. Vengerskaya, *Gig. Tr. Prof. Zabol.,* **10,** (11), 56-60, 1966.

170.Neligan, R.E., Hydrocarbons in the Los Angeles atmosphere, *Arch. Environ. Health,* **5,** 581-591, 1962.

171.Neligan, R.E., Paper presented at Princeton/BTL Conference on Air Pollution, Princeton, N.J., Dec. 10, 1974.

172.Nichols, R., Ethylene production during senescence of flowers, *J. Hort, Sci.,* **41,** 279-290, 1966.

173.Nieboer, H. and J. van Ham, Peroxyacetyl nitrate (PAN) in relation to ozone and some meteorological parameters at Delft in the Netherlands, *Atmos. Environ.,* **10,** 115-120, 1976.

174.Nishida, K., T. Honda and T. Tsuji, The effect of ozone deodorization equipment for the sewage odor, *Akushu no Kenkya,* **4** (20), 24-30, 1975. (Ozone Chemistry and Technology Index No. 581, 1975).

175.Obremski, R.J., Design notes-dioxane, *Poll. Eng.,* **7,** (9), 22, 1975.

176.O'Donnell A. and A. Dravnieks, *Chemical species in engine exhaust and their contributions to exhaust odors,* Report No. IITRI. C6183-5, IIT Research Institute (Chicago, IL), 1970.

177.Okada, S., Collected data on allowable and minimum perceptible concentrations of odor substances, *J. Water Waste (Japan),* **13,** 1136-1142, 1971.

178.Okita, T., M. Watanabe and S. Kiyono, Analysis of HC of low grade in atmosphere, Paper presented at 24th Ann. Mtg., *Japan Chem. Soc. (Tokyo)*, March, 1971. (APTIC No. 29482).

179.Okuno, T., Measurement of polycyclic hydrocarbons in air, *J. Japan Soc. Air Poll.*, **5**, 74, 1970. (APTIC No. 32929).

180.Okuno, T., M. Tsuji and K. Takada, *Problems of public nuisance caused by bad odors in Hyogo Prefecture*, Environmental Sci. Inst., Hyogo Prefacture, Kobe (Japan), Report 2, 18-24, 1971.

181.Okuno, T., M. Tsuji, Y. Shintani and H. Watanabe, On the pyrolysis of PCB, *J. Japan Soc. Air Poll.*, **8**, 351, 1973. (APTIC No. 60236).

182.Olsen, D. A. and J.L. Haynes, *Air Pollution aspects of Organic Carcinogens*, Litton Systems, Inc. (Bethesda, Maryland), 1969. (NTIS Document PB 188090).

183.O'Mara, M.M., The combustion products from synthetic and natural products. Part 1. Wood, *J. Fire Flammability*, **5**, 34-53, 1974.

184.O'Mara, M.M., L.B. Crider and R.L. Daniel, Combustion products from vinyl chloride monomer, *Am. Ind. Hyg. Assoc. J.*, **32**, 153-156, 1971.

185.Pasco, L.I. and F.Kh. Yunusov, Air pollution in waste-water cleaning premises, *Bezop. Tr. Prom.*, **15**, 46-48, 1971. (APTIC No. 39402).

186.Penkett, S.A., F.J. Sandalls and J.E. Lovelock, Observations of peroxyacetyl nitrate (PAN) in air in Southern England, *Atmos. Environ.*, **9**, 139-140, 1975.

187.Pfaff, R.O., Swift Agricultural Chemicals, Inc., Birmingham, Alabama, Environ. Engg., Inc. (Gainesville, Fla., Office of Air Quality Planning and Standards), Contract Rpt. 73-FRT-6, 1973.

188.Pierce, R.C. and M. Katz, Dependency of polynuclear aromatic hydrocarbon content on size distribution of atmospheric aerosols, *Environ. Sci. Technol.*, **9**, 347-353, 1975.

189.Pierce, R.C. and M. Katz, Determination of atmospheric isomeric polycyclic arenes by thin-layer chromatography and fluorescence spectroscopy, *Anal. Chem.*, **47**, 1743-1748, 1975.

190.Pierce, R.C. and M. Katz, Chromatographic isolation and spectral analysis of polycyclic quinones: application to air pollution analysis, *Environ. Sci. Technol.*, **10**, 45-51, 1976.

191.Polgar, L.G., R.A. Duffee and L.J. Updyke, Odor characteristics of mixture of sulfur compounds emitted from the viscose process, Paper presented at 68th Annual Meeting, Air Pollution Control Assoc., Boston, Mass., 1975.

192.Prakash, C.B. and F.E. Murray, Studies on air emissions from the combustion of wood-waste, *Combust. Sci. Technol.*, **6**, 81-88, 1972.

193.Puchner, F., Air pollution problems of Dorog, Hungary, *Tatabanyai Szenbanyak Muszaki-Kozgazdasagi Kozl*, **2-3**, 74-76, 1974.

194.Que Hee, S.S., R.G. Sutherland and M. Vetter, GLC analysis of 2,4-D concentrations in air samples from central Saskatchewan in 1972, *Environ. Sci. Technol.*, **9**, 62-66, 1975.

195.Randall, C.W., Bacterial air pollution from activated sludge units-field and tracer studies, *College of Engg. (Texas Univ. Austin, Texas)*, Ph.D. Dissertation, 1966.

196.Rasch, R., Lead in waste seen in connection with waste incineration, *Aufbereitungs Technik. (W. Ger.)*, **15**, 234-237, 1974. (APTIC No. 62515).

197.Rasmussen, R.A., *Qualitative analysis of the hydrocarbon emission from veneer dryers*, Washington State Univ. (Pullman, Wash.), Research Grant AP 1232, 17, 1970. (APTIC No. 32180).

198.Rasmussen, R.A., What do the hydrocarbons from trees contribute to air pollution?, *J. Air Poll. Contr. Assoc.*, **22**, 537-543, 1972.

199.Rasmussen, R.A., Emission of biogenic hydrogen sulfide, *Tellus*, **26**, 254-260, 1974.

200.Raymond, A. and G. Guiochon, Gas chromatographic analysis of $C_8 - C_{18}$ hydrocarbons in Paris air, *Environ. Sci. Technol.*, **8**, 143-148, 1974.

201.Rondia, D., The solution of a hygienic problem in steel works. Exposure of workers to a fog containing 3,4-benzopyrene, *Arch. Maladies Profess. Med. Trav. Securite Sociale (Paris)*, **25**, 403-406, 1964. (APTIC No. 33279).

202.Rosen, A.A., R.T. Skeel and M.B. Ettinger, Relationship of river water odor to specific organic contaminants, *J. Water Poll. Control Fed.*, **35**, 777-782, 1963. (APTIC No. 07089).

203.Ryosaka, E., Research on the source of odor development and the residential reaction in Hokkaido, *J. Pollution Control (Japan)*, **4**, (4), 209-220, 1968.

204.Saijo, T., T. Tsujimoto and T. Takahashi, Odor pollution of Kashima District, *J. Japan Soc. Air Poll.*, **6**, 222, 1971. (APTIC No. 36845).

205.Sakai, T. and E. Ito, Studies on vehicle exhaust(III) investigation on hydrogen cyanide and formaldehyde in the exhaust emissions, *Rep. Environ. Pollut. Res. Inst. (City of Nagoya)*, **2**, 39-42, 1973. (APTIC No. 69016).

206.Saltzman, B.E. and J.E. Cuddeback, Air pollution, *Anal. Chem.*, **47**, 1R-15R, 1975.

207.Sandberg, D.V., S.G. Pickford and E.F. Darley, Emissions from slash burning and the influence of flame retardant chemicals, *J. Air Poll. Contr. Assoc.*, **25**, 278-281, 1975.

208.Saunders, R.A., J.R. Griffith and F.E. Saalfeld, Identification of some organic smog components based on rain water analysis, *J. Biomedical Mass Spectrometry*, **1**, 192-194, 1974.

209.Sawicki, E., T.R. Hauser, W.C. Elbert, F.T. Fox and J.E. Meeker, Polynuclear aromatic hydrocarbon composition of the atmosphere in some large American cities, *Am. Ind. Hyg. Assoc. J.*, **23**, 137-144, 1962.

210.Schlegel, H.G., Production, modification, and consumption of atmospheric trace gases by microorganisms, *Tellus*, **26**, 11-20, 1974.

211.Schuler, M. and L. Borla, PVC and air hygiene, *Chem. Rundschau (Solothurn)*, **25**, (2), 17-18, 1972. (APTIC No. 44163).

212.Schuphan, W. Potential and actual hazards of fertilizer and pesticide use for the environment, *Schriftenreihe Ver Wasser-Boden-Lufthyg (Berlin)*, No. 34, 35-50, 1971. (APTIC No. 57230).

213.Schuetzle, D., Computer Controlled High Resolution Mass Spectrometric Analysis of Air Pollutants, *University of Washington*, Ph.D. Dissertation, 1972.

214.Schuetzle, D., D. Cronn, A.C. Crittenden and R.J. Charlson, Molecular composition of secondary aerosol and its possible origin, *Environ. Sci. Technol.*, **9**, 838-845, 1975.

215.Sciamanna, A.F. and A.S. Newton, *A survey of the occurrence of dimethyl mercury in the atmosphere*, Atomic Energy Comm. Contract W-7405-eng-48, Report TID-4500-R61, Lawrence Berkeley Lab., Berkeley, Cal., 410-413, 1974.

216.Seba, D.B. and J.M. Prospero, Pesticides in the lower atmosphere of the northern equatorial Atlantic Ocean, *Atmos. Environ.*, **5**, 1043-1050, 1971.

217.Seizinger, D.E. and B. Dimitriades, *Oxygenates in automotive exhaust gas,* Bureau of Mines Report of Investigation 7675, U.S. Dept. of the Interior (Washington, D.C.), 1972.

218.Sforzolini, G., G. Scassellati and G. Saldi, Further research on the polycyclic hydrocarbons of cigarette smoke. Comparison of the inhaled smoke and that taken from the ambient atmosphere, *Bell. Soc. Ital. Biol. Sper. (Naples)*, **37**, 769-771, 1961. (APTIC No. 31630.

219.Shaw, A.C. and K.T. Waldock, Vanillin analysis and odor evaluation by gas chromatography, *Pulp Paper Mag. Can.*, **68** (3), T-118 to T-122, 1967.

220.Shendrikar, A.D. and P.W. West, Determination of selenium in the smoke from trash burning, *Environ. Lett.*, **5**, 35-39, 1973.

221.Shigeta, Y., Examples of bad odor measurement designated by the bad odor control law, and sensitivity test, Preprint, Bad Odor Pollution Study Group (Japan), p. 46-54, 1974. (APTIC No. 69235).

222.Sibbitt, D.J., R.H. Moyer and G.H. Milly, Emission of mercury from latex paints, *Am. Chem. Soc., Div. Water, Air, Waste Chem., Gen. Papers,* **12** (1), 20-26, 1972.

223.Siegert, H., H.H. Oelert and J. Zajontz, Rapid gas chromatographic analysis for individual hydrocarbons in the exhaust of diesel engines, *Motortech. Z. (Stuttgart),* **35** (4), 101-106, 1974. (APTIC No. 59888).

224.Simmonds, P.G., S.L. Kerrin, J.E. Lovelock and F.H. Shair, Distribution of atmospheric halocarbons in the air over the Los Angeles basin, *Atmos. Environ.,* **8**, 209-216, 1974.

225.Sohn, H.Y. and J. Szekely, On the oxidation of cyanides in the stack region of the blast furnace, Steel Ind. Environ, New York State Univ. (Buffalo), *2nd C.C. Furnas Mem. Conf.,* p. 249-264, 1973. (APTIC No. 69318).

226.Spicer, C.L., Nonregulated photochemical pollutants derived from nitrogen oxides, in *Scientific Seminar on Automotive Pollutants,* EPA 600/9-75-003, Environ. Prot. Agency, Washington, D.C., 1975.

227.Spindt, R.S., G.J. Barnes and J.H. Somers, The characterization of odor components in diesel exhaust gas, Paper presented at Midyear Meeting, Society of Auto. Engrs. Int. (Montreal, Can.), June 7, 1974.

228.Spynu, Ye.I., L.N. Ivanova and A.V. Bolotnyy, Environmental pollution due to organophosphorus pesticides, *Gigiena i Sanit.,* **10**, 75-79, 1973. (APTIC No. 57347).

229.Staiff, D.C., G.E. Quinby, D.L. Spencer and H.G. Starr, Jr., *Polychlorinated biphenyl emission from fluorescent lamp ballasts,* Environmental Protection Agency, Wenctchee, Wash., 1973. (APTIC No. 57527).

230.Stanley, C.W., J.E. Barney, II, M.R. Helton and A.R. Yobs, Measurement of atmospheric levels of pesticides, *Environ. Sci. Technol.,* **5**, 430-435, 1971.

231.Stephens, E.R., Chemistry of atmospheric oxidants, *J. Air Poll. Contr. Assoc.,* **19**, 181-185, 1969.

232.Stephens, E.R., *Hydrocarbons in polluted air,* Coordinating Research Council (New York, New York), CRC Project CAPA 5-68, 1973.

233.Stephens, E.R., Observations of peroxyacetyl nitrate (PAN) in air in Southern England, *Atmos. Environ.,* **9**, 461, 1975.

234.Stoiber, R.E., D.C. Leggett, T.F. Jenkins, R.P. Murrmann and W.I. Rose, Organic compounds in volcanic gas from Santiaguito Volcano, Guatamala, *Geol. Soc. Amer. Bull.,* **82**, 2299-2302, 1971.

235.Stoiber, R.E. and A. Jepsen, Sulfur dioxide contributions to the atmosphere by volcanoes, *Science,* **182**, 577-578, 1973.

236.Lewis, B.G., C.M. Johnson, and C.C. Delwiche, Release of volatile selenium compounds by plants, *J. Agr. Food Chem.,* **14** , 638-640, 1966.

237.Suzuki, F., An estimation of PCT's in incinerators, *J. Japan Soc. Air Pollution,* **9**, 440, 1974. (APTIC No. 70941).

238.Suzuki, Y., K. Nishiyama, M. Oe and F. Kametani, Studies on the prevention of public nuisance by the exhaust gases from the Kraft pulp mill. Part 1. (Analysis of exhaust gases), *Tohoku J. Exp. Med. (Tokyo),* **11** (2), 120-126, 1964. (APTIC No. 96240).

239.Tachikawa, R., Mechanism of BHC pollution, *Kagaku Asahi (Japan),* **30** (12), 45-51, 1970. (APTIC No. 30376).

240.Takahashi, T., A few observations on the generation mechanism of obnoxious odor, *Odor Res. J. (Japan),* **1** (4), 33-38, 1971.

241.Takahashi, M., M. Yamazaki, Y. Sigeta, Y. Sakaida and H. Nagasawa, Studies on the offensive odor and odor substances originating from manufacture process fish-meal, bone-meal, and feather meal, *Human Hyg. (Japan),* **38** (3), 122-136, 1973. (APTIC No. 53493).

242.Tanaka, A., M. Hori and Y. Kobayashi, Preprint , *Japan Soc. for Safety Eng. (Tokyo),* #28, 1973. Cited by Saltzman and Cuddeback(206).

243.Tanimoto, M. and H. Uehara, Detection of acrolein in engine exhaust with microwave cavity spectrometer of Stark voltage sweep·type, *Environ. Sci. Technol.,* **9,** 153-154, 1975.

244.Tatsukawa, R., A new environmental pollutant-polychlorinated biphenyls (PCB), *J. Pollution Control (Japan),* **7,** 419-425, 1971. (APTIC No. 29984).

245.Tatsukawa, R. and W. Tadaaki, Pesticide residues in air. Part 2. Air pollution by BHC, *J. Japan Soc. Air Poll.,* **5,** 92, 1970. (APTIC No. 28776).

246.Tazieff, H., Volcanism and atmospheric conditions, *Proc. Inst. Symp. Env. Meas.,* Bechman Instruments S.A. (Geneva), 131-132, 1973.

247.Tebbens, B.D., Gaseous pollutants in the air, in *Air Pollution, I,* (A.C. Stern, ed.), New York: Academic Press, p. 23-46, 1968.

248.Thomas, G.H., *Environ. Health Perspect.,* **3,** 23. 1973. Cited by Giam, et al.(77).

249.Tichatschke, J., Studies of the emission from refuse incinerators, *Mitt. Ver. Grosskesselbesitzer,* **51** (3), 219-223, 1971.

250.Trieff, N.M. and V.M.S. Ramanujam, Oxidation of aromatic amine air pollutants using chloramine-T and hypochlorous acid, Paper presented at 68th Annual Meeting, Air Poll. Cont. Assoc., Boston, MA, 1975. June 16, 1975.

251.Tseudrovskaya, V.A., *Gigiena i Sanit.,* **1** , 62, 1973. Cited by Saltzman and Cuddeback(206).

252.Tsifrinovich, A.N. and N.I. Lulova, Composition of organosulfur trace impurities in gases, *Neftepererab. Neftekhim.,* **6,** 33-35, 1972. (APTIC No. 54825).

253.Tsuji, M. and T. Okuno, The GC analysis of carbonyl sulfide emitted by the viscose plant, *Proc. 13th Symp., Japan Soc. Air Poll.,* p. 119, 1972. (APTIC No. 49266).

254.Tsuji, M., T. Okuno and N. Takada, On the concentration of amine and aldehyde compounds from fish-meal plants, *J. Japan Soc. Air Poll.,* **6,** 226, 1971. (APTIC No. 36850).

255.Tsuji, M., T. Okuno and K. Takada, Investigation of the components of bad odor from fish meal factories, *Rpt. Public Nuisance Res. Inst. (Hyogo Prefect.),* **3,** 18-20, 1972. (APTIC No. 48285).

256.Vandegrift, A.E., L.J. Shannon, E.W. Lawless, P.G. Gorman, E.E. Sallee and M. Reichel, *Particulate pollutant system study. Volume III. Handbook of emission properties,* Midwest Research Institute (Kansas City, Mo.), Contract Report CPA 22-69-104, 1971.

257.Vanderpol, A.H., F.D. Carsey, D.S. Covert, R.J. Charlson and A.P. Waggoner, Aerosol chemical parameters and air mass character in the St. Louis region, *Science,* **190,** 570, 1975.

258.Vol'fson, U.Ya. and A. F. Sudak, Chromatographic determination of carbon disulfide, carbon dioxide, and sulfur dioxide micro- contaminants in air, *Ind. Lab (USSR),* **36,** 1322-1323, 1970.

259.Wang, C.C., L.I. Davis, Jr., C.H. Wu, S. Japar, H. Niki, and B. Weinstock, Hydroxyl radical concentrations measured in ambient air, *Science, 189,* 797-800, 1975.

260.Watanabe, I. and T. Okita, Distribution and estimation of main sources of ambient light hydrocarbons in Tokyo, *J. Japan Soc. Air Poll.,* **8,** 710-728, 1973. (APTIC No. 65743).

261.Wessler, M.A., Mass spectrographic analysis of exhaust products from an air aspirating diesel fuel burner, *Dept. of Mech. Eng. (Purdue Univ.),* Ph.D. Thesis, 1968.

262.Wilby, F.V., Variation in recognition odor threshold of a panel, *J. Air Poll. Contr. Assoc.,* **19,** 96-100, 1969.

263.Williams, I.H., Gas chromatographic techniques for the identification of low concentrations of atmospheric pollutants, *Anal. Chem.,* **37,** 1723-1732, 1965.

264.Williams, F.W. and J.E. Johnson, Atmospheric contamination with a cleaning solvent, *Chem. Res. in Nuclear Sub. Atmos. Purification,* Prog. Rpt. 7037, Naval Research Lab., Washington, D.C., 1970.

265.Willis, G.H., J.F. Parr and S. Smith, Pesticides in air. Volatilizaion of soil-applied DDT and DDD from flooded plots, *Pesticides Monitoring J.,* **4,** 204-208, 1971.

266.Willis, G.H., J.F. Parr, S. Smith and B.R. Carroll, Volatilization of dieldrin from fallow soil as affected by different soil water regimes, *J. Environ. Quality,* **1,** 193-196, 1972.

267.Wilson, D.F. and B.F. Hrutfiord, Formation of volatile organic compounds in the Kraft pulping process, *TAPPI,* **54,** 1094-1098, 1971.

268.Wilson, H.H. and L.D. Johnson, *Characterization of air pollutants emitted from brick plant kilns,* Environmental Protection Agency (Research Triangle Park, N.C.), 1973.

269.Wilson, W.E., Jr., W.E. Schwartz and G.W. Kinzer, *Haze formation-its nature and origin,* Battelle Columbia Laboratories, Columbus, Ohio, 1972. (NTIS Document No. PB 212609).

270.Winkler, H-D. and K. Welzel, Studies of the formaldehyde emission from the production of wooden boards, *Wasser Luft Betrieb.,* **16,** 213-215, 1972. (APTIC No. 43912).

271.Wohlers, H.C. and G.B. Bell, Stanford Research Institute, Project No. SU-1816, 1956.

272.Yamaguchi Prefecture (Japan) Research Inst. of Health, Result of PCP estimation in the air around a factory of agricultural chemicals in Ogori-town, *Ann. Rpt. Yamaguchi Prefect. Res. Inst. Health,* **13,** 77-78, 1971. (APTIC No. 37289).

273.Yanagisawa, S. A report of investigation on actual conditions of air pollution in and around Osaka international airport, (Osaka kokusaikuko shuken ni okeru taiki osen jittai chosa hokokusho), 1971. (APTIC No. 61300).

274.Yavorovskaya, S.F. and L.P. Anvayer, *Gig. Sanit.,* **8,** 64, 1973.

275.Yokohama National Univ., Kanagawa-Ken Taiki Osen Chosa Kenkyu Hokuku, **14,** 152, 1972 (cited in Saltzman and Cuddeback (206)).

276.Yule, W.N., A.F.W. Cole and I. Hoffman, A survey for atmospheric contamination following forest spraying with fenitrothion, *Bull. Environ. Contamination Toxicol.,* **6,** 289-296, 1971.

277.Zdrazil, J. and F. Picha, Cancerogenic substances-3, 4-benzopyrene in moulding sand mixtures and foundry dust, *Prakovni Lekar. (Prague),* **15,** 207-211, 1963. (APTIC No. 33297).

278.Zielinski, M., M. Zamfirescu and M. Decusara, Air pollution by nitric oxide from fertilizer factories, *Wiss. Z. Humboldt Univ. Berlin Math. Naturw. Reihe,* **19,** 523-525, 1970.

279.Bates, C.E. and L.D. Scheel, Processing emissions and occupational health in the ferrous foundry industry, *Am. Ind. Hyg. Assoc. J.,* **35,** 452-462, 1974.

280.Blosser, E.R. and W.M. Henry, *Identification and estimation of ions, molecules, and compounds in particulate matter collected from ambient air,* Battelle Columbus Laboratories, Columbus, Ohio, 1971. (NTIS Document No. PB 201738).

281.Environmental Protection Agency, *Air Quality Criteria for Carbon Monoxide,* Report No. AP-62, Washington, D.C., 1970.

282.Environmental Protection Agency, *Air Quality Criteria for Nitrogen Oxides,* Report AP-84 (Washington, D.C.), 1971.

283.Environmental Protection Agency, *Control of photochemical oxidants-technical basis and implications of recent findings,* Research Triangle Park (N.C.), 1975.

284.Hafstad, L.R., Automobiles and air pollution, in *Universities, National Laboratories and Man's Environment,* (D. Jared, ed.), U.S. Atomic Energy Commission, Oak Ridge, Tenn., p. 122, 1969.

285.Hangebrauck, R.P., D.J. von Lehmden and J.E. Meeker, *Sources of polynuclear hydrocarbons in the atmosphere,* U.S. Department of Health, Education, and Welfare, Cincinnati, Ohio, 1967. (NTIS Document No. PB 174706).

286.Jepsen, A.F., Measurements of mercury vapor in the atmosphere, *ACS Adv. Chem. Ser.,* **123,** 81-95, 1973.

287.D. D. Davis, W. Heaps, and T. McGee, Direct measurements of natural tropospheric levels of OH via an aircraft borne tunable dye laser, *Geophys. Res. Lett.,* **3** ,331-333, 1976.

288.Popov, V.A., *Hyg. Sanit,* **35** (5), 178-182, 1970.

289.Environmental Protection Agency, *Air Quality Data-1973 Annual Statistics*, EPA 450/2-74-015, Research Triangle Park, N.C., 1974.

289.Robinson, E. and R.C. Robbins, *Sources, Abundance and Fate of Gaseous Atmospheric Pollutants*, Stanford Research Institute, Project PR-6755, 1968. (NTIS Document N71-25147).

290.Robinson, E. and R.C. Robbins, *Sources, Abundance and Fate of Gaseous Atmospheric Pollutants Supplement*, Stanford Research Institute, Project PR-6755, 1971. (NTIS Document ZZ18194).

291.Schmidt, U., Molecular hydrogen in the atmosphere, *Tellus*, **26**, 78-89, 1974.

292.Schulten, H.R. and U. Schurath, Aerosol analysis by field desorption mass spectrometry combined with a new sampling technique, *Atmos. Environ.*, **9**, 1107-1112, 1975.

293.Seiler, W. and C. Junge, Carbon monoxide in the atmosphere, *J. Geophys. Res.*, **75**, 2217-2226, 1970.

294.Valley, S.L., ed., *Handbook of Geophysics and Space Environments*, New York: McGraw-Hill, 1965.

295.Watt, A.D., Placing atmospheric CO_2 in perspective, *IEEE Spectrum*, **8** (11), 59-72, 1971.

296.Bates, D.R. and P.B. Hays, Atmospheric nitrous oxide, *Planet. Space Sci.*, **15**, 189-197, 1967.

297.Hahn, J., The North Atlantic Ocean as a source of atmospheric N_2O, *Tellus*, **26**, 160-168, 1974.

298.Hoffmann, D., C. Patrianakos, K.D. Brunnemann and G.B. Gori, Chromatographic determination of vinyl chloride in tobacco smoke, *Anal. Chem.*, **48**, 47-50, 1976.

299.Lahue, M.D., J.B. Pate and J.P. Lodge, Jr., Atmospheric nitrous oxide concentrations in the humid tropics, *J. Geophys. Res.*, **75**, 2922-2926, 1970.

300.Schutz, K., C. Junge, R. Beck and B. Albrecht, Studies of atmospheric N_2O, *J. Geophys. Res.*, **75**, 2230-2246, 1970.

301.Koyama, T., Gaseous metobolism in lake sediments and paddy soils and the production of atmospheric methane and hydrogen, *J. Geophys. Res.*, **68**, 3971-3973, 1963.

302.Babich, H. and G. Stotzky, Air pollution and microbial ecology, *CRC Critical Rev. Environ. Control*, **4** (3), 353-421, 1974.

303.Farwell, S.O., E. Robinson, W.J. Powell and D.F. Adams, Survey of airborne 2, 4-D in South-Central Washington, *J. Air Poll. Contr. Assoc.*, **26**, 224-230, 1976.

304.Woolley, W.D., Decomposition products of PVC for studies of fires, *British Polymer J.*, **3**, 186-193, 1971.

305.Woodwell, G.M., R.A. Houghton, and N.R. Tempel, Atmospheric CO_2 at Brookhaven, Long Island, New York: Patterns of variation up to 125 meters, *J. Geophys. Res.*, **78**, 932-940, 1973.

306.Robinson, E., R.A. Rasmussen, H.H. Westberg and M.W. Holdren, Nonurban nonmethane low molecular weight hydrocarbon concentrations related to air mass identification, *J. Geophys. Res.*, **78**, 5345-5351, 1973.

307.Pitts, J.N., Jr., D. Grosjean, B. Shortridge, G. Doyle, J. Smith, T. Mischke and D. Fitz, The nature, concentration and size distribution of organic particulates on the eastern part of the Southern California air basin, Paper presented at Centennial Meeting, Amer. Chem. Soc., New York, New York, Apr. 5, 1976.

308.Rappaport, S.M. and D.A. Fraser, Gas chromatographic-mass spectrometric identification of volatiles released from a rubber stock during simulated vulcanization, *Anal. Chem.*, **48**, 476-481, 1976.

309.Smythe, R.J., The application of high resolution gas chromatography and mass spectrometry to analysis of engine exhaust emissions, *Univ. of Waterloo (Waterloo, Ontario)*, Ph.D Dissertation, 1973.

310.Hauser, T.R. and J.N. Pattison, Analysis of aliphatic fraction of air particle matter, *Environ. Sci. Technol.,* **6**, 549-555, 1972.

311.Boyer, K.W., Analysis of automobile exhaust particulates, *Univ. Of Illinois (Urbana-Champaign),* Ph.D Dissertation, 1974.

312.Hoffmann, D. and E.L. Wynder, Studies on gasoline engine exhaust, *J. Air Poll. Contr. Assoc.,* **13**, 322-327, 1963.

313.Alperstein, M., and R.L. Bradow, Exhaust gas emissions related to engine combustion, *Soc. Automot. Eng. J.,* **4** 85(8):52-53, 1968.

314.Smythe, R.J. and F.W. Karasek, The analysis of diesel exhausts for low molecular weight carbonyl compounds, *J. Chromatog,* **86**, 228-231, 1973.

315.Larsen, R.I., and W.C. Nelson, Preprint, 1967. 1967.

316.Environmental Protection Agency, *Air Quality Criteria for Sulfur Dioxide,* Report AP-50, Washington, D.C., 1969.

317.Chass, R.L. and R.E. George, Contaminant emissions from the combustion of fuels, *J. Air Poll. Contr. Assoc.,* **10**, 34-43, 1960.

318.Chaimedes, W. Private communication, 1976.

319.Gay, B.W., Jr. and J.J. Bufalini, Hydrogen peroxide in the urban atmosphere, *ACS Adv. Chem.,* **113**, 255-263, 1972.

320.Nash, T. Nitrous acid in the atmosphere and laboratory experiments on its photolysis, *Tellus,* **26**, 175-179, 1974.

321.Lovelock, J.E., R.J. Maggs and R.J. Wade, Halogenated hydrocarbons in and over the Atlantic, *Nature,* **241**, 194-196, 1973.

322.Lovelock, J.E., Natural halocarbons in the air and in the sea, *Nature,* **256**, 193-194, 1975.

323.Little, A.D., Inc., *Preliminary economic impact assessment of possible regulatory action to control atmospheric emissions of selected halocarbons,* EPA 450/3-75-073, Cambridge, Mass., 1975.

324.Froelich, A.A., Private communication, 1970.

325.Osag, T.R. and G.B. Crane, *Control of odors from inedible-rendering plants,* EPA450/1-74-006, Environmental Protection Agency, Research Triangle Park, N.C., 1974.

326.Noller, C.R., *Textbook of Organic Chemistry,* Philadelphia, PA: W.B. Saunders Co., 2nd ed., 1958.

327.White, R.K., Gas chromatogrylic analysis of dairy animal wastes, Ph.D. Dissertation, *Ohio State University,* (1970), *(Dissert. Abst. Int. B.),* **31** (2), 643, 1970.

328.Rosehart, R.G. and R. Chu., Methods for identification of arsenic compounds, *Water Air Soil Poll.,* **4**, 395-398, 1975.

329.Lonneman, W.A., J.J. Bufalini and R.L. Seila, PAN and oxidant measurement in ambient atmospheres, *Environ. Sci. Technol.,* **10**, 374-380, 1976.

330.Vogh, J.W., Nature of odor components in diesel exhaust, *J. Air Poll. Cont. Assoc.,* **19**, 773-777, 1969.

331.Toyosawa, S., Y. Umezawa, T. Shirai, S. Yanagisawa, T. Yukawa and S. Nakamura, Analysis of gaseous products from rubber wear, *J. Japan Soc. Air Poll.,* **9**, 200, 1974. (APTIC No. 71608).

332.Calcraft, A.M., R.J.S. Green and T.S. McRoberts, Burning of plastics. Part 1. Smoke formation, *J. Plastics Inst.,* **42**, 200-208, 1974.

333.Sato, Y., I. Mizoguchi, K. Makino and H. Yagyu, On the correlation between light chlorinated hydrocarbons and other pollutants in Tokyo atmosphere, *Ann. Rpt. Tokyo Metro. Lab. Public Health,* **25**, 371-379, 1974. (APTIC No. 70587).

334.Morita, M., H. Mimura and T. Nishizawa, Environmental pollution by p-Dichlorobenzene, *Japan. Soc. Public Health (Fukushima),* p. 311, 1974. (APTIC No. 70517).

335.Miller, C.W. and T.M. Shafik, Concentrations of propoxur in air following repeated indoor applications, *Bull. World Health Org. (Geneva),* **51** (1), 41-44, 1974.

336.Zeier, U. von and M. Semlitsch, Microstructure studies on magnetic spheroids found in Alpine snow and foundry dust samples in Switzerland, *Suizer Tech. Rev. (Switz.)*, **4**, 268-274, 1970.

337.Egger, A.M. and E. Widmer, Environmental protection measures in petrochemical plant as exemplified by the gasoline cracking plant of Lonza AG, *Chimia*, **28**, 674-678, 1974.

338.Numata, H. and S. Takano, A case of environmental pollution by coal-tar pitch, *J. Japan Soc. Air Poll.*, **9**, 540, 1974. (APTIC No. 72056).

339.Schmeltz, I., J. Tosk and D. Hoffmann, Formation and determination of naphthalenes in cigarette smoke, *Anal. Chem.*, **48**, 645-650, 1976.

340.Wallcave, L., D.L. Nagel, J.W. Smith and R.D. Waniska, Two pyrene derivatives of widespread environmental distribution: cyclopenta(cd) pyrene and acepyrene, *Environ. Sci. Technol.*, **9**, 143-145, 1975.

341.Davies, I.W., R.M. Harrison, R. Perry, D. Ratnayaka and R.A. Wellings, Municipal incinerator as a source of polynuclear aromatic hydrocarbons in environment, *Environ. Sci. Technol.*, **10**, 451-453, 1976.

342.Pitts, J. N., Jr., B. J. Finlayson-Pitts, and A. M. Winer, Optical systems unravel smog chemistry, *Environ. Sci. Technol.*, **11**, 568-573, 1977.

343.King, R.B., J.S. Fordyce, A.C. Antoine, H.F. Leibecki, H.E. Neustadter and S. M. Sidik, *Extensive 1-year survey of trace elements and compounds in the airborne suspended particulate matter in Cleveland, Ohio*, National Aeronautics and Space Administration, Washington, D.C., NASA TN D-8110, 1976.

344.Gibbard, S. and R. Schoental, Simple semi-quantitative estimation of sinapyl and certain related aldehydes in wood and in other materials, *J. Chromatog.*, **44**, 396-398, 1969.

345.Moore, H.E. and S.E. Poet, Background levels of ^{226}Ra in the lower troposphere, *Atmos. Environ.*, **10**, 381-383, 1976.

346.Chau, Y.K., P.T.S. Wong and H. Saitoh, Determination of tetraalkyl lead compounds in the atmosphere, *J. Chromat. Sci.*, **14**, 162-164, 1976.

347.Bidleman, T.F. and C.P. Rice, TLC analysis of carbaryl insecticide on sprayed foliage, *J. Chem. Ed.*, **53**, 173, 1976.

348.Temple, P.J. and S.N. Linzon, Boron as a phytotoxic air pollutant, *J. Air. Poll. Contr. Assoc.*, **26**, 498-499, 1976.

349.Pauling, L., *General Chemistry,* 2nd ed., San Francisco: W.H. Freeman and Co., 1954.

350.Anonymous, Zinc oxide emissions controlled by dry collection system, *Pollution Eng.*, **8** (5), 26, 1976.

351.Anonymous, Planning permits economical air pollution control and product recovery, *Pollution Eng.*, **8** (5), 22, 1976.

352.Finch, S.P., III, E.R. Stephens and M.A. Price, Paper presented at Pacific Conference on Chemistry and Spectroscopy, San Francisco, Cal., Oct. 16, 1974.

353.Pierotti, D. and R.A. Rasmussen, Combustion as a source of nitrous oxide in the atmosphere, *Geophys. Res. Lett.*, **3**, 265-267, 1976.

354.Hartstein, A.M. and D.R. Forshey, *Coal mine combustion products: neoprenes, polyvinyl chloride compositions, urethane foam, and wood,* Bureau of Mines Report of Investigations 7977, U.S. Dept. of the Interior, Washington, D.C., 1974.

355.American Petroleum Institute, *Bulletin on photochemical evaporation loss from storage tanks,* API Bull. 2523, New York, NY, 1969.

356.Sternling, C.V. and J.O.L. Wendt, *Kinetic mechanisms governing the fate of chemically bound sulfur and nitrogen in combustion,* EPA650/2-74-017, *Environmental Protection Agency (Washington, D.C.),* 1972. (NTIS Document No. PB 230895).

357.Pearson, C.D., The determination of trace mercaptans and sulfides in natural gas by a gas chromatography-flame photometric detector technique, *J. Chromatog. Sci.*, **14**, 154-158, 1976.

358.Pellizzari, E.D., J.E. Bunch, R.E. Berkley and J. McRae, Determination of trace hazardous organic vapor pollutants in ambient atmospheres by gas chromatography/mass spectrometry/computer, *Anal. Chem.*, **48**, 803-807, 1976.

359.Katzman, H. and W.F. Libby, Hydrocarbon emissions from jet engines operated at simulated high-altitude supersonic flight conditions, *Atmos. Environ.*, **9**, 839-842, 1975.

360.Candeli, A., G. Morozzi, A. Paolacci and L. Zoccolillo, Analysis using thin layer and gas-liquid chromatography of polycyclic aromatic hydrocarbons in the exhaust products from a European car running on fuels contaning a range of concentrations of these hydrocarbons, *Atmos. Environ.*, **9**, 843-849, 1975.

361.Lao, R.C., R.S. Thomas and J.L. Monkman, Computerized gas chromatographic-mass spectrometric analysis of polycyclic aromatic hydrocarbons in environmental samples, *J. Chromatog.*, **112**, 681-700, 1975.

362.Lao, R.C., R.S. Thomas, H. Oja and L. Dubois, Application of a gas chromatograph-mass spectrometer-data processor combination to the analysis of the polycyclic aromatic hydrocarbon content of airborne pollutants, *Anal. Chem.*, **45**, 908-915, 1973.

363.Cautreels, W. and K. Van Cauwenberghe, Determination of organic compounds in airborne particulate matter by gas chromatography-mass spectrometry, *Atmos. Environ.*, **10**, 447-457, 1976.

364.Kunen, S.M., M.F. Burke, E.L. Bandurskii and B. Nagy, Preliminary investigations of the pyrolysis products of insoluble polymer-like components of atmospheric particulates, *Atmos. Environ.*, **10**, 913-916, 1976.

365.Arbesman, P., U.S. EPA monitoring results for hazardous substances in New Jersey, Memorandum, State of New Jersey Department of Environmental Protection (Trenton, New Jersey), July 16, 1976.

366.Creac'h, P.V. and G. Point, Mise en evidence, dans l'atmosphere d'acide borique gazeux provement de l'evaporation de l'eau de mer, *C.R. Acad. Sci. (Paris)*, **Ser. B263**, 89-91, 1966.

367.Johnson, D.L. and R.S. Braman, Distribution of atmospheric mercury species near the ground, *Environ. Sci. Technol.*, **8**, 1003-1009, 1974.

368.Fine, D.H., D.P. Rounbehler, N.M Belcher and S.S. Epstein, N-Nitroso compounds: Detection in ambient air, *Science*, **192**, 1328-1330, 1976.

369.Anonymous, *Investigation reports on the environmental atmosphere at a chemical plants region in the Kita Ward*, Kita Ward Office (Tokyo, Japan), Construction and Public Nuisance Div., April, 1975. (APTIC No. 78264).

370.Katou, T. and R. Nakagawa, Determination of alkyl leads by combined system of gas chromatography and atomic absorption spectrophotometry, *Bull. Inst. Environ. Sci. Technol. (Yokohama Nat'l. Univ.)*, **1**, 19-24, 1975.

371.Uchimura, R., Pollution caused by petrochemical complex: plants for butadiene and aromatic hydrocarbon, *Tech. Hum. Being (Japan)*, **4**, 112-117, 1975. (APTIC No. 78806).

372.Axtmann, R.C., Emission control of gas effluents from geothermal power plants, *Environ. Lett.*, **8**, 135-146, 1975.

373.Singh, H.B., L.J. Silas and L.A. Cavanagh, Distribution, sources and sinks of atmospheric halogenated compounds, *J. Air Poll. Contr. Assoc.*, **27**, 332-336, 1977.

374.Gaddo, P.P., F. Corazzari and L. Giacomelli, Characterization of hydrocarbons pollution in the urban area of Turin, Paper presented at 69th Annual Meeting, Air Poll. Contr. Assoc., Portland, OR, June 28, 1976.

375.Fox, M.A. and S.W. Staley, Determination of polycyclic aromatic hydrocarbons in atmospheric particulate matter by high pressure liquid chromatography coupled with fluorescence techniques, *Anal. Chem.*, **48**, 992-998, 1976.

376.Bellar, T.A. and J.E. Sigsby, Jr., Direct gas chromatographic analysis of low molecular weight substituted organic compounds in emissions, *Environ. Sci. Technol.*, **4**, 150-156, 1970.

377.Pellizzarri, E.D., J.E. Bunch, J.T. Bursey, R.E. Berkley, E. Sawicki and K. Krost, Estimation of N-nitrosodimethylamine levels in ambient air by capillary gas-liquid chromatography/mass spectrometry, *Anal. Lett.*, **9**, 579-594, 1976.

378.Fine, D.H., D.P. Rounbehler, E. Sawicki, K. Krost and G.A. De Marrais, N-Nitroso compounds in the ambient community air of Baltimore, Maryland, *Anal. Lett.*, **9**, 595-604, 1976.

379.Gardner, J., *Sulfuric Acid Emissions from ESB Battery Plant-Forming Room*, Report 77-BAT-5, York Research Co., Stamford, Conn., 1977. (APTIC No. 102953.)

380.Giannovario, J.A., R.L. Grob and P.W. Rulon, Analysis of trace pollutants in the air by means of cryogenic gas chromatography, *J. Chromatog.*, **121**, 285-294, 1976.

381.Murayama, T., Blue flame, white flame, and odor in compressed and ignited engine, *Internal Combust. Engine (Japan)*, **14** (2), 50-59, 1975. (ATPIC No. 73769).

382.Nishida, K., T. Honda and Y. Miyake, On the exhausted odor component by spraying paints, *Environ. Conserv. Eng. (Japan)*, **4** (4), 240-247, 1975. (APTIC No. 73893).

383.Gordon, S.J. and S.A. Meeks, A study of the gaseous pollutants in the Houston, Texas area, Paper presented at 79th National Meeting, Amer. Inst. of Chem. Engrs., Houston, Tex., Mar. 16-20, 1975.

384.Saijo, T. and T. Nozaki, Fact-finding survey for offensive odor, Ibaraki-ken Research Cent. (Japan), *Ann. Rpt. on Environ. Pollution*, **6** p. 83-93, 1973. (APTIC No. 74213).

385.Hoshika, Y., S. Kadowaki, I. Kojima, K. Koike and K. Yoshimoto, The gas chromatographic analysis of sulfur compounds and aliphatic amines in the exhaust gas from a fishmeal dryer and cooker, *J. Japan Oil Chem. Soc.*, **24** (4), 27, 1975. (APTIC No. 77123).

386.Uchiyama, M., Y. Shimada, I. Takahashi, Y. Saito and F. Hayashi, Studies on the photochemical air pollution: Halogen gas in the automobile exhaust gas, *Gunma Prefect. Res. Cent. Environ. Sci. (Japan)*, Ann. Rpt. No. 6, p. 134, 1974. (APTIC No. 76917).

387.Neiser, J., M. Kaloc and V. Masek, Pyridine and its homologs in the atmosphere of coke oven plants, *Czech. Sh.Ved. Pr. Vys. Sk. Banske Ostrave. (Czechoslovakia)*, **19** (1), 287-309, 1973. (APTIC No. 79013).

388.Braman, R.S. and D.L. Johnson, Ambient forms of mercury in air, *2nd Ann. Trace Contam. Conf. Proc.*, National Science Foundation., Washington., D.C., p. 75-78, 1974. (NTIS No. LBL-3217).

389.Kagawa, F., K. Inazawa and H. Saruyama, Detection of cyanogen in atmosphere and rain, Paper presented at 1st Meeting, Soc. for General Research, Environ. Science (Japan), June, 1975. (APTIC No. 74803).

390.Nagai, K., K. Naka, B. Matsuoayashi, M. Takatsuka and M. Murata, Methods of measurements and examples of measurement on odorous substances (lower aldehydes and styrene) in the petrochemical complex area, *Odor Research J. (Japan)*, **4** (19), 23-32, 1975. (APTIC No. 74561).

391.Kunz, C.O. and C.J. Paperiello, Xenon-133: Ambient activity from nuclear power stations, *Science*, **192**, 1235-1237, 1976.

392.Davis, D.D., R.T. Watson, T. McGee, W. Heaps, J. Chang and D. Wuebbles, Tropospheric residence times for several halocarbons based on chemical degradation via OH radicals, Paper ENVT-63 presented at Amer. Chem. Soc. Centennial Meeting, New York, New York, April 8, 1976.

393.Chau, Y.K., P.T.S. Wong, B.A. Silverberg, P.L. Luxon and G.A. Bengert, Methylation of selenium in the aquatic environment, *Science*, **192**, 1130-1131, 1976.

394.Singh, H.B., Phosgene in the ambient air, *Nature*, 264, 428-429, 1976.

395.Franconeri, P. and L. Kaplan, Determination and evaluation of stack emissions from municipal incinerators, *J. Air Poll. Contr. Assoc.*, **26**, 887-888, 1976.

396.Johnstone, R.A.W. and J.R. Plimmer, The chemical constituents of tobacco and tobacco smoke, *Chem. Rev.*, **59**, 885-936, 1959.

397. Yamaguchi, M. and N. Shimojo, Methyl-mercury derivatives in engine exhaust and its synthesis, *J. Japan Soc. Air Poll.,* **10,** 413, 1975. (APTIC No. 79929).

398. Suzuki, F., M. Mimakami, I. Ogawa, S. Oote, M. Hoshino, A. Matsura and K. Sato, Researches on iodine emitted from iodine manufacturing plants in Chiba, *J. Japan Soc. Air Poll.,* **9,** 439, 1974. (APTIC No. 73348).

399. Kunen, S.M., *University of Utah,* Ph.D Dissertation, 1977.

400. Crutzen, P.J., The possible importance of CSO for the sulfate layer of the stratosphere, *Geophys. Res. Lett.,* **3,** 73-76, 1976.

401. Jones, J.H., J.T. Kummer, K. Otto, M. Shelef and E.E. Weaver, Selective catalytic reaction of hydrogen with nitric oxide in the presence of oxygen, *Environ. Sci. Technol.,* **5,** 790-798, 1971.

402. Cavanagh, L.A. and R.E. Ruff, Sources and sampling of pollutants for geothermal steam areas, *1975 Environ. Sensing and Assessment,* Inst. of Elect. and Electronic Engrs., New York, NY, IEEE Pub. 75-CH 1004-1 ICESA, p. 18-5.1 to p. 18-5.5, 1976.

403. Environmental Protection Agency, *Air Pollution Aspects of Sludge Incineration,* EPA 625/4-75-009, Washington, D.C., 1975.

404. Council on Environmental Quality, *Carcinogens in the Environment,* Sixth Annual Report, Washington, D.C., 1975.

405. Brunneman, K.D., H.C. Lee and D. Hoffmann, Chemical studies on tobacco smoke. XLVII. On the quantitative analysis of catechols and their reduction, *Anal. Lett.,* **9,** 939-955, 1976.

406. Severson, R.F., M.E. Snook, R.F. Arrendale and O.T. Chortyk, Gas chromatographic quantitation of polynuclear aromatic hydrocarbons in tobacco smoke, *Anal. Chem.,* **48,** 1866-1872, 1976.

407. Dmitrieva, V.N., O.V. Meshkova and V.D. Bezuglyi, Determination of low contents of inorganic impurities in effluents from polymer production, *J. Anal. Chem. (USSR),* **30** (7), Part 2, 1181-1183, 1975.

408. Lee, M.L., M. Novotny and K.D. Bartle, Gas chromatography/mass spectrometric and nuclear magnetic resonance determination of polynuclear aromatic hydrocarbons in airborne particulates, *Anal. Chem.,* **48,** 1566-1572, 1976.

409. Ishizaka, T. and K. Isono, Distribution of aerosols in urban atmospheres. I., Paper presented at Japan Met., Soc. Spring Meeting (Tokyo), May 21-23, 1975. (APTIC No. 74751).

410. Dimitriades, B., *Photochemical Oxidants in the Ambient Air of the United States,* EPA-600/3-76-017, Environmental Protection Agency, Research Triangle Park, N.C., 1976.

411. Kato, T. and Y. Hanai, GC and GC-MS environmental analysis of photochemical smog component, Paper presented at 24th Lecture Meeting, Japan Soc. for Anal. Chem., (Sapporo, Japan), Oct. 1-5, 1975. (APTIC No. 81207).

412. Kozyr, N.P., A.Y. Chebanov and B.P. Titomer, Properties of gases and dust generated by titanium slag smelting in closed ore-smelting furnaces, *Tsvetn. Metal. (USSR),* **11,** 52-53, 1974. (APTIC No. 71642).

413. Kalpasanov, Y. and G. Kurchatova, A study of the statistical distribution of chemical pollutants in air, *J. Air Poll. Contr. Assoc.,* **26,** 981-985, 1976.

414. Conkle, J.P., W.W. Lackey, C.L. Martin and R.L. Miller, Organic compounds in turbine combustor exhaust, *1975 Environ. Sensing and Assessment,* Inst. of Elect. and Electronic Engrs., New York, N.Y., IEEE Pub. 75-CH 1004-1 ICESA, p. 27-2.1 to 27-2.11, 1976.

415. Hurn, R.W., F.W. Cox and J.R. Allsup, *Effects of Gasoline Additives on Gaseous Emissions. Part II.* EPA 600/2-76-026, Energy Research and Development Admin., Bartlesville, Okla., 1976.

416. Kaneko, M., Measurement of cyanide concentration in automobile exhaust gases, stack gases and environment, *Kankyo Sozo (Japan),* **3** (7), 2-8, 1973. (APTIC No. 53231).

417. Kanamaru, G., H. Shima, T. Matsui and T. Hashizume, An example of atmospheric pollution by fluorides, *Mie Prefect. Pub. Nuisance Center Ann. Rpt.,* **1,** 134-137, 1973. (APTIC No. 52664).
418. Dietzmann, H.E., *Protocol to Characterize Gaseous Emissions as a Function of Fuel and Additive Composition,* EPA 600/2-75-048, Southwest Research Institute, San Antonio, Texas, 1975.
419. Litton Systems, Inc., *Air Pollution Aspects of Ethylene,* Bethesda, Maryland, 1969. (NTIS Document PB-188069).
420. Barger, W.R., and W.D. Garrett, Surface active organic material in air over the Mediterranean and over the eastern equatorial Pacific, *J. Geophys. Res.,* **81** , 3151-3157, 1976.
421. Holzer, G., J. Oro, and W. Bertsch, Gas chromatographic-mass spectrometric evaluation of exhaled tobacco smoke, *J. Chromatog.,* **126** , 771-785, 1976.
422. Rains, B.A., M.J. DePrimo, and I.L. Groseclose, *Odors Emitted from Raw and Digested Sewage Sludge,* EPA-670/2-73-098, Environmental Protection Agency, Washington, D.C., 1973.
423. Mullins, B.J., Jr., D.E. Solomon, G.L. Austin, and L.M. Kacmarcik, *Atmospheric Emissions Survey of the Gas Processing Industry,* EPA-450/3-75-076, Environmental Protection Agency, Washington, D.C., 1975.
424. Committee on Biologic Effects of Environmental Pollutants, *Particulate Polycyclic Organic Matter,* National Academy of Sciences, Washington, D.C., 1972.
425. National Assn. of Secondary Material Industries, Inc., *Studies of dislocation factors: No. II. The secondary material industries and environmental problems,* New York, NY, 1968.
426. Muenzer, M., and K. Heder, Results of the medical and technical inspection of chemical dry cleaning plants, *Zbl. Arbeitsmedizin.,* **22** (5),133-138,1972.
427. Merz,O., Lacquers and surface coatings and their susceptible emissions, *Staub, Reinhaltung Luft,* **31,** 395-396,1971.
428. Guenther,R., *A study of the substances liberated from binding agents at the drying of lacquers with respect to air pollution,* Ph.D. Dissertation, Karlsruhe Univ. (West Germany), 1971.
429. Hirose,K., S.Yamanaka, and S. Takada, Monthly variations of hydrocarbon concentrations in the urban atmosphere ,*J. Japan Soc. Air. Poll.,* **10** (4),586,1975.
430. Lambert, G., P.Bristeau, and G. Polian, Emission and enrichemnts of radon daughters from Etna volcano magma, *Geophys. Res. Lett.,* **3,**.724-726,1976.
431. Crittenden, B.D., and R. Long, Diphenylene oxide and cyclopentacenaphthylene(s) in flame soots, *Environ. Sci. Technol.,* **7** ,742-744,1973.
432. Scott Research Laboratories, *Diesel exhaust composition and odor: Progress report for year 1965,* CRC Project Rpt. CD-9-61, New York, NY, 1966.
433. Scott Research Laboratories, *Diesel exhaust composition and odor: Progress report for year 1964,* CRC Project Rpt. No. CD-9-61, New York, NY, 1965.
434. Walsh, W.H., *Analysis of diesel engine exhaust hydrocarbons,* M.S. Thesis, Penn. State Univ., 1963.
435. Landen, E.W., and J.M. Perez, Some diesel exhaust reactivity information derived by gas chromatography ,SAE Paper No. 740530, Soc. of Automotive Engineers, Inc., New York, NY, 1974.
436. Smith, M.S., A.J. Francis, and J.M. Duxbury, Collection and analysis of organic gases from natural ecosystems: application to poultry manure, *Environ. Sci. Technol.,* **11** , 51-55,1977.
437. Banwart, W.L., and J.M. Bremner, Paper presented at Div. of Soil Microbiol. and Biochem., 66th Annual Meeting, ASA, Chicago, IL, 1974.
438. Diebel, R.H., in *Agriculture and the quality of our environment,* N.C. Brady, Ed., Am. Assoc. for the Advancement of Science, Washington, D.C., 1967.
439. Committee on Medical and Biologic Effects of Environ. Pollutants, *Vapor-phase organic pollutants,* National Academy of Sciences, Washington, D.C., 1976.

440.Rasmussen, R.A., *Terpenes: their analysis and fate in the atmosphere,* Ph.D. Dissertation, Washington Univ., St. Louis, Mo., 1964.
441.Epstein, J., G.T. Davis, L. Eng, and M.M. Demek, Potential hazards associated with spray drying operations, *Environ. Sci.Technol.,* **11,** 70-75,1977.
442.Suprenant, N.F., and M.I. Bornstein, *New source classification codes for processes which cause hydrocarbon and organic emissions,* EPA-450/3-75-067, GCA Corporation, Bedford, Mass., 1975.
443.Cautreels, W., and K. Van Cauwenberghe, Extraction of organic compounds from airborne particulate matter, *Water, Air, Soil Poll.,* **6** ,103-110, 1976.
444.Kircher,D.S., and D.P. Armstrong, *An Interim Report on Motor Vehicle Emission Estimation,* EPA-450/2-73-003, Environmental Protection Agency, Research Triangle Park, N.C., 1973.
445.Baker, R.R., The formation of oxides of carbon by the pyrolysis of tobacco, *Beit. zur Tabak.,* **8,** (1),16-27, 1975 (SHB 75-0462).
446.Ceschini,P., and D. Chem, Effect of sampling conditions on the composition of the volatile phase of cigarette smoke, *Beit. zur Tabak.,* **7** (5),294-301, 1974 (SHB 75-0156).
447.Klus, H., and H. Kuhn, The determination of nitrophenols in tobacco smoke condensate (preliminary results), *Fach. Mitt. der Austria Tabakwerke A.G.,* **15,** 275-288, 1974 (SHB 75-0164).
448.Spears, A.W., Effect of manufacturing variables on cigarette smoke composition, *CORESTA Information Bulletin* Special Issue: 65-78, 1974 (SHB 75-0175).
449.Sander, J., F. Schweinsberg, J. LaBar, G. Burkle, and E. Schweinsberg, Nitrate and nitrosable amino compounds in carcinogenesis, *GANN Monograph on Cancer Research,* **17** ,145-160, 1975 (SHB 76-0016).
450.Tso, T.C., J.L. Sims, and D.E. Johnson, Some agronomic factors affecting N-dimethylnitrosamine content in cigarette smoke, *Beir. zur Tabak.* **8** (1),34-38, 1975 (SHB 75-0483).
451.Walters, D.B., W.J. Chamberlain, and O.T. Chortyk, *Cis* and *trans* fatty acids in cigarette smoke condensate, *Anal. Chim. Acta,* **77** ,309-311,1975.
452.Mestres, R., S. Illes, C. Espinoza, and C. Chevallier, Detection and assay of pesticide residues in tobaccos, *Trav. de la Soc. de Pharm. de Montpellier,* **34** (3),255-266, 1974 (SHB 76-0313).
453.Burton, H.R., and G. Childs, Jr., The thermal degradation of tobacco.VI. Influence of extraction on the formation of some major gas phase constituents, *Beit. zur Tabak.,* **8** (4), 174-180, 1975 (SHB 76-0473).
454.Hoshika, Y., and Y. Iida, Gas chromatographic determination of sulphur compounds in town gas, *J. Chromatog.,* **134** , 423-432, 1977.
455.Jones, P.W., Analysis of nonparticulate organic compounds in ambient atmospheres, Paper presented at 67th Annual Meeting, Air Poll. Cont. Assoc., Denver, CO, 1974.
456.Moore, B.J., *Analyses of Natural Gases, 1975,* Bureau of Mines Inform. Circ. 8717, U.S. Dept. of the Interior, Washington, D.C., 1976.
457.Environmental Health Research Center, *Proposed Standard and Health Effects of Ambient Chlorine,* Rept. IIEQ 76-07, Chicago, IL, 1976. (NTIS PB-258983).
458.Kunen, S.M., K.J. Voorhees, A.C. Hill, F.D. Hileman, and D.N. Osborne, Chemical analysis of the insoluble carbonaceous components of atmospheric particulates with pyrolysis/gas chromatography/mass spectrometry techniques, Paper 77-36.4, 70th Annual Meeting , Air Poll. Cont. Assoc., Montreal, Ont., Can., June, 1977.
459.Rupp, W.H., Air pollution sources and their control, in *Air Pollution Handbook,* ed. P.L. Magill, F.R. Holden, and C. Ackley, New York: McGraw-Hill, p. 1-7, 1956.
460.Dillard, J.G., R.D. Seals, and J.P. Wightman, A study of aluminum-containing atmospheric particulate matter, paper presented at Spring Annual Meeting, American Chemical Society, New Orleans, LA, 1977.

461.Gay, B.W., Jr., R.R. Arnts, and W.A. Lonneman, Naturally emitted hydrocarbons, their atmospheric chemistry and effects on rural and urban oxidant formation, paper presented at Spring Annual Meeting, American Chemical Society, New Orleans, LA, 1977.

462.Berkley, R.E., E.D. Pellizzari, and J.T. Bursey, Analysis of trace levels of organic vapors in ambient air by glass capillary gc/ms: industrial pollutants in the Los Angeles basin, paper presented at Spring Annual Meeting, American Chemical Society, New Orleans, La., 1977.

463.Maddox, W.L., and G. Mamantov, Analysis of cigarette smoke by fourier transform infrared spectrometry, *Anal. Chem.,* **49,** 331-336, 1977.

464.Moore, R.E., Volatile compounds from marine algae, *Acc. Chem. Res.,* **10,** 40-47, 1977.

465.Mayrsohn, H., J. H. Crabtree, M. Kuramoto, R. D. Sothern, and S. H. Mano, Source reconciliation of atmospheric hydrocarbons 1974, *Atmos. Environ.,* **11** , 189-192, 1977.

466.Shultz, J. L., T. Kessler, R. A. Friedel, and A. G. Sharkey, Jr., High-resolution mass spectrometric investigation of heteroatom species in coal-carbonization products, *Fuel,* **51** , 242-246, 1972.

467.Nozaki, Y., D. J. McMaster, L. K. Benninger, D. M. Lewis, W. C. Graustein, and K. K. Turekian, Atmospheric Pb-210 fluxes determined from soil profiles, *EOS-Trans. AGU,* **58** , 396, 1977.

468.Birkenheuer, D. L., and B. L. Davis, X-ray quantitative analysis of hi-volume aerosol samples, *EOS-Trans. AGU,* **58** , 396, 1977.

469.Stephens, E. R., and F. R. Burleson, Unpublished data cited by Winer, et al. (1033).

470.Ryan, J.A., and N.R. Mukherjee, Sources of atmospheric gaseous chloring, *Rev. Geophys. Space Phys.,* **13** ,650-658, 1975.

471.Morgan, E. D., and R. C. Tyler, Microchemical methods for the identification of volatile pheromones, *J. Chromatog.,* **134** , 174-177, 1977.

472.Rasmussen, R. A., R. B. Chatfield, and M. W. Holdren, Hydrocarbon species in rural Missouri air, paper presented at symposium on "The Non-Urban Tropospheric Composition", Hollywood, FL, Nov. 11, 1976.

473.Mason, B. *Principles of Geochemistry* 2nd ed., New York: John Wiley, 1958.

474.Murrell, J. T., and R. B. Channell, Fragrence analyses of *Trillium luteum* and *Trillium cuneatum, J. Tenn. Acad. Sci.,* **48** , 101-103, 1973.

475.Lodge, J. P., Jr., J. B. Pate, B. E. Ammons, and G. A. Swanson, The use of hypodermic needles as critical orifices in sampling, *J. Air Poll. Cont. Assoc.,* **16** , 197-200, 1966.

476.Rahn, K. A., *The Chemical Composition of the Atmospheric Aerosol,* Technical Report, Graduate School of Oceanography, University of Rhode Island, Kingston, RI, July, 1976.

477.Dams, R., J. A. Robbins, K. Rahn, and J. W. Winchester, Nondestructive neutron activation analysis of pollution particulates, *Anal. Chem.,* **42** , 861-867, 1970.

478.King, R. B., J. S. Fordyce, A. C. Antoine, H. F. Leibecki, H. E. Neustadler, and S. M. Sidik, Elemental composition of airborne particulates and source identification: an extensive one-year survey, *J. Air Poll. Cont. Assoc.,* **26** , 1073-1084, 1976.

479.Moore, H. E., E. A. Martell, and S. E. Poet, Sources of polonium-210 in the atmosphere, *Environ. Sci. Technol,* **10** , 586-591, 1976.

480.Grosjean, D., K. Van Cauwenberghe, J.P. Schmid, P.E. Kelley, and J.N. Pitts, Jr., Identification of C_3-C_{10} aliphatic dicarboxylic acids in airborne particulate matter, *Environ. Sci. Technol.,* in press, 1978.

481.Volchok, H. L., and B. Krajewski, Radionuclides and lead in surface air, *Health and Safety Laboratory Fallout Program,* Quarterly Summary Report, p. C-1 to C-9 and C-102 to C-107, New York, NY, June 1, 1972.

482.Thomas, C.W., Preliminary finding of the concentration of ^{99}Tc in environmental samples, *Pacific Northwest Laboratory Annual Report for 1972* , BNWL-1951, Pt. 2, UC-48, 1973.

483.Junge, C. *Air Chemistry and Radioactivity* , New York: Academic Press, 1963.

484.Krey, P. W., private communication, 1976.

485.Koide, M., J. J. Griffin, and E. D. Goldberg, Records of plutonium fallout in marine and terrestrial samples, *J. Geophys. Res.,* **80** , 4153-4162, 1975.

486.Black, F. M., L. E. High, and A. Fontijn, Chemiluminescent measurement of reactivity weighted ethylene-equivalent hydrocarbons, *Environ. Sci. Technol.,* **11** , 597-601, 1977.

487.Kaiser, K. L. E., and J. Lawrence, Polyelectrolytes: potential chloroform precursors, *Science,* **196** , 1205-1206, 1977.

488.Dong, M. W., D. C. Locke, and D. Hoffmann, Characterization of aza-arenes in basic organic portion of suspended particulate matter, *Environ. Sci. Technol.,* **11** , 612-618, 1977.

489.Robertson, D. E., E. A. Crecelius, J. S. Fruchter, and J. D. Ludwick, Mercury emissions from geothermal power plants, *Science,* **196** , 1094-1097, 1977.

490.Billings, C. E., A. M. Sacco, W. R. Matson, R. M. Griffin, W. R. Coniglio, and R. A. Hartley, Mercury balance on a large pulverized coal-fired furnace, *J. Air Poll. Cont. Assoc.,* **23** , 773-777, 1973.

491.Maroulis, P. J., and A. R. Bandy, Estimate of the contribution of biologically produced dimethyl sulfide to the global sulfur cycle, *Science,* **196** , 647-648, 1977.

492.Iverson, R. E., Air pollution in the aluminum industry, *J. Metals,* **25** , 19-23, 1973.

493.Stahly, E. E., Removal of metal carbonyls from tobacco smoke, Patent U.S. 3, 246, 654, *Chem. Abstr.,* **65** , 9358, 1966.

494.Wilkniss, P. E., J. W. Swinnerton, D. J. Bressan, R. A. Lamontagne, and R. E. Larson, CO, CCl_4, Freon-11, CH_4 and Rn-222 concentrations at low altitude over the Arctic ocean in January, 1974, *J. Atmos. Sci.,* **32** , 158-162, 1975.

495.Palmer, T. Y., Combustion sources of atmospheric chlorine, *Nature,* **263** , 44-46, 1976.

496.Loper, G. M., and A. M. Lapidi, Photoperiodic effects on the emanation of volatiles from alfalfa *(Medicago sativa L.)* florets, *Plant. Physiol.,* **49** , 729-732, 1971.

497.Kingston, B. H., Selection of "naturals" in modern perfumery, *Soap, Perfum., Cosmetics,* **44** , 553-558, 1971.

498.Saxena, K. N., and S. Prabha, Relationship between the olfactory sensilla of *Papileo demoleus L.* larvae and their orientation response to different odours, *J. Entomol.,* **A50** , 119-126, 1975.

499.Holman, R. T., and W. H. Heimermann, Identification of components of orchid fragrances by gas chromatography-mass spectrometry, *Bull. Amer. Orchid Soc.,* **42** , 678-682, 1973.

500.Dodson, C. H., R. L. Dressler, H. G. Hills, R. M. Adams, and N. H. Williams, Biologically active compounds in orchid fragrances, *Science,* **164** , 1243-1249, 1969.

501.Williams, N. H., and C. H. Dodson, Selective attraction of male euglossine bees to orchid floral fragrances and its importance in long distance pollen flow, *Evolution,* **26** , 84-95, 1972.

502.Waller, G. D., G. M. Loper, and R. L. Berdel, A bioassay for determining honey bee response to flower volatiles, *Environ. Entomol.,* **2** , 255-259, 1973.

503.Opdyke, D. L. J., Monographs on fragrance raw materials, *Food Cosmet. Toxicol.,* **11** , 95-115, 1973.

504.Opdyke, D. L. J., Monographs on fragrance raw materials, *Food Cosmet. Toxicol.,* **11** , 477-495, 1973.

505.Opdyke, D. L. J., Monographs on fragrance raw materials, *Food Cosmet. Toxicol.,* **11** , 855-876, 1973.

506.Opdyke, D. L. J., Monographs on fragrance raw materials, *Food Cosmet. Toxicol.,* **11** , 1011-1081, 1973.

507.Opdyke, D. L. J., Monographs on fragrance raw materials, *Food Cosmet. Toxicol.,* **12** , 385-405, 1974.

508.Opdyke, D. L. J., Monographs on fragrance raw materials, *Food Cosmet. Toxicol.,* **12** , 517-537, 1974.

509.Opdyke, D. L. J., Monographs on fragrance raw materials, *Food Cosmet. Toxicol.,* **12** , 703-736, 1974.

510.Opdyke, D. L. J., Monographs on fragrance raw materials, *Food Cosmet. Toxicol.,* **12** , 807-1016, 1974.

511.Opdyke, D. L. J., Monographs on fragrance raw materials, *Food Cosmet. Toxicol.,* **13** , 91-112, 1975.

512.Opdyke, D. L. J., Monographs on fragrance raw materials, *Food Cosmet. Toxicol.,* **13** , 449-457, 1975.

513.Opdyke, D. L. J., Monographs on fragrance raw materials, *Food Cosmet. Toxicol.,* **13** , 545-554, 1975.

514.Opdyke, D. L. J., Monographs on fragrance raw materials, *Food Cosmet. Toxicol.,* **13** , 681-923, 1975.

515.Opdyke, D. L. J., Monographs on fragrance raw materials, *Food Cosmet. Toxicol.,* **14** , 307-338, 1976.

516.Opdyke, D. L. J., Monographs on fragrance raw materials, *Food Cosmet. Toxicol.,* **14** , 443-481, 1976.

517.Opdyke, D. L. J., Monographs on fragrance raw materials, *Food Cosmet. Toxicol.,* **14** , 601-633, 1976.

518.Hocking, G. M., Plant flavor and aromatic values in medicine and pharmacy, in *Current Topics in Plant Science* , New York: Academic Press, 1969.

519.Hagen, D. F., and G. W. Holiday, The effects of engine operating and design variables on exhaust emissions, *SAE TP-6* , Society of Automotive Engineers, New York, NY, p. 206, 1964.

520.Lee, R. E., Jr., and F. V. Duffield, EPA's catalyst research program: environmental impact of sulfuric acid emissions, *J. Air Poll. Cont. Assoc.,* **27** , 631-635, 1977.

521.Akagi, H., and A. Kobayashi, Analysis of plasticizers and hydrogen chloride during processing of polyvinyl chloride compounds, *J. Japan Soc. Air Poll.,* **10** , 374, 1975, (APTIC No. 84207).

522.Banwart, W. L., and J. M. Bremner, Identification of sulfur gases evolved from animal manures, *J. Environ. Qual.,* **4** , 363-366, 1975.

523.Hoshika, Y., I. Kozima, K. Koike, and K. Yoshimoto, Analysis of the sulfur compounds in the exhaust gases from two corn starch factories, *J. Japan Oil Chem. Soc.,* **24** , 317-318, 1975.

524.Stephens, E. R., and F. R. Burleson, Distribution of light hydrocarbons in ambient air, *J. Air Poll. Contr. Assoc.,* **19** , 929-936, 1969.

525.Bertsch, W., R. C. Chang, and A. Zlatkis, The determination of organic volatiles in air pollution studies: characterization of profiles, *J. Chromatog. Sci.,* **12** , 175-182, 1974.

526.Papa, L. J., D. L. Dinsel, and W. C. Harris, Gas chromatographic determination of C_1 to C_{12} hydrocarbons in automotive exhaust, *J. Gas Chromatog.,* **6** , 270-279, 1968.

527.Jacobs, E. S., Rapid gas chromatographic determination of C_1 to C_{10} hydrocarbons in automotive exhaust gas, *Anal. Chem.,* **38** , 43-48, 1966.

528.Ketseridis, G., J. Hahn, R. Jaenicke, and C. Junge, The organic constituents of atmospheric particulate matter, *Atmos. Environ.,* **10** , 603-610, 1976.

529.Soldano, B. A., P. Bien, and P. Kwan, Air-borne organo-mercury and elemental mercury emissions with emphasis on central sewage facilities, *Atmos. Environ.,* **9** , 941-944, 1975.

530.Shendrikar, A. D., and P. W. West, Air sampling methods for the determination of selenium, *Anal. Chim. Acta,* **89** , 403-406, 1977.

531.Arctander, S., *Perfume and Flavor Materials of Natural Origin* , Privately published, Elizabeth, New Jersey, 1960.
532.Cronn, D.R., R.J. Charlson, R.L. Knights, A.L. Crittenden, and B.R. Appel, A survey of the molecular nature of primary and secondary components of particles in urban air by high-resolution mass spectrometry, *Atmos. Environ.,* 11 , 929-937, 1977.
533.Wakayama, S., S. Namba, and M. Ohno, Odorous constituents of lilac flower oil, *Nippon Kagaku Zasshi,* 92 , 256-259, 1971.
534.Schmeltz, I., and D. Hoffmann, Nitrogen-containing compounds in tobacco and tobacco smoke, *Chem. Rev.,* 77 , 295-311, 1977.
535.Levaggi, D. A., and M. Feldstein, The collection and analysis of low molecular weight carbonyl compounds from source effluents, *J. Air Poll. Contr. Assoc.,* 19 , 43-45, 1969.
536.Hendel, F. J., Aerothermochemistry of the terrestrial atmosphere, *CRC Crit. Rev. Environ. Control,* 3 , 129-152, 1973.
537.Kimura, K., H. Kaji, N. Nakano, and T. Ito, Measurement of trace hydrocarbons in automobile exhaust, *Proc. Lecture Meeting, Soc. Auto. Engr. Japan* , p. 609-614, 1975 (APTIC No. 79610).
538.McLane, J. E., R. B. Finkelman, and R. R. Larson, *Minerological examination of particulate matter from the fumeroles of Sherman Crater, Mt. Baker, Washington (State),* Preprint, Geological Survey, Reston, VA, 2 p., 1975 (APTIC No. 82574).
539.Michalak, L., Air pollution by a cement plant at Chelm Lubelski, *Ochrona Powietrza (Warsaw)* , no. 3, 81-86, 1975 (APTIC No. 79697).
540.Terajima, T., Offensive odor, *Mokuzai Koryo,* 30 , 508-510, 1975 (APTIC No. 82209).
541.Blosser, E. R., L. J. Hillenbrand, J. Lathouse, W. R. Pierson, and J. W. Butler, Sampling and analysis for sulfur compounds in automobile exhaust, *Proc. IMR Symp.* , 389-400, (National Bureau of Standards Pub. 422), 1976.
542.Chamberlain, W. J., Polar lipid materials in cigarette smoke condensate, *Tobacco,* 178 (#23), 51-53, 1976.
543.Trijonis, J. C., and K. W. Arledge, *Utility of Reactivity Criteria in Organic Emission Control Strategies for Los Angeles* , Contract Report No. 68-02-1735, TRW Environ. Systems, Redondo Beach, CA, 1975.
544.Carter, J. A., and W. R. Musick, Platinum metals in air particulates near a catalytic converter test site as measured by isotope dilution SSMS, in *Proc. Platinum Res. Rev. Conf.,* Catal. Res. Program, Rougemont, NC, 1975.
545.Hrutfiord, B. F., L. N. Johanson, and J. L. McCarthy, *Steam Stripping Odorous Substances from Kraft Effluent Streams,* EPA-R2-73-196, Environmental Protection Agency, Washington, D. C., 1973. (NTIS PB-221335).
546.Dong, M., D. Hoffman, D. C. Locke, and E. Ferrand, The occurrence of caffeine in the air of New York City, *Atmos. Environ.,* 11, 651-653, 1977.
547.Rasmussen, R., private communication, 1977.
548.Tabor, E. C., T. E. Hauser, J. P. Lodge, and R. H. Burttschell, Characteristics of the organic particulate matter in the atmosphere of certain American cities, *Arch. Ind. Health.,* 17, 58-63, 1958.
549.Meijer, G. M., and H. Niebour, Determination of peroxybenzoyl nitrate (PBzN) in ambient air, *Ver. Deutsch. Ingen.,* 270, 55-56, 1977.
550.Hoshika, Y., Simple and rapid gas-liquid-solid chromatographic analysis of trace concentrations of acetaldelyde in urban air, *J. Chromatogr.,* 137, 455-460, 1977.
551.Tausch, H., and G. Stehlik, Bestimmung polycyclischer aromaten in russ mittels gaschromatographie-massenspektrometerie, *Chromatographia,* 10, 350-357, 1977.
552.Nicholas, H. J., Miscellaneous volatile plant products, in *Phytochemistry,* (L. P. Miller, ed.), v. 2, p. 381-399, 1973

553.Krstulovic, A. M., D. M. Rosie, and P. R. Brown, Distribution of some atmospheric polynuclear aromatic hydrocarbons, *Amer. Lab.,* 9 (#7), 11-18, 1977.

554.Mansfield, C. T., B. T. Hodge, R. B. Hege, Jr., and W. C. Hamlin, Analysis of formaldehyde in tobacco smoke by high performance liquid chromatography, *J. Chromatog. Sci.,* 15, 301-302, 1977.

555.Ito, Y. Offensive smell in tap water, Paper presented at Chem. Soc. Japan 34th Ann. Meeting, Tokyo, 1976 (APTIC No. 101209).

556.Grimmer, G., Analysis of automobile exhaust condensates, in *Air Pollut. Cancer Man,* Lyon, World Health Org., p. 29-39, 1977.

557.Suzuki, R., M. Ito, M. Noma, I. Moritani, Y. Watanabe, T. Nakaya, and N. Saito, Determination of PCB in dust, ash and combustion gas from city wastes incinerators, *Rept. Environ. Pollut. Res. Cent. Aichi Pref.* (Japan), 2, 43-49 (1974) (APTIC No. 77236).

558.Nicolet, M., The properties and constitution of the upper atmosphere, in *Physics of the Upper Atmosphere,* J. A. Ratcliffe, ed., New York: Academic Press, p. 17, 1960.

559.Little, A. D., Inc. *Preliminary Economic Input of Possible Regulatory Action to Control Atmosphere Emissions of Selected Halocarbons,* EPA Contract No. 68-02-1349, Task 8, EPA-450/3-75-073, Research Triangle Park, NC, 1975.

560.Panel on Atmospheric Chemistry, *Halocarbons: Effects on Stratospheric Ozone,* National Academy of Sciences, Washington, D. C., 1976.

561.Lapp, T. W., H. M. Gadberry, R. R. Wilkinson, and T. Weast, *Chemical Technology and Economics in Environmental Perspectives. Task III - Chlorofluorocarbon Emission Control in Selected End-Use Applications,* EPA-560/1-76-009, Environmental Protection Agency, Washington, D. C., 1976.

562.Lapp, T. W., G. J. Hennon, H. M. Gadberry, I. C. Smith, and K. Lawrence, *Chemical Technology and Economics in Environmental Perspectives; Task I - Technical Alternatives to Selected Chlorofluorocarbon Uses,* EPA-560/1-76-002, Environmental Protection Agency, Washington, D. C., 1976.

563.Federal Task Force on Inadvertent Modification of the Stratosphere, *Fluorocarbons and the Environment,* Council on Environmental Quality and Federal Council for Science and Technology, Washington, D. C., 1975.

564.Amato, W. S., B. Bandyopadhyay, B. E. Kurtz, and R. H. Fitch, Development of a low emissions process for ethylene dichloride production, *Int. Conf. on Photochem. Oxidant,* EPA-600/3-77, 001b, Environmental Protection Agency, Research Triangle Park, NC, 1977.

565.Bernstein, L. J., K. K. Kearby, A.K.S. Roman, J. Vardi, and E. E. Wigg, Application of catalysts to automotive NO_x emissions control, *Tech. Paper 710014,* Soc. Automot. Engineers, Warrenton, PA, 1971.

566.Ciccioli, P., G. Bertoni, E. Brancaleoni, R. Fratarcangeli, and F. Bruner, Evaluation of organic pollutants in the open air and atmospheres in industrial sites using graphetized carbon black traps and gas chromatographic-mass spectrometric analysis with specific detectors, *J. Chromatogr.,* 126, 757-770, 1976.

567.Turner, B. C., and D. E. Glotfelty, Field air sampling of pesticide vapors with polyurethane foam, *Anal. Chem.,* 49, 7-10, 1977.

568.Plimmer, J. R., Photochemistry of organochlorine insecticides, *Proc. 2nd Int. Conf. on Pesticide Chem.,* 1, 413-432, 1972.

569.Arnts, R. R., A. Appleby, D. Lillian, and H. B. Singh, Vapor phase photodecomposition of ortho-chlorobiphenyl, Paper ENVT-68, 172nd Meeting, American Chemical Society, San Francisco, Cal., 1976.

570.Lunde, G., J. Gether, N. Gjos, and M.-B. S. Lande, Organic micropollutants in precipitation in Norway, *Atmos. Environ.,* 11, 1007-1014, 1977.

571.Rubel, F. N., Incinerator emissions, in *Incineration of Solid Wastes,* Noyes Data Corp., Park Ridge, NJ, p. 49-67, 1974.

572.Singmaster, J. N. III, and D. G. Crosby, Volatilization of hydrophobic pesticides from water, Paper PEST-6, 173rd Meeting, American Chemical Society, New Orleans, LA, 1977.

573.Miller, C. W., and M. T. Shafik, Occurrence of propoxur in air following intradomicilliary spraying, Paper PEST-26, 173rd Meeting, American Chemical Society, New Orleans, LA, 1977.

574.Taylor, A. W., D. E. Glotfelty, B. C. Turner, R. E. Silver, H. P. Freeman, and A. Weiss, Volatilization of dieldrin and heptachlor residues from field vegetation, Paper PEST-7, 173rd Meeting, American Chemical Society, New Orleans, LA, 1977.

575.Turner, B. C., D. E. Glotfelty, A. W. Taylor, and D. R. Watson, Volatilization of microencapsulated and conventionally applied CIPC in the field, Paper PEST-8, 1973rd Meeting, American Chemical Society, New Orleans, LA, 1977.

576.Caro, J. H., B. A. Bierl, H. P. Freeman, D. E. Glotfelty, and B. C. Turner, Disparlure: volatilization rates of two microencapsulated formulations from a grass field, Paper PEST-9, 173rd Meeting, American Chemical Society, New Orleans, LA, 1977.

577.Abbott, D. C., R. B. Harrison, J. O'G. Tatton, and J. Thomson, Organochlorine pesticides in the atmosphere, *Nature,* **211**, 259-261, 1966.

578.Miles, J. W., L. E. Fetzer, and G. W. Pearce, Collection and determination of trace quantities of pesticides in air, *Environ. Sci. Technol.,* **4**, 420-425, 1970.

579.Schmeltz, I., J. Tosk, G. Jacobs, and D. Hoffmann, Redox potential and quinone content of cigarette smoke, *Anal. Chem.,* **49**, 1924-1929, 1977.

580.Cheney, J. L., and C. R. Fortune, The present relationships of sulfuric acid concentration to acid dew point for flue gases, *Anal. Lett.,* **10**, 797-816, 1977.

581.Meguerian, G. H., and C. R. Lang, NO_x reduction catalysts for vehicle emission control, Tech. Paper 710291, Soc. Automot. Engineers, Warrenton, PA, 1971.

582.Louw, C. W., J. F. Richards, and P. K. Faure, The determination of volatile organic compounds in city air by gas chromatography combined with standard addition, selective subtraction, infrared spectrometry and mass spectrometry, *Atmos. Environ.,* **11**, 703-717, 1977.

583.Woodhead, S., and E. Bernays, Changes in release rates of cyanide in relation to palatability of *Sorghum* to insects, *Nature,* **270**, 235-236, 1977.

584.Oaks, D. M., H. Hartmann, and K. P. Dimick, Analysis of sulfur compounds with electron capture/hydrogen flame dual channel gas chromatography, *Anal. Chem.,* **36**, 1560-1565, 1964.

585.Duce, R. A., W. H. Zoller, and J. L. Moyers, Particulate and gaseous halogens in the Antarctic atmosphere, *J. Geophys. Res.,* **78**, 7802-7811, 1973.

586.Savenko, V. S., Is boric acid evaporation from the sea water the major boron source in the atmosphere?, *Okeanolognya,* **17**, 445-448, 1977.

587.Evans, C. S., C. J. Asher, and C. M. Johnson, Isolation of dimethyl diselenide and other volatile selenium compounds from *Astragalus racemosus* (pursh.), *Aust. J. Biol. Sci.,* **21**, 13-20, 1968.

588.Winer, A. M., A. C. Lloyd, K. R. Darnall, R. Atkinson, and J. N. Pitts, Jr., Rate constants for the reaction of OH radicals with *n*-propyl acetate, *sec*-butyl acetate, tetrahydrofuran, and peroxyacetyl nitrate, *Chem. Phys. Lett.,* **51**, 221-226, 1977.

589.Schuetzle, D., A. L. Crittenden, and R. J. Charlson, Application of computer controlled high resolution mass spectrometry to the analysis of air pollutants, *J. Air Poll. Control. Assoc.,* **23**, 704-709, 1973.

590.Drivas, P. J., and F. H. Shair, Dispersion of a crosswind line source of tracer released from an urban highway, *Atmos. Environ.,* **8**, 1155-1164, 1974.

591.Holzer, G., H. Shanfield, A. Zlatkis, W. Bertsch, P. Juarez, H. Mayfield, and H. M. Liebich, Collection and analysis of trace organic emissions from natural sources, *J. Chromatogr.,* **142,** 755-764, 1977.

592.Hurn, R. W., Mobile Combustion sources, in *Air Pollution, III,* (A.C. Stern, ed.) 2nd Ed., New York: Academic Press, p. 55-95, 1968.

593.Krey, P. W., R. J. Lagomarsino, and L. E. Toonkel, Gaseous halogens in the atmosphere in 1975, *J. Geophys. Res.,* **82,** 1753-1766, 1977.

594.Fujii, M., M. Matsuoka, K. Nakamura, K. Oikawa, K. Honma, and Y. Hashimoto, Air pollutants in exhaust gas of jet-aircraft engine, *Japan J. Public Health,* **23** (4), 299-300, 1976. (APTIC No. 103711.)

595.Wofsy, S. C., M. B. McElroy, and Y. C. Yung, The chemistry of atmospheric bromine, *Geophys. Res. Lett.,* **2,** 215-218, 1975.

596.Fishbein, J., Chromatographic and biological aspects of DDT and its metabolites, *J. Chromatog.,* **98** , 177-251, 1974.

1001. Cahn, R. S., *Introduction to Chemical Nomenclature,* Fourth ed., New York: Halstead Press, NY, 1974.
1002. Bach, W., Global air pollution and climate, *Rev. Geophys. Space Phys.,* **14** , 429-474, 1976.
1003. Heicklen, J., *Atmospheric Chemistry,* New York: Academic Press, 1976.
1004. Graedel, T. E., and N. Schwartz, Air quality reference data for the design and testing of materials and components, *Materials Performance,* **16** (8),17-25,1977.
1005. Green, A. E. S., T. Sawada, and E. P. Shettle, The middle ultraviolet reaching the ground, *Photochem. Photobiol.,* **19** , 251-259, 1974.
1006. Watson, R. T., G. Machado, B. Conaway, S. Wagner, and D. D. Davis, A temperature dependent kinetics study of the reaction of OH with CH_2ClF, $CHCl_2F$, $CHClF_2$, CH_3CCl_3, CH_3CF_2Cl, and $CF_2ClCFCl_2$, *J. Phys. Chem.,* **81** , 256-262, 1977.
1007. Graedel, T. E., L. A. Farrow and T. A. Weber, Kinetic studies of the photochemistry of the urban troposphere, *Atmos. Environ.,* **10** , 1095-1116, 1976.
1008. *Air Quality Criteria Document for Photochemical Oxidants,* Environmental Protection Agency, Research Triangle Park, NC, (in press), 1977.
1009. Leighton, P. A. *Photochemistry of Air Pollution,* New York: Academic Press, 1961.
1010. Graedel, T. E., and D. L. Allara, The kinetic ozone photochemistry of natural and perturbed nonurban tropospheres, *Int'l. Conf. on Photochem. Oxidant,* v. 1, EPA-600/3-77-001a, Environmental Protection Agency, Research Triangle Park, N. C., p. 467-473, 1977 (NTIS PB264232).
1011. Sze, N. D., Anthropogenic CO emissions: implications for the atmospheric $CO-OH-CH_4$ cycle, *Science,* **195** , 673-675, 1977.
1012. Klemperer, W., private communication, 1976.
1013. Noxon, J., paper presented at symposium on "The Non-Urban Tropospheric Composition", Hollywood, Fla., Nov. 12, 1976.
1014. Graedel, T. E., The homogeneous chemistry of atmospheric sulfur, *Rev. Geophys. Space Phys.,* **15** ,421-428,1977.
1015. Crutzen, P. J., The possible importance of CSO for the sulfate layer of the stratosphere, *Geophys. Res. Lett.,* **3** , 73-76, 1976.
1016. Graedel, T. E., Sulfur dioxide, sulfate aerosol, and urban ozone, *Geophys. Res. Lett.,* **3** , 181-184, 1976.
1017. Davis, D. D. and G. Klauber, Atmospheric gas phase oxidation mechanisms for the molecule SO_2 , *Int. J. Chem. Kinetics Symp.,* **1** , 543-556, 1975.
1018. Weiss, R.E., A.P. Waggoner, R.J. Charlson, and N.C. Ahlquist, Sulfate aerosol: its geographical extent in the midwestern and southern United States, *Science,* **195** ,979-981,1977.
1019. Altshuller, A. P., Atmospheric sulfur dioxide and sulfate: Distribution of concentration at urban and nonurban sites in United States, *Environ. Sci. Technol.,* **7** , 709-712, 1973.
1020. Prahm, L. P., U. Torp, and R. M. Stern, Deposition and transformation rates of sulphur oxides during atmospheric transport over the Atlantic, *Tellus,* **28** , 355-372, 1976.
1021. Hampson, R. F., Jr., and D. Garvin, eds., *Chemical Kinetic and Photochemical Data for Modelling Atmospheric Chemistry,* NBS Technical Note 866, National Bureau of Standards, Washington, D. C., 1975.
1022. Demerjian, K. L., J. A. Kerr, and J. G. Calvert, The mechanism of photochemical smog formation, *Adv. Environ. Sci. Technol.,* **4** , 1-262, 1974.
1023. Levy, H., II, Photochemistry of the troposphere, *Adv. Photochem.,* **9** , 369-524, 1974.

1024. Ford, H. W., and N. Endow, Rate constants at low concentrations. IV. Reactions of atomic oxygen with various hydrocarbons, *J. Chem. Phys.,* **27** , 1277-1279, 1957.

1025. Darnall, K. R., A. M. Winer, A. C. Lloyd, and J. N. Pitts, Jr., Relative rate constants for the reaction of OH radicals with selected C_6 and C_7 alkanes and alkenes at $305\pm2K$, *Chem. Phys. Lett.,* **44** , 415-418, 1976.

1026. Alcock, W. G., and B. Mile, Gas-phase reactions of alkylperoxy and alkoxy radicals. Part I. The photoinitiated oxidation of 2,3-dimethylbutane, *Comb. Flame,* **29** , 133-144, 1977.

1027. Carter, W. P. L., K. R. Darnall, A. C. Lloyd, A. M. Winer, and J. N. Pitts, Jr., Evidence for alkoxy radical isomerization in photo-oxidations of C_4-C_6 alkanes under simulated atmospheric conditions, *Chem. Phys. Lett.,* **42** , 22-27, 1976.

1028. Campbell, I. M., D. F. McLaughlin, and B. J. Handy, Rate constants for reactions of hydroxyl radicals with alcohol vapours at 292°K , *Chem. Phys. Lett.,* **38** , 362-364, 1976.

1029. Kanofsky, J. R., D. Lucas, F. Pruss, and D. Gutman, Direct identification of the reactive channels in the reactions of oxygen atoms and hydroxyl radicals with acetylene and methylacetylene, *J. Phys. Chem.,* **78** , 311-316, 1974.

1030. Harker, A. B., and C. S. Burton, A study of the mechanism and kinetics of the reaction of $O(^3P)$ atoms with propane, *Int. J. Chem. Kinetics,* **7** , 907-917, 1975.

1031. Japar, S. M., and H. Niki, Gas phase reactions of the nitrate radical with olefins, *J. Phys. Chem.,* **79** , 1629-1632, 1975.

1032. Herron, J. T., and R. E. Huie, Reactions of $O_2(^1\Delta_g)$ with olefins and their significance in air pollution, *Environ. Sci. Technol.,* **4** , 685-686, 1970.

1033. Winer, A. M., A. C. Lloyd, K. R. Darnall, and J. N. Pitts, Jr., Relative rate constants for the reaction of the hydroxyl radical with selected ketones, chloroethenes , and monoterpene hydrocarbons, *J. Phys. Chem.,* **80** , 1635-1639, 1976.

1034. Japar, S. M., C. H. Wu, and H. Niki, Rate constants for the reaction of ozone with olefins in the gas phase, *J. Phys. Chem.,* **78** , 2318-2320, 1974.

1035. Ashford, R. D., and E. A. Ogryzlo, Temperature dependence of some reactions of singlet oxygen with olefins in the gas phase, *J. Am. Chem. Soc.,* **97** , 3604-3607, 1975.

1036. Atkinson, R., R. A. Perry, and J. N. Pitts, Jr., Rate constants for the reaction of OH radicals with 2-Methyl-2-butene over the temperature range 297-425K, *Chem. Phys. Lett.,* **38** , 607-610, 1976.

1037. Husain, L., P.E. Coffey, R.E. Meyers, and R.T. Cederwall, Ozone transport from stratosphere to troposphere, *Geophys. Res. Lett.,* **4** , 363-365, 1977.

1038. Toby, F. S., and S. Toby, Reaction between ozone and allene in the gas phase, *Int. J. Chem. Kinetics,* **6** , 417-428, 1974.

1039. Arrington, C. A., Jr., and D. A. Cox, Arrhenius parameters for the reaction of oxygen atoms, $O(^3P)$, with propyne, *J. Phys. Chem.,* **79** , 2584-2586, 1975.

1040. Lissi, E. A., G. Massiff, and A. Villa, Addition of methoxy radicals to olefins, *Int. J. Chem. Kinetics,* **7** , 625-631, 1975.

1041. Nakamura, K., and S. Koda, Reaction of atomic oxygen with cyclopentadiene, *Int. J. Chem. Kinetics,* **9** , 67-81, 1977.

1042. Michael, J.V., D.A. Whytock, J.H. Lee, W.A. Payne, and L.J. Stief, Absolute rate constant for the reaction of atomic chlorine with hydrogen peroxide vapor over the temperature range 265-400 K, *J. Chem. Phys.,* **67** , 3533-3536, 1977.

1043. Weaver, J., J. Meagher, and J. Heicklen, Photo-oxidation of CH_3CHO vapor at 3130Å , *J. Photochem.* **6** , 111-126, 1976/77.

1044. Cadle, R. D., S. S. Lin, and R. F. Hausman, Jr., The reaction of $O(^3P)$ with propionaldehyde and acrolein in a fast-flow system, *Chemosphere,* **1** , 15-20, 1972.

1045. Cadle, R. D., H. H. Wickman, C. B. Hall, and K. M. Eberle, The reaction of atomic oxygen with formaldehyde, crotonaldehyde, and dimethyl sulfide, *Chemosphere,* **3** , 115-118, 1974.

1046. Morris, E. D., Jr., and H. Niki, Reaction of the nitrate radical with acetaldehyde and propylene, *J. Phys. Chem.,* **78** , 1337-1338, 1974.

1047. Lee, J. H., and R. B. Timmons, Kinetics and mechanism of the gas-phase reaction of $O(^3P)$ atoms with acetone, *Int. J. Chem. Kinetics,* **9** , 133-139, 1977.

1048. LeFevre, H. F., J. F. Meagher, and R. B. Timmons, The kinetics of the reactions of $O(^3P)$ atoms with dimethyl ether and methanol, *Int. J. Chem. Kinetics,* **4** , 103-116, 1972.

1049. Lloyd, A. C., K. R. Darnall, A. M. Winer, and J. N. Pitts, Jr., Relative rate constants for the reactions of OH radicals with isopropyl alcohol, diethyl and di- n -propyl ether at $305 \pm 2K$, *Chem. Phys. Lett.,* **42** , 205-209, 1976.

1050. Lee, J. H., R. B. Timmons, and L. J. Stief, Absolute rate parameters for the reaction of ground state atomic oxygen with dimethyl sulfide and episulfide, *J. Chem. Phys.,* **64** , 300-305, 1976.

1051. Boden, J. C., and B. A. Thrush, Kinetics of reactions involving CN emission. IV. Study of the reactions of CN by electronic absorption spectroscopy, *Proc. Roy. Soc.,* **A305** , 107-123, 1968.

1052. Davies, P. B., and B. A. Thrush, Reactions of oxygen atoms with hydrogen cyanide, cyanogen chloride and cyanogen bromide, *Trans. Faraday Soc.,* **64** , 1836-1843, 1968.

1053. Bonanno, R. J., R. B. Timmons, L. J. Stief, and R. B. Klemm, The kinetics and mechanisms of the reactions of $O(^3P)$ atoms with CH_3CN and CF_3CN , *J. Chem. Phys.,* **66** , 92-98, 1977.

1054. Hanst, P. L., J. W. Spence, and M. Miller, Atmospheric chemistry of N-nitroso dimethylamine, *Environ. Sci. Technol.,* **11** , 403-405, 1977.

1055. Kuntz, R. L., S. L. Kopczynski, and J. J. Bufalini, Photochemical reactivity of benzaldehyde- NO_x and benzaldehyde-hydrocarbon- NO_x mixtures, *Environ. Sci. Technol.,* **7** , 1119-1123, 1973.

1056. Atkinson, R., R.A. Perry, and J.N. Pitts, Jr., Rate constants for the reaction of the OH radical with CH_3SH and CH_3NH_2 over the temperature range $299-426^oK$, *J. Chem. Phys.* **66** ,1578-1581, 1977.

1057. Hamilton, E. J., Jr., and C. A. Naleway, Theoretical calculation of strong complex formation by the HO_2 radical: $HO_2 \cdot H_2O$ and $HO_2 \cdot NH_3$, *J. Phys. Chem.,* **80** , 2037-2040, 1976.

1058. Schofield, K., Evaluated chemical kinetic rate constants for various gas phase reactions, *J. Phys. Chem. Ref. Data,* **2** , 25-84, 1973.

1059. Becker, K. H., M. A. Inocencio, and U. Schurath, The reaction of ozone with hydrogen sulfide and its organic derivatives, *Int. J. Chem. Kinetics Symp.,* **1** , 205-220, 1975.

1060. Toby, S., F. S. Toby, and B. A. Kaduk, Chemiluminescence in the gas-phase reactions of ozone with cyanoacetylene, carbon disulfide, and thiophene, *J. Photochem.,* **6** , 297-300, 1976/77.

1061. Davis, D. D., J. Prusazcyk, M. Dwyer, and P. Kim, A stop-flow time-of-flight mass spectrometry kinetics study. Reaction of ozone with nitrogen dioxide and sulfur dioxide, *J. Phys. Chem.,* **78** , 1775-1779, 1974.

1062. Campbell, I. M., and K. Goodman, Rate constants for reactions of hydroxyl radicals with nitromethane and methyl nitrite vapours at 292K, *Chem. Phys. Lett.,* **36** , 382-384, 1975.

1063. Committee on Medical and Biologic Effects of Environmental Pollutants, *Ozone and Other Photochemical Oxidants,* National Academy of Sciences, Washington, D.C., 1977.

1064. Niki, H., P.D. Maker, C.M. Savage, and L.P. Breitenbach, Fourier transform IR spectroscopic observation of pernitric acid formed via $HOO+NO_2 \rightarrow HOONO_2$, *Chem. Phys. Lett.,* **45** ,564-566, 1977.

1065. Atkinson, R., R. A. Perry, and J. N. Pitts, Jr., Kinetics of the reactions of OH radicals with CO and N_2O , *Chem. Phys. Lett.,* **44** , 204-207, 1976.

1066. Margitan, J. J., M. S. Zahniser, F. Kaufman, and J. G. Anderson, Laboratory kinetic studies of reactions of atmospheric interest using resonance fluorescence spectroscopy, Paper R7, 8th Int. Conf. Photochem., Edmonton, Alta., Aug. 7-13, 1975.

1067. Takacs, G. A., and G. P. Glass, Reactions of hydrogen atoms and hydroxyl radicals with hydrogen bromide, *J. Phys. Chem.,* **77** , 1060-1064, 1973.

1068. Singleton, D.L., and R.J. Cvetanović Determination by the phase shift technique of the temperature dependence of the reactions of $O(^3P)$ with HCl, HBr, and HI, Paper presented at 12th Inf. Conf. Photochem., Gaithersberg, Md., June 28, 1976.

1069. Park, C., Reaction rates for $O_3+HCl \rightarrow O+O_2+HCl$, $Cl+O_3 \rightarrow ClO+O_2$, and $HCl+O \rightarrow OH+Cl$ at elevated temperatures, *J. Phys. Chem.,* **81** , 499-504, 1977.

1070. Huie, R. E., J. T. Herron, and D. D. Davis, Rates of reaction of atomic oxygen with C_2H_3F , C_2H_3Cl , C_2H_3Br , $1,1-C_2H_2F_2$, and $1,2-C_2H_2F_2$, *Int. J. Chem. Kinetics,* **4** , 521-527, 1972.

1071. Furuyama, S., and N. Ebara, A kinetic study of the reaction of $O(^3P)$ with toluene, *Int. J. Chem. Kinetics,* **7** , 689-698, 1975.

1072. Westenberg, A. A., and N. deHaas, Reaction rates of $O+CH_3Br$ and $O+CH_3Cl$, *J. Chem. Phys.,* **62** , 4477-4479, 1975.

1073. Howard, C. J., and K. M. Evenson, Rate constants for the reactions of OH with CH_4 and fluorine, chlorine,and bromine substituted methanes at 296K, *J. Chem. Phys.* **64** , 197-202, 1976.

1074. Toby, F. S., and S. Toby, Kinetics of the reaction of ozone with tetrafluoroethene, *J. Phys. Chem.,* **80** , 2313-2316, 1976.

1075. Bogan, D.J., R.S. Sheinson, R.G. Gann, and F.W. Williams, Formaldehyde $(A^1A_2 \rightarrow X^1A_1)$ chemiluminescence in the gas phase reaction of $O_2(a^1\Delta_g)$ plus ethyl vinyl ether, *J. Amer. Chem. Soc.,* **97** , 2560-2562, 1975.

1076. Sidebottom, H. W., C. C. Badcock, G. E. Jackson, J. G. Calvert, G. W. Reinhardt, and E. K. Damon, Photooxidation of sulfur dioxide, *Environ. Sci. Technol.,* **6** , 72-79, 1972.

1077. Rao, T. N., S. S. Collier, and J. G. Calvert, Primary photophysical processes in the photo- chemistry of sulfur dioxide at 2875 $\overset{o}{A}$, *J. Amer. Chem. Soc.,* **91** , 1609-1615, 1969.

1078. Rao, T.N., S.S. Collier, and J.G. Calvert, The quenching reactions of the first excited singlet and triplet states of sulfur dioxide with oxygen and carbon dioxide, *J. Amer. Chem. Soc.,* **91** ,1616-1621, 1969.

1079. Demerjian, K. L., J. G. Calvert, and D. L. Thorsell, A kinetic study of the chemistry of the $SO_2(^3B_1)$ reactions with cis- and trans-2-butene, *Int. J. Chem. Kinetics,* **6** , 829-848, 1974.

1080. Smith, P. P., and L. D. Spicer, Characterization of products from the photochemically induced reactions between SO_2 and aliphatic hydrocarbons, *Chemosphere,* **4** , 131-136, 1975.

1081. Meagher, J.F., and J. Heicklen, Reaction of HO with C_2H_4, *J. Phys. Chem.,* **80** , 1645-1652, 1976.

1082. Hansen, D. A., and J. N. Pitts, Jr., Chemiluminescent emissions from the gas phase reactions of ozone with acetylene and allene, *Chem. Phys. Lett.,* **35** , 569-573, 1975.

1083. Lin, M. C., R. G. Shortridge, and M. E. Umstead, The dynamics of reactions of $O(^3P)$ atoms with allene and methylacetylene, *Chem. Phys. Lett.,* **37** , 279-284, 1976.

1084. Walter, T. A., J. J. Bufalini, and B. W. Gay, Jr., Mechanism for olefin-ozone reactions, *Environ. Sci. Technol.,* **11** , 382-386, 1977.

1085. Lloyd, A. C., K. R. Darnall, A. M. Winer, and J. N. Pitts, Jr., Relative rate constants for reaction of the hydroxyl radical with a series of alkanes, alkenes, and aromatic hydrocarbons, *J. Phys. Chem.,* **80** , 789-794, 1976.

1086. Olszyna, K. J., R. G. DePena, and J. Heicklen, Kinetics of particle growth VII. NH_4NO_3 from the NH_3-O_3 reaction in the presence of air and water vapor, *Int. J. Chem. Kinetics,* **8** , 357-379, 1976.

1087. Schwartz, W., P.W. Jones, C.J. Riggle, and D.F. Miller, *Chemical Characterization of Model Aerosols,* Battelle Memorial Inst.; Columbus, OH, 1974. (NTIS Document PB-238 557).

1088. Rasmussen, R. A., Private communication, 1974.

1089. Akimoto, H., M. Hoshino, G. Inoue, M. Okuda, and N. Washida, Photo-oxidation of toluene- NO_2 - O_2 - N_2 system in gas phase, Paper presented at 12th Inf. Conf. Photochem., Gaithersburg, MD, 1976.

1090. Fatiadi, A. J. Effects of temperature and of ultraviolet radiation on pyrene adsorbed on garden soil, *Environ. Sci. Technol.,* **1** , 570-572, 1967.

1091. Tebbens, B. D., M. Mukai, and J. F. Thomas, Fate of arenes incorporated with airborne soot: effect of irradiation, *J. Amer. Ind. Hyg. Assoc.,* **32** , 365-372, 1971.

1092. Jaffe, S. Photooxidation of ethylene oxide and propionaldehyde in the presence of NO_2 and light, Proc. 2nd Int. Clean Air Congress, (H. Englund, ed.) New York: Academic Press, p.316-324, 1971.

1093. Glasson, W. A., and J. M. Heuss, Synthesis and evaluation of potential atmospheric eye irritants, *Environ. Sci. Technol.,* **11** , 395-398, 1977.

1094. Heuss, J. M., and W. A. Glasson, Hydrocarbon reactivity and eye irritation, *Environ. Sci. Technol.,* **2** , 1109-1116, 1968.

1095. Vaish, S.P., R.D. McAlpine, and M. Cocivera, Photo-CIDNP studies of 2-butanone and 3-pentanone in solution, *Can. J. Chem.,* **52** , 2978-2984, 1974.

1096. Rowland, F. S., and M. J. Molina, Chlorofluoromethanes in the environment, *Rev. Geophys. Space Phys.,* **13** , 1-35, 1975.

1097. Spence, J. W., P. L. Hanst, and B. W. Gay, Jr., Atmospheric oxidation of methyl chloride, methylene chloride, and chloroform, *J. Air Poll. Cont. Assoc.,* **26** , 994-996, 1976.

1098. Gay, B.W., Jr., P.L. Hanst, J.J. Bufalini, and R.C. Noonan, Atmospheric oxidation of chlorinated ethylenes, *Environ. Sci. Technol.,* **10** , 58-67, 1976.

1099. Hesstvedt, E., O. Hov, and I. S. A. Isaksen, *Photochemical Lifetime of Vinylchloride in the Atmosphere,* Report No. 17, Institutt for Geofysikk, Univ. Oslo, Norway, 1976.

1100. Rabani, J., D. Klug-Roth, and A. Henglein, Pulse radiolytic investigations of $OHCH_2O_2$ radicals, *J. Phys. Chem.,* **78** , 2089-2093, 1974.

1101. Sanhueza, E., and J. Heicklen, The $Hg6(^3P)$-sensitized photo-oxidation of C_2Cl_3H, *J. Photochem.,* **4** , 161-169, 1975.

1102. Nojima, K., K. Fukaya, S. Fukui, and S. Kanno, The formation of glyoxals by the photochemical reaction of aromatic hydrocarbons in the presence of nitrogen monoxide, *Chemosphere,* **5** , 247-252, 1974.

1103. Niki, H., P. D. Maker, C. M. Savage, and L. P. Breitenbach, Fourier transform IR spectroscopic observation of propylene ozonide in the gas phase reaction of ozone-*cis*-2-butene-formaldehyde, *Chem. Phys. Lett.,* **46** , 327-330, 1977.

1104. Graedel, T. E., L. A. Farrow, and T. A. Weber, Urban kinetic chemical calculations with altered source conditions, *Atmos. Environ.,* **12** , (in press), 1978.

1105. Chameides, W. L., and R. J. Cicerone, Nonmethane hydrocarbons and the photochemistry of the troposphere and stratosphere, *EOS-Trans. AGU,* **58** , 463, 1977.

1106. Ehhalt, D. H., The atmospheric cycle of methane, *Tellus,* **26** , 58-70, 1974.

1107. Blank, B., A. Henne, and H. Fischer, Photochemische primarreaktionen α-verzweigter aliphatischer ketone und aldehyde in losung, *Helv. Chim. Acta,* **57** , 920-936, 1974.

1108. Martinez, J.R., K.T. Tran, A.C. Lloyd, and G.M. Hidy, *Development of an Atmospheric Model for Sulfate Formation,* Document No. P-1534, Environmental Research and Technology, Inc., Santa Barbara, CA, 1977.

1109. Ho, S. Y., R. A. Gorse, and W. A. Noyes, Jr., Recent work on the photochemistry of acetone in the gaseous phase, *J. Phys. Chem.,* **79** , 1632-1634, 1975.

1110. Smith, R. M., and J. G. Calvert, The vapor-phase photolysis of trifluoroacetophenone and mixtures of trifluoroacetophenone and trifluoroacetone, *J. Am. Chem. Soc.,* **78** , 2345-2351, 1956.

1111. Charlton, J. L., R. G. Smerchanski, and C. E. Burchill, The aqueous photochemistry of sodium 9,10-anthraquinone- 2-sulfonate, *Can. J. Chem.,* **54** , 512-515, 1976.

1112. Grimsrud, E. P., H. H. Westberg, and R. A. Rasmussen, Atmospheric reactivity of monoterpene hydrocarbons, NO_x photo-oxidation and ozonolysis, *Int. J. Chem. Kinetics Symp.,* **1** , 183-195, 1975.

1113. Brosset, C., Sulphate aerosols, Paper ENVT-7, 171st Meeting, American Chemical Society, New York, NY, 1976.

1114. Graedel, T. E., and D. L. Allara, The kinetic photochemistry of natural tropospheres, (In preparation), 1978.

1115. Grosjean, D., and S. K. Friedlander, Gas-particle distribution factors for organic and other pollutants in the Los Angeles atmosphere, *J. Air Poll. Contr. Assoc.,* **25,** 1038-1044, 1975.

1116. Heisler, S. L., and S. K. Friedlander, Gas-to-particle conversion in photochemical smog: aerosol growth laws and mechanisms for organics, *Atmos. Environ.,* **11,** 157-168, 1977.

1117. Went, F. W., On the nature of Aitken condensation nuclei, *Tellus,* **18,** 549-556, 1966.

1118. Gorse, R. A., and D. H. Volman, Photochemistry of the gaseous hydrogen peroxide-carbon monoxide system. II. Rate constants for hydroxyl radical reactions with hydrocarbons and for hydrogen atom reactions with hydrogen peroxide, *J. Photochem.,* **3,** 115-122, 1974.

1119. Burrows, J. P., G. W. Harris, and B. A. Thrush, Rates of reaction of HO_2 with HO and O studied by laser magnetic resonance, *Nature,* **267,** 233-234, 1977.

1120. Hack, W., K. Hoyermann, and H. G. Wagner, The reaction $NO + HO_2 \rightarrow NO_2 + OH$ with $OH + H_2O_2 \rightarrow HO_2 + H_2O$ as an HO_2 source, *Int. J. Chem. Kinetics Symp.,* **1,** 329-339, 1975.

1121. Greiner, N. R., Hydroxyl radical kinetics by kinetic spectroscopy. VI. Reactions with alkanes in the range 300-500°K, *J. Chem. Phys.,* **53,** 1070-1076, 1970.

1122. Perry, R.A., R. Atkinson, and J.N. Pitts, Jr., Kinetics and mechanism of the gas phase reaction of OH radicals with aromatic hydrocarbons over the temperature range 296-473 K, *J. Phys. Chem.,* **81** , 296-304, 1977.

1123. Atkinson, R., and J. N. Pitts, Jr., Temperature dependence of the absolute rate constants for the reaction of $O(^3P)$ atoms with a series of aromatic hydrocarbons over the range 299-392°K, *J. Phys. Chem.,* **79**, 295-297, 1975.

1124. Colussi, A. J., D. L. Singleton, R. S. Irwin, and R.J. Cvetanović, Absolute rates of oxygen (^3P) atom reactions with benzene and toluene, *J. Phys. Chem.,* **79**, 1900-1903, 1975.

1125. Graedel, T. E., The kinetic photochemistry of the marine atmosphere, submitted for publication, 1978.

1126. Crutzen, P. J., I. S. A. Isaksen, and J. R. McAfee, The impact of the chlorocarbon industry on the ozone layer, *J. Geophys. Res.,* **83** , 345-363, 1978.

1127. Cicerone, R. J., R. S. Stolarski, and S. Walters, Stratospheric ozone destruction by man-made chlorofluoromethanes, *Science,* **185**, 1165-1167, 1974.

1128. Wofsy, S. C., M. B. McElroy, and N. D. Sze, Freon consumption: implications for atmospheric ozone, *Science,* **187**, 535-537, 1975.

1129. Panel on Atmospheric Chemistry, *Halocarbons: Effects on Stratospheric Ozone,* National Academy of Sciences, Washington, D. C., 1976.

1130. Finlayson, B. J., and J. N. Pitts, Jr., Photochemistry of the polluted troposphere, *Science,* **192**, 111-119, 1976.

1131. Ingold, K. U., Rate constants for free radical reactions, in *Free Radicals, Vol. I,* ed. J. K. Kochi, New York: John Wiley, p. 37-112, 1973.

1132. Hendry, D. G., and R. A. Kenley, The role of peracetyl nitrate (PAN) in smog formation, Paper PHYS-194, 173rd Meeting, American Chemical Society, New Orleans, LA, 1977.

1133. Cox, R. A., R. G. Derwent, and P. M. Holt, Relative rate constants for the reactions of OH radicals with H_2, CH_4, CO, NO, and HONO at atmospheric pressure and 296K, *J. C. S.-Faraday I,* **72**, 2031-2043, 1976.

1134. Mill, T., and D. G. Hendry, Estimation procedures for persistence of organic compounds in the environment, paper presented at 172nd Meeting, American Chemical Society, San Francisco, CA, 1976.

1135. Jayanty, R.K.M., R. Simonaitis, and J. Heicklen, H_2 formation in the reaction of $O(^1D)$ with CH_4, *Int. J. Chem. Kinetics,* **8** , 107-110, 1976.

1136. Bhatia, K., Hydroxyl radical induced oxidation of nitrobenzene, *J. Phys. Chem.,* **79** , 1032-1038, 1975.

1137. Colussi, A.J., and R.J. Cvetanović Reaction of excited oxygen atoms, $O(^1D_2)$, with cyclobutane, *J. Phys. Chem.,* **79** , 1891-1899, 1975.

1138. Blumer, M., Polycyclic aromatic hydrocarbons in nature, *Scient. Amer.,* **234** (3), 35-45, 1976.

1139. Commins, B.T., Formation of polycyclic aromatic hydrocarbons during pyrolysis and combustion of hydrocarbons, *Atmos. Environ.,* **3** , 565-572, 1969.

1140. Dillemuth, F. J., D. R. Skidmore, and C. C. Shubert, The reaction of ozone with methane, *J. Phys. Chem.,* **64**, 1496-1499, 1960.

1141. Morrissey, R. J., and C. C. Shubert, The reactions of ozone with propane and ethane, *Combust. Flame,* **7**, 263-268, 1963.

1142. Shubert, C. C., and R. N. Pease, The oxidation of lower paraffin hydrocarbons. I. Room temperature reaction of methane, propane, *n*-butane, and isobutane with ozonized oxygen, *J. Amer. Chem. Soc.,* **78**, 2044-2048, 1956.

1143. Heicklen, J., The reaction of ozone with perfluoroolefins, *J. Phys. Chem.,* **70**, 477-480, 1966.

1144. Bufalini, J. J., and A. P. Altshuller, Kinetics of vapor-phase hydrocarbon-ozone reactions, *Can. J. Chem.,* **43**, 2243-2250, 1965.

1145. Wei, Y.K., and R.J. Cvetanović A study of the vapor phase reaction of ozone with olefins in the presence and absence of molecular oxygen, *Can. J. Chem.*, **41**, 913-925, 1963.

1146. Cadle, R. D., and C. Schadt, Kinetics of the gas phase reaction of olefins with ozone, *J. Amer. Chem. Soc.*, **74**, 6002-6004, 1952.

1147. Cleveland, W. S., T. E. Graedel, and B. Kleiner, Urban formaldehyde: observed correlation with source emission and photochemistry. *Atmos. Environ.*, **11**, 357-360, 1977.

1148. Niki, H., P. Maker, C. Savage, and L. Breitenbach, IR fourier-transform spectroscopic studies of atmospheric reactions, Paper presented at 12th Inf. Conf. Photochem., Gaithersberg, MD., June 28, 1976.

1149. Davidson, J. A., C. M. Sadowski, H. I. Schiff, G. E. Streit, C. J. Howard, D. A. Jennings, and A. L. Schmeltekopf, Absolute rate constant determinations for the deactivation of $O(^1D)$ by time resolved decay of $O(^1D) \rightarrow O(^3P)$ emission, *J. Chem. Phys.*, **64**, 57-62, 1976.

1150. Howard, C. J., and K. M. Evenson, Rate constants for the reactions of OH with ethane and some halogen substituted ethanes at 296K, *J. Chem. Phys.*, **64**, 4303-4306, 1976.

1151. Garvin, D., and Hampson, R. F., *Chemical kinetics data survey. VII. Tables of rate and photochemical data for modeling of the stratosphere (revised)*, National Bureau of Standards, Washington, D. C., NBSIR 74-430, 1974.

1152. Lloyd, A. C., Evaluated and estimated kinetic data for gas phase reactions of the hydroperoxyl radical, *Int. J. Chem. Kinetics*, **6**, 169-228, 1974.

1153. Michaud, P., G. Paraskevopoulos, and R. J. Cvetanović, Relative rates of the reactions of $O(^1D_2)$ atoms with alkanes and cycloalkanes, *J. Phys. Chem.*, **78**, 1457-1461, 1974.

1154. Schubert, C. C., and R. N. Pease, Reaction of paraffin hydrocarbons with ozonized oxygen: possible role of ozone in normal combustion, *J. Chem. Phys.*, **24**, 919-920, 1956.

1155. Bass, A.M., and A.H. Laufer, Extinction coefficients of azomethane and dimethyl mercury in the near ultra-violet, *J. Photochem.*, **6** , 465-470, 1974.

1156. Atkinson, R., and J. N. Pitts, Jr., Absolute rate constants for the reaction of $0(^3P)$ atoms with selected alkanes, alkenes, and aromatics as determined by a modulation technique, *J. Phys. Chem.*, **78**, 1780-1784, 1974.

1157. Havel, J. J., Atomic oxygen. I. The reactions of allenes with oxygen (^3P) atoms, *J. Amer. Chem. Soc.*, **96**, 530-533, 1974.

1158. Atkinson, R., R. A. Perry, and J. N. Pitts, Jr., Absolute rate constants for the reaction of OH radicals with allene, 1,3-butadiene, and 3-methyl-1-butene over the temperature range 299-424°K, *J. Chem. Phys.*, **67**, 3170-3174, 1977.

1159. Davis, D. D., S. Fischer, R. Schiff, R. T. Watson, and W. Bollinger, A kinetics study of the reaction of OH radicals with two C_2 hydrocarbons: C_2H_4 and C_2H_2, *J. Chem. Phys.*, **63**, 1707-1712, 1975.

1160. Atkinson, R., and J. N. Pitts, Jr., Absolute rate constants for the reaction of $0(^3P)$ atoms with a series of olefins over the temperature range 298-439°K, *J. Chem. Phys.*, **67**, 38-43, 1977.

1161. Ravishankara, A. R., S. Wagner, S. Fischer, G. Smith, R. Schiff, R. T. Watson, G. Tesi, and D. D. Davis, A kinetics study of the reaction of OH radicals with aromatics and olefins, Preprint, 1977.

1162. Atkinson, R., and J. N. Pitts, Jr., Absolute rate constants for the reaction of $0^3(P)$ atoms with allene, 1,3-butadiene, and vinyl methyl ether over the temperature range 297-439°K, *J. Chem. Phys.*, **67**, 2492-2495, 1977.

1163. Morris, E. D., Jr., and H. Niki, Reactivity of hydroxyl radicals with olefins, *J. Phys. Chem.*, **75**, 3640-3641, 1971.

1164. Atkinson, R., and J. N. Pitts, Jr., Rate constants for the reaction of OH radicals with propylene and the butenes over the temperature range 297-425°K, *J. Chem. Phys.*, **63**, 3591-3595, 1975.

1165. DeMore, W. B., Arrhenius constants for the reactions of ozone with ethylene and acetylene, *Int. J. Chem. Kinetics*, **1**, 209-220, 1969.

1166. Singleton, D.L., S. Furuyama, R.J. Cvetanović, and R. S. Irwin, Temperature dependence of the rate constants for the reactions $0(^3P)$ + 2,3-dimethyl-2-butene and $0(^3P)$ + NO + M determined by a phase shift technique, *J. Chem. Phys.*, **63**, 1003-1007, 1975.

1167. Sehested, K., H. Corfitzen, H. C. Christensen, and E. J. Hart, Rates of reaction of 0^-, OH, and H with methylated benzenes in aqueous solution. Optical spectra of radicals, *J. Phys. Chem.*, **79**, 310-315, 1975.

1168. Morris, E. D., Jr., D.H. Stedman, and H. Niki, Mass spectrometric study of the reactions of the hydroxyl radical with ethylene, propylene, and acetaldehyde in a discharge-flow system, *J. Am. Chem. Soc.*, **93**, 3570-3572, 1971.

1169. Christy, M. I., and M. A. Voisey, Reaction of acetyl radicals-2-Reaction of acetaldehyde with nitrogen dioxide in the presence of nitric oxide, *Trans. Faraday Soc.*, **63**, 2702-2707, 1967.

1170. Hudson, R. D., ed., *Chlorofluoromethanes and the Stratosphere*, NASA Ref. Pub. 1010, National Aeronautics and Space Administration, Greenbelt, MD, 1977.

1171. Braslavsky, S., and J. Heicklen, The gas-phase reaction of O_3 with H_2CO, *Int. J. Chem. Kinetics*, **8**, 801-808, 1976.

1172. Daby, E. E., D.H. Stedman, and H. Niki, Mass spectrometric studies of the reactions of formaldehyde and acetaldehyde with atomic oxygen in a discharge-flow system, Paper PHYS-122, 160th Meeting, American Chemical Society, Chicago, IL, 1970.

1173. Perry, R. A., R. Atkinson, and J. N. Pitts, Jr., Kinetics and mechanism of the gas phase reaction of OH radicals with methoxybenzene and *o* -cresol over the temperature range 299-435K, *J. Phys. Chem.*, **81**, 1607-1611, 1977.

1174. Takezaki, Y., S. Mori, and H. Kawasaki, The reaction of oxygen atoms with dimethyl ether, *Bull. Chem. Soc. Japan*, **39**, 1643-1650, 1966.

1175. Perry, R. A., R. Atkinson, and J. N. Pitts, Jr., Rate constants for the reaction of OH radicals with dimethyl ether and vinyl methyl ether over the temperature range 299-427°K, *J. Chem. Phys.*, **67**, 611-614, 1977.

1176. Molina, L.T., S.D. Schinke, and M.J. Molina, Ultraviolet absorption spectrum of hydrogen peroxide vapor, *Geophys. Res. Lett.*, **4**, 580-582, 1977.

1177. Salter, L. F., and B. A. Thrush, Reaction of oxygen atoms with methyl and ethyl nitrates, *JCS-Faraday Trans.-I*, **73**, 1098-1103, 1977.

1178. Slagle, I. R., R. E. Graham, and D. Gutman, Direct identification of reactive routes and measurement of rate constants in the reactions of oxygen atoms with methanethiol, ethanethiol, and methylsulfide, *Int. J. Chem. Kinetics*, **8**, 457-458, 1976.

1179. Slagle, I. R., J. R. Gilbert, and D. Gutman, Kinetics of the reaction between oxygen atoms and carbon disulfide, *J. Chem. Phys.*, **61**, 704-709, 1974.

1180. Atkinson, R., R.A. Perry, and J.N. Pitts, Jr., Rate constants for the reaction of OH radicals with COS, CS_2, and CH_3SCH_3 over the temperature range 299-430 K, *Chem. Phys. Lett.*, **54**, 14-18, 1978.

1181. Walker, J. C. G., Stability of atmospheric oxygen, *Amer. J. Sci.*, **274**, 193-214, 1974.

1182. Delwiche, C. C., The nitrogen cycle, *Scient. Amer.*, **223** (#3), 137-146, 1970.
1183. Robinson, E. and R. C. Robbins, Gaseous nitrogen compound pollutants from urban and natural sources, *J. Air Poll. Contr. Assoc.*, **20**, 303-306, 1970.
1184. Bolin, B., The carbon cycle, *Scient. Amer.*, **223** (#3), 125-132, 1970.
1185. Junge, C. E., M. Schidlowski, R. Eichmann, and H. Pietrek, Model calculations for the terrestrial carbon cycle: carbon isotope geochemistry and evolution of photosynthetic origin, *J. Geophys. Res.*, **80**, 4542-4552, 1975.
1186. Kellogg, W. W., R. D. Cadle, E. R. Allen, A. L. Lazrus, and E. A. Martell, The sulfur cycle, *Science*, **175**, 587-596, 1972.
1187. Friend, J. P., The global sulfur cycle, in *Chemistry of the Lower Atmosphere*, S. I. Rasool, ed., New York: Plenum Press, p. 177-201, 1973.
1188. Penner, J. E., M. B. McElroy, and S. C. Wofsy, Sources and sinks for atmospheric H_2: a current analysis with projections for the influence of antropogenic activity, *Planet. Space Sci.*, **25**, 521-540, 1977.
1189. Singh, H. B., F. L. Ludwig, and W. B. Johnson, *Ozone in Clean Remote Tropospheres: Concentration and Variabilities*, SRI Project 5661 Final Report, Stanford Research Institute, Menlo Park, CA, 1977.
1190. McElroy, M. B., J. W. Elkins, S. C. Wofsy, and Y. L. Yung, Sources and sinks for atmospheric N_2O, *Rev. Geophys. Space Phys.*, **14**, 143-150, 1976.
1191. Liu, S. C., R. J. Cicerone, T. M. Donahue, and W. L. Chameides, Sources and sinks of atmospheric N_2O and the possible ozone reduction due to industrial fixed nitrogen fertilizers, *Tellus*, **29**, 251-263, 1977.
1192. McConnell, J. C., Atmospheric ammonia, *J. Geophys. Res.*, **78**, 7812-7821, 1973.
1193. Bortner, M. H., R. H. Kummler, and L. S. Jaffe, Carbon monoxide in the Earth's atmosphere, *Water, Air, and Soil Poll.*, **3**, 17-52, 1974.
1194. Seiler, W., The cycle of carbon monoxide in the atmosphere, *Int. Conf. on Sensing and Assessment*, IEEE Cat #75-CH 1004-1, Inst. of Elec. Electronic Engrs., NeYrk, NY, p. 35-4-1 to 35-4-9, 1976.
1195. Gerakis, P. A., Man-made disruptions of the carbon dioxide and oxygen concentration in the atmosphere and the role of plants, *J. Environ. Qual.*, **3**, 299-304, 1974.
1196. Woodwell, G. M., The carbon dioxide question, *Scient. Amer.*, **238** (1), 34-43, 1978.
1197. Wofsy, S. C., Interactions of CH_4 and CO in the Earth's atmosphere, *Ann. Rev. Earth Planet. Sci.*, **4**, 441-469, 1976.
1198. Singh, H. B., D. P. Fowler, and T. O. Peyton, Atmospheric carbon tetrachloride: another man-made pollutant, *Science*, **192**, 1231-1233, 1976.
1199. Singh, H. B., Preliminary estimation of average tropospheric HO concentrations in the Northern and Southern Hemispheres, *Geophys. Res. Lett.*, **4**, 453-456, 1977.
1200. Lovelock, J. E., Methyl chloroform in the troposphere as an indicator of OH radical abundance, *Nature*, **267**, 32, 1977.
1201. Fletcher, I. S., and D. Husain, Absolute reaction rates of oxygen (2^1D_2) with halogenated paraffins by atomic absorption spectroscopy in the vacuum ultraviolet, *J. Phys. Chem.*, **80**, 1837-1840, 1976.
1202. Cox, R. A., R. G. Derwent, A. E. J. Eggleton, and J. E. Lovelock, Photochemical oxidation of halocarbons in the troposphere, *Atmos. Environ.*, **10**, 305-308, 1976.
1203. Manning, R. G., and M. J. Kurylo, Flash photolysis resonance fluorescence investigation of the temperature dependencies of the reactions of $Cl(^2P)$ atoms with CH_4, CH_3Cl, CH_3F^t, and C_2H_6, *J. Phys. Chem.*, **81**, 291-296, 1977.
1204. Dillemuth, F. J., B. D. Lalancette, and D. R. Skidmore, Reaction of ozone with 1,1-difluoroethane and 1,1,1-trifluoroethane, *J. Phys. Chem.*, **80**, 571-575, 1976.

1205. Atkinson, R., and J. N. Pitts, Jr., Rate constants for the reaction of $O(^3P)$ atoms with $CH_2=CHF$, $CH_2=CHCl$, and $CH_2=CHBr$ over the temperature range 298-422°K, *J. Chem. Phys.*, **67**, 2488-2491, 1977.

1206. Jones, D. S., and S. J. Moss, Arrhenius parameters for reactions of oxygen atoms with the fluorinated ethylenes, *Int. J. Chem. Kinetics*, **6**, 443-452, 1974.

1207. Howard, C.J., Rate constants for the gas-phase reactions of OH radicals with ethylene and halogenated ethylene compounds, *J. Chem. Phys.*, **65** , 4771-4777, 1976.

1208. Cloud, P., and A. Gibor, The oxygen cycle, *Scient. Amer.*, **223** (#3), 111-123, 1970.

1209. Jesson, J. P., P. Meakin, and L. C. Glasgow, The fluorocarbon-ozone theory - II. Tropospheric lifetimes - an estimate of the tropospheric lifetime of CCl_3F, *Atmos. Environ.*, **11**, 499-508, 1977.

1210. Committee on Medical and Biologic Effects of Environment Pollutants, *Selenium,* Natural Academy of Sciences, Washington, D. C., p. 41-50, 1976.

1211. Atkinson, R., D. A. Hansen, and J. N. Pitts, Jr., Rate constants for the reaction of the OH radical with H_2 and NO(M = Ar and N_2), *J. Chem. Phys.*, **62**, 3284-3288, 1975.

1212. Hampson, R. F. (ed.) *Chemical kinetics data survey. VI. Photochemical and rate data for twelve gas phase reactions of interest for atmospheric chemistry,* NBSIR 73-207, National Bureau of Standards, Washington, D. C., 1973.

1213. Perry, R. A., R. Atkinson, and J. N. Pitts, Jr., Rate constants for the reactions $OH + H_2S \rightarrow H_2O + SH$ and $OH + NH_3 \rightarrow H_2O + NH_2$ over the temperature range 297-427°K, *J. Chem. Phys.*, **64**, 3237-3239, 1976.

1214. Whytock, D. A., R. B. Timmons, J. H. Lee, J. V. Michael, W. A. Payne, and L. J. Stief, Absolute rate of the reaction of $O(^3P)$ with hydrogen sulfide over the temperature range 263 to 495 K, *J. Chem. Phys.*, **65**, 2052-2055, 1976.

1215. Castleman, A. W., Jr., and I. N. Tang, Kinetics of the association reaction of SO_2 with the hydoxyl radical, *J. Photochem.*, **6**, 349-354, 1966/77.

1216. Westenberg, A. A., and N. de Haas, Rate of the reaction $O + SO_2 + M \rightarrow SO_3 + M$, *J. Chem. Phys.*, **63**, 5411-5415, 1975.

1217. Bell, J., A. W. Castleman, Jr., R. Davis, and I. N. Tang, Association reactions involved in H_2SO_4 aerosol formation, Paper presented at 68th Annual Meeting, Air Pollution Contr. Assoc., Boston, MA, June 20, 1975.

1218. Overend, R., and G. Paraskevopoulos, The question of a pressure effect in the reaction OH + CO at room temperature, *Chem. Phys. Lett.*, **49**, 109-111, 1977.

1219. Countess, R. J., and J. Heicklen, Kinetics of particle growth. II. Kinetics of the reaction of ammonia with hydrogen chloride and the growth of particulate ammonia chloride, *J. Phys. Chem.*, **77**, 444-447, 1973.

1220. Hellman, T. M., and F. H. Small, Characterization of the odor properties of 101 photochemicals using sensory methods, *J. Air Poll. Contr. Assoc.*, **24**, 979-982, 1974.

1221. Clyne, M. A. A., P. B. Monkhouse, and L. W. Townsend, Reactions of O^3P_J atoms with halogens: the rate constants for the elementary reactions O + BrCl, O + Br_2, and O + Cl_2, *Int. J. Chem. Kinetics*, **8**, 425-449, 1976.

1122. Engleman, V. S., *Survey and Evaluation of Kinetic Data on Reactions in Methane/Air Combustion,* EPA-600/2-76-003, Environmental Protection Agency, Research Triangle Park, NC, 1976.

1223. Howard, C. J., and K. M. Evenson, Kinetics of the reaction of HO_2 with NO, *Geophys. Res. Lett.*, **4**, 437-440, 1977.

1224. Anastasi, C., and I. W. M. Smith, Rate measurements of reactions of OH by resonance absorption. Part 5. Rate constants for $OH + NO_2(+M) \rightarrow HNO_3(+M)$ over a wide range of temperature and pressure, *JCS Faraday Trans. II*, **72**, 1459-1468, 1976.

1225. Howard, C. J., New laboratory rate measurments, *Upper Atmos. Prog. Bull.*, 77-3, Federal Aviation Admin., Washington, D. C., 1977.
1226. Baulch, D. L., D. D. Drysdale, D. G. Horne, and A. C. Lloyd, *Evaluated Kinetic Data for High Temperature Reactions, Vol. I. Homogeneous Gas Phase Reactions of the* H_2O_2 *System,* London: Butterworths, 1972.
1227. Chang, J. S., and F. Kaufman, Upper limits of the rate constants for the reactions of $CFCl_3(F-11)$, $CF_2Cl_2(F-12)$, and N_2O with OH. Estimates of corresponding lower limits to their tropospheric lifetimes, *Geophys. Res. Lett.*, **4**, 192-194, 1977.
1228. Herron, J. T., and R. E. Huie, Rate constants for the reactions of atomic oxygen ($O(^3P)$) with organic compounds in the gas phase, *J. Phys. Chem. Ref. Data*, **2**, 467-518, 1973.
1229. Pitts, J. N., Jr., Photochemical air pollution: singlet molecular oxygen as an environmental oxidant, *Adv. Environ. Sci. Technol.*, **1**, 289-337, 1969.
1230. Bogan, D. J., *Kinetics of the reaction of ethylene oxide with* $O(^3P)$ *atoms.*, Ph.D. Dissertation, Carnegie-Mellon University, 1973.
1231. Calvert, J. G., and J. N. Pitts, Jr., *Photochemistry*, New York: John Wiley, 1966.
1232. Houston, P. L., and C. B. Moore, Formaldehyde photochemistry: appearance rate, vibrational relaxation, and energy distribution of the CO product, *J. Chem. Phys.*, **65**, 757-770, 1976.
1233. Altshuller, A. P., I. R. Cohen, and T. C. Purcell, Photooxidation of propional-dehyde at low partial pressure of aldehyde, *Can. J. Chem.*, **44**, 2973-2979, 1966.
1234. Glasson, W. A., and C. S. Tuesday, The atmospheric thermal oxidation of nitric oxide in the presence of dienes, *Environ. Sci. Technol.*, **4**, 752-757, 1970.
1235. Watson, R. T., Rate constants for reactions of ClO_x of atmospheric interest, *J. Phys. Chem. Ref. Data*, **6**, 871-917, 1977.
1236. Pruss, F. J., Jr., I. R. Slagle, and D. Gutman, Determination of branching ratios for the reaction of oxygen atoms with ethylene, *J. Phys. Chem.*, **78**, 663-665, 1974.
1237. Graedel, T. E., L. A. Farrow, and T. A. Weber, The influence of aerosols on the chemistry of the troposphere, *Int. J. Chem. Kinetics Symp.*, **1**, 581-594, 1975.
1238. Orel, A. E., and J. H. Seinfeld, Nitrate formation in atmospheric aerosols, *Environ. Sci. Technol.*, **11**, 1000-1007, 1977.
1239. Kenley, R.A., J.E. Davenport, and D.G. Hendry, Hydroxyl radical reactions in the gas phase. Products and pathways for the reaction of OH with toluene, *J. Phys. Chem.*, **82** ,1095-1096, 1978.
1240. Bornstein, R. D., and D. S. Johnson, Urban-rural wind velocity differences, *Atmos. Environ.*, **11**, 597-604, 1977.
1241. Maroulis, P. J., A. L. Torres, and A. R. Bandy, Atmospheric concentrations of carbonyl sulfide in the Southwestern and Eastern United States, *Geophys. Res. Lett.*, **4**, 510-512, 1977.
1242. Umstead, M. E., and M. C. Lin, The dynamics of CO production from the reaction of $O(^3P)$ with 1- and 2-butyne, *Chem. Phys.*, **25**, 353-359, 1977.
1243. Glotfelty, D. E., B. C. Turner, and A. W. Taylor, Transport and reactions of pesticides in the atmosphere, Paper PEST-3, 172nd Meeting, American Chemical Society, San Francisco, CA, 1976.
1244. Zepp, R. G., N. L. Wolfe, L. V. Azarraga, and R. H. Cox, Photochemical transformation of the 1,1-diaryl-2,2-dichloroethylenes, DDE and DMDE, by sunlight, Paper ENVT-69, 172nd Meeting, American Chemical Society, San Francisco, CA, 1976.
1245. Sanhueza, E., I. C. Hisatsune, and J. Heicklen, Oxidation of haloethylenes, *Chem. Rev.*, **76**, 801-826, 1976.

1246. Mackay, D., and P. J. Leinonen, Rate of evaporation of low-solubility contaminants from water bodies to atmosphere, *Environ. Sci. Technol.,* **9,** 1178-1180, 1975.

1247. Lee, J. H., J. V. Michael, W. A. Payne, L. J. Stief, and D. A. Whytock, Absolute rate of the reaction of $Cl(^2P)$ with molecular hydrogen from 200-500 K, *JCS-Faraday. I,* **73,** 1530-1536, 1977.

1248. Sprung, J. L., Tropospheric oxidation of H_2S, *Adv. Environ. Sci. Technol.,* **7,** 263-278, 1977.

1249. Thiemens, M. W. and S. E. Schwartz, The fate of HS radical under atmospheric conditions, Paper presented at 13th Inf. Conf. Photochem., Clearwater, Fla., 1978.

1250. Benson, S. W., Thermochemistry and kinetics of sulfur-containing molecules and radicals, *Chem. Rev.,* **78** , 23-35, 1978.

1251. Singleton, D. L., R. S. Irwin, and R. J. Cvetanović, Arrhenius parameters for the reaction of $O(^3P)$ atoms with several aldehydes and the trend in aldehydic $C-H$ bond dissociation energies, *Can. J. Chem.,* **55,** 3321-3327, 1977.

1252. Sloane, T. M., Reaction product identification from $O(^3P)$ + benzene, toluene, and 1,3,5-trimethylbenzene collisions in crossed molecular beams, *J. Chem. Phys.,* **67,** 2267-2274, 1977.

1253. Kaiser, E. W., and S. M. Jagar, The kinetics of gas phase reactions of nitrous acid with ozone, *Upper Atmos. Prog. Bull.,* **77-5,** National Aeronautics and Space Administration-Federal Aviation Admin., Washington, D. C., 1977.

1254. Kaiser, E. W., and S. M. Japar, The kinetics of the gas phase reaction of $O(^3P)$ with N_2O_5, *Upper Atmos. Prog. Bull.,* **77-5,** National Aeronautics and Space Administration - Federal Aviation Admin., Washington, D. C., 1977.

1255. Pierrard, J. M., Photochemical decomposition of lead halides from automobile exhuast, *Environ. Sci. Technol.,* **3,** 48-51, 1969.

1256. Robbins, J. A., and F. L. Snitz, Bromine and chlorine loss from lead halide automobile exhaust particulates, *Environ. Sci. Technol.,* **6,** 164-169, 1972.

1257. Stahl, W. H., ed., *Compilation of Odor and Taste Threshhold Values Data,* ASTM Pub. D5-48, American Soc. for Testing and Materials, Philadelphia, PA, 1973.

1258. Atkinson, R., R.A. Perry, and J.N. Pitts, Jr., Rate constants for the reactions of the OH radical with $(CH_3)_2NH$, $(CH_3)_3N$, and $C_2H_5NH_2$ over the temperature range $298-426°K$, *J. Chem. Phys.,* **68** , 1850-1853, 1978.

1259. Schulten, H.-R., and U. Schurath, Analysis of aerosols from the ozonolysis of 1-butene by high-resolution field desorption mass spectrometry, *J. Phys. Chem.,* **79,** 51-57, 1975.

1260. Falk, H. L., I. Markul, and P. Kotin, Aromatic hydrocarbons. IV. Their fate following emission into the atmosphere and experimental exposure to washed air and synthetic smog, *Arch. Ind. Health,* **13,** 13-17, 1956.

1261. Barofsky, D. F., and E. J. Baum, Field desorption mass analysis of the photooxidation products of adsorbed polycyclic aromatic hydrocarbons, Paper 76-20.5, 69th Annual Meeting, Air Poll. Contr. Assoc., (Portland, OR), 1976.

1262. Herron, J. T., and R. E. Huie, Stopped-flow studies of the mechanisms of ozone-alkene reactions in the gas phase. Ethylene, *J. Amer. Chem. Soc.,* **99,** 5430-5435, 1977.

1263. Martinez, R. I., R. E. Huie, and J. T. Herron, Mass spectrometric detection of dioxirane, H_2COO, and its decomposition products, H_2 and CO, from the reaction of ozone with ethylene, *Chem. Phys. Lett.,* **51,** 457-459, 1977.

1264. Kurylo, M. J., Rate constants for the reaction of OH with CS_2 and COS, Paper presented at 13th Inf. Conf. Photochem., Clearwater, FL, 1978.

1265. Scott, J. D., G. C. Causley, and B. R. Russell, Vacuum ultraviolet absorption spectra of dimethyl sulfide, dimethyl selenide, and dimethyl telluride, *J. Chem. Phys.,* **59,** 6577-6586, 1973.

1266. Atkinson, R., R.A. Perry, and J.N. Pitts, Jr., Rate constants for the reaction of OH radicals with COS, CS_2 and CH_3SCH_3 over the temperature range 299-430 K, *Chem. Phys. Lett.*, **54** , 14-18, 1978.

1267. Pratt, P.F., et al., Effect of increased nitrogen fixation on stratospheric ozone, *Climatic Change,* **1** , 109-135, 1977.

1268. Galloway, J.N., G.E. Likens, and E.S. Edgerton, Hydrogen ion speciation in the acid precipitation of the northeastern United States, *Water, Air, Soil Poll.,* **6** , 423-433, 1976.

1269. Benson, S.W., *The Foundations of Chemical Kinetics,* New York: McGraw-Hill, 1960.

1270. Benson, S.W., *Thermochemical Kinetics,* New York: John Wiley, 1968.

1271. Stern, A.C., ed. *Air Pollution,* Third Ed.(five vols.), New York: Academic Press, 1976-77.

1272. Likens, G.E., Acid precipitation, *Chem. Eng. News,* **54** (48), 29-44, 1976.

1273. Appel, B.R., S.M. Wall, and R.L. Knights, *Characterization of Carbonaceous Materials in Atmospheric Aerosols by High Resolution Mass Spectrometric Thermal Analysis,* Report 182, Air and Industrial Hygiene Lab., State of California Dept. of Health, Berkeley, CA, 1975.

APPENDIX A
SOURCES OF ATMOSPHERIC COMPOUNDS

This appendix is derived from the tables in Chapters 2-9. It presents the source information contained therein in a format cross-indexed by source name instead of by compound. Each source name in this appendix is followed by one or more species numbers. For "air conditioning", for example, three species numbers are listed. Reference to Table 8.1 indicates that the compounds emitted into the atmosphere as a result of air conditioner operation and maintenance are fluorodichloromethane (8.1-10), fluorotrichloromethane (8.1-11), and dichlorodifluoromethane (8.1-12). Similar analyses may be performed for sources of interest to a particular user.

As with other data in this book, that in this appendix should not be regarded as complete or perfectly accurate. It represents, rather, the information available when this compilation was performed. Sources that have been studied intensively for one reason or another will thus be more completely described than those which have received less attention. Nonetheless, considerable chemical emissions information is available from this listing. Some of the implications of these data from a broader perspective are discussed in Section 10.2.

acetyl. mfr.
    5.5-37

air cond.
    8.1-10   8.1-12   8.1-13

algae
    7.2-1

aluminum mfr.

2.4-1	2.4-3	2.4-11	2.4-16	2.4-17	2.4-18	2.4-20	2.4-21
2.5-1	2.5-4	2.5-9	2.5-11	2.5-13	2.5-24	3.7-18	3.7-41
3.7-44	3.7-47	3.7-64	3.7-68	3.7-74	3.7-86	3.7-89	3.7-96
3.7-101	3.7-111	6.4-46	8.1-1	8.1-17			

amine mfr.
    6.2-6

ammonia mfr.
    2.2-2

animal secretion
  4.7-21   5.11-15

animal waste

2.2-2	2.3-1	4.1-1	4.1-2	4.1-5	4.1-7	4.1-8	4.1-11
4.1-22	4.1-26	4.1-28	4.5-1	4.5-8	4.5-9	5.1-2	5.1-4
5.1-6	5.1-10	5.1-12	5.1-17	5.5-1	5.5-2	5.5-8	5.5-9
5.5-10	5.5-12	5.5-13	5.5-14	5.5-53	5.9-1	5.9-2	5.9-3
5.9-6	5.9-8	5.9-9	5.9-12	6.2-2	6.2-3	6.2-7	6.2-8
6.2-13	6.2-15	6.2-16	6.2-18	6.2-23	6.2-24	6.4-35	6.4-36
7.1-1	7.1-2	7.1-3	7.2-1	7.2-3	7.2-7	7.2-16	7.2-17

auto body comb.
  5.1-2

auto

2.1-7	2.2-2	2.2-5	2.2-6	2.3-1	2.3-7	2.3-8	2.3-11
2.4-4	2.4-5	2.4-13	2.4-14	2.4-23	2.5-7	2.5-8	3.1-1
3.1-2	3.1-3	3.1-4	3.1-5	3.1-6	3.1-7	3.1-8	3.1-10
3.1-11	3.1-13	3.1-14	3.1-15	3.1-16	3.1-17	3.1-18	3.1-19
3.1-20	3.1-21	3.1-22	3.1-29	3.1-30	3.1-31	3.1-32	3.1-33
3.1-34	3.1-38	3.1-40	3.1-44	3.1-45	3.1-46	3.1-47	3.1-49
3.1-57	3.1-58	3.1-65	3.1-69	3.1-86	3.1-88	3.1-90	3.1-92
3.1-93	3.1-94	3.1-96	3.1-97	3.1-98	3.1-99	3.1-101	3.1-102
3.1-104	3.1-106	3.1-108	3.1-110	3.1-112	3.1-115	3.2-1	3.2-2
3.2-3	3.2-4	3.2-5	3.2-7	3.2-8	3.2-9	3.2-10	3.2-12
3.2-13	3.2-16	3.2-18	3.2-19	3.2-21	3.2-22	3.2-23	3.2-24
3.2-26	3.2-27	3.2-28	3.2-31	3.2-33	3.2-34	3.2-38	3.2-39
3.2-40	3.2-41	3.2-42	3.2-44	3.2-46	3.2-47	3.2-49	3.2-50
3.2-53	3.2-57	3.2-58	3.2-60	3.2-61	3.2-63	3.2-65	3.2-68
3.2-71	3.2-72	3.2-73	3.2-74	3.2-75	3.2-80	3.2-81	3.2-90
3.4-7	3.4-8	3.4-14	3.4-19	3.4-20	3.4-22	3.4-23	3.4-27
3.4-31	3.4-36	3.4-37	3.4-38	3.5-1	3.5-2	3.5-3	3.5-4
3.5-5	3.5-6	3.5-7	3.5-8	3.5-11	3.5-13	3.5-14	3.5-15
3.5-16	3.5-17	3.5-19	3.5-30	3.5-31	3.5-33	3.5-34	3.5-38
3.5-40	3.5-67	3.6-1	3.6-2	3.6-16	3.6-25	3.6-27	3.7-1
3.7-9	3.7-13	3.7-18	3.7-20	3.7-21	3.7-37	3.7-38	3.7-40
3.7-41	3.7-47	3.7-50	3.7-62	3.7-64	3.7-74	3.7-76	3.7-79
3.7-80	3.7-83	3.7-86	3.7-94	3.7-96	3.7-97	3.7-101	3.7-102
3.7-103	3.7-104	3.7-106	3.7-109	3.7-111	3.7-122	3.7-126	3.7-130
3.7-130	3.7-132	3.7-133	3.7-136	4.1-1	4.1-2	4.1-5	4.1-7
4.1-8	4.1-10	4.1-11	4.1-12	4.1-13	4.1-28	4.2-1	4.2-2
4.2-3	4.2-4	4.4-1	4.4-2	4.4-3	4.4-4	4.4-5	4.4-6
4.4-15	4.4-21	4.4-22	4.5-1	4.5-4	4.5-6	4.5-7	4.5-10
4.5-11	4.5-18	4.6-1	4.6-2	4.6-3	4.8-7	4.8-27	4.8-29
4.8-30	4.8-34	4.8-38	4.8-57	4.8-64	4.8-67	4.8-68	5.4-1
5.4-9	5.5-1	5.8-15	5.9-1	5.9-2	5.9-6	5.9-13	5.9-43
5.10-5	5.12-1	5.12-2	5.12-3	5.12-4	5.12-5	5.12-6	5.12-7
5.12-8	5.12-9	5.12-10	5.12-11	5.12-13	5.12-14	5.12-15	5.12-16
5.12-19	5.12-30	5.12-37	5.12-38	5.12-39	5.16-1	5.16-2	5.16-8
5.16-15	5.16-19	5.16-24	5.17-1	5.17-29	6.1-1	6.1-3	6.1-6
6.3-1	6.3-3	6.4-56	7.2-17	9.1-14			

bacteria
9.1-12   9.1-13

battery mfr.
2.3-8

brewing
3.1-57   3.1-65   3.1-69   3.1-74   3.1-77   3.1-79   3.1-81   3.1-84
3.2-75   3.2-82   3.2-86   3.2-88   3.2-90   3.2-92   3.2-93   3.2-95
3.6-1   3.6-2   4.3-1   4.3-2   5.12-1   5.12-4   5.12-5   5.12-10
5.12-33

brick mfr.
2.3-8   2.4-3

building mtls.
3.1-4   3.1-10   3.1-11   3.1-29   3.1-44   3.1-57

building resin
4.1-1   4.7-3   4.8-24   5.10-1   5.14-6   8.1-56   3.5-21

calc. carbide mfr.
3.2-1

casting resin
4.1-1   4.1-2

cement mfr.
2.3-8   2.3-19   2.5-9   2.5-10   2.5-11   2.5-13   2.5-15   2.5-16
2.5-24

ceramics mfr.
2.4-3

charcoal mfr.
4.1-1   4.1-2   5.9-1

chemical mfr.
2.4-7   3.1-29   3.5-1   3.5-2   3.5-13   3.5-21   4.5-1   4.5-18
4.5-19   5.5-13   8.1-2   8.1-3   8.1-4   8.1-5   8.1-18   8.1-20
8.1-27   8.1-28   8.1-29   8.1-30   8.1-51   8.2-1

chlorine mfr.
2.4-1   2.5-2

coal comb
6.4-53   2.5-2   3.5-57   3.7-1   3.7-2   3.7-3   3.7-4   3.7-5
3.7-10   3.7-11   3.7-13   3.7-14   3.7-18   3.7-19   3.7-21   3.7-23
3.7-29   3.7-32   3.7-39   3.7-41   3.7-42   3.7-43   3.7-44   3.7-46
3.7-47   3.7-49   3.7-62   3.7-63   3.7-64   3.7-65   3.7-66   3.7-69
3.7-76   3.7-79   3.7-86   3.7-87   3.7-89   3.7-93   3.7-94   3.7-95
3.7-96   3.7-97   3.7-98   3.7-101   3.7-106   3.7-114   3.7-130   3.7-131

drug mfr.
   6.4-36

dry cleaning
   8.1-24   8.1-29   8.1-30

explosives mfr.
   2.2-10   2.3-11   6.2-30

fat subst. plant
   5.9-1   5.9-2   5.9-3   5.9-8   5.9-15   5.9-23

fertilizer mfr.
   2.2-2   2.2-8   2.4-3   2.4-7   2.4-9   2.4-20   2.5-10   2.5-11
   2.5-13   2.5-13   2.5-16   2.5-24   6.2-18   8.2-9

fiberglass mfr.
   2.5-46

fish meal mfr.
   2.2-2   2.3-1   4.1-2   4.1-5   4.1-7   4.1-8

fish oil mfr.
   3.5-1   3.5-2   3.5-3   3.5-4   3.5-5   4.2-1   4.5-1   5.16-15

fish processing
   6.2-2   6.2-3   6.2-7   6.2-10   6.2-15   6.2-18   7.1-3   7.2-16
   7.2-17

fluorescent lamps
   8.3-17

foaming agent
   2.5-8   8.1-3   8.1-12   8.1-13   8.1-23   8.1-25

food processing
   6.2-19   6.2-25   6.4-35   6.4-36   7.4-1   7.4-4

forest fire
   2.1-7   2.4-4   3.1-1   3.1-2   3.1-3   3.1-4   3.1-5   3.1-6
   3.1-7   3.1-10   3.1-11   3.1-13   3.1-14   3.1-29   3.2-1   3.2-2
   3.2-3   3.2-5   3.2-7   3.2-8   3.2-16   3.2-18   3.2-21   3.2-22
   3.2-23   3.2-39   3.4-14   3.5-1   3.5-2   3.5-3   4.1-1   4.1-2
   4.1-5   4.1-7   4.1-8   4.2-1   4.3-1   4.3-2   4.5-1   4.5-4
   4.5-9   4.5-10   4.5-16   5.1-1   5.1-2   5.5-1   5.5-8   5.5-28
   5.9-1   5.9-2   5.12-33   5.12-36   5.16-15   5.16-16   5.16-19   7.2-17
   8.1-2   8.1-18

foundry
   2.2-2   2.3-6   2.5-1   2.5-10   2.5-11   2.5-13   2.5-13   2.5-16
   2.5-18   2.5-19   2.5-20   2.5-21   2.5-23   2.5-24   2.5-26   2.5-27
   2.5-32   3.1-1   3.2-1   3.2-2   5.12-1   6.1-1

fruit ripening
    3.2-2

fuel oil comb.
    2.3-8

fumigating agent
    8.1-5   8.1-6

furnace soot
    2.3-11

gasoline vapor
    3.1-2   3.1-3   3.1-4   3.1-5   3.1-6   3.1-7   3.1-10   3.1-11
    3.1-13   3.1-14   3.1-16   3.1-17   3.1-20   3.1-21   3.1-22   3.1-29
    3.1-31   3.1-32   3.1-38   3.1-40   3.1-44   3.1-45   3.1-46   3.1-49
    3.1-57   3.1-58   3.1-65   3.2-7   3.2-8   3.2-16   3.2-19   3.2-21
    3.2-22   3.2-23   3.2-26   3.2-38   3.2-39   3.2-40   3.4-7   3.4-8
    3.4-15   3.4-20   3.5-1   3.5-2   3.5-3   3.5-4   3.5-5   3.5-6
    3.5-7   3.5-8   3.5-13   3.5-14   3.5-15   3.5-16   3.5-17   3.5-30
    3.5-33   9.1-17   9.1-18   9.1-19   9.1-20   9.1-21

geothermal steam
    2.1-7   2.2-1   2.2-2   2.3-1   2.5-2   2.5-8   2.5-46   3.1-1
    3.1-2

glass fibre mfr.
    5.12-1   5.12-3   5.12-4

glue vapor
    2.2-2   3.5-21   5.5-9   8.1-45

gold mfr.
    2.5-30

H_2SO_4 mfr.
    2.3-7   2.3-8   2.3-11

HCHO mfr.
    4.1-1   4.1-2

HCl mfr.
    2.4-4

HF mfr.
    2.4-3

HNO_3 mfr.
    2.2-10

industrial mfr.
  2.4-28

industrial
2.3-4	2.3-23	2.4-24	2.5-39	2.5-44	4.1-8	4.5-4	4.5-34
4.8-25	4.8-33	4.8-35	4.8-36	5.1-6	5.1-21	5.2-1	5.4-8
5.5-28	5.6-3	5.9-4	5.9-7	5.9-10	5.9-12	5.9-13	5.9-26
5.9-27	5.9-39	5.10-1	5.11-3	5.12-34	5.12-51	5.14-3	5.14-5
5.14-6	5.14-8	5.16-1	6.1-2	6.1-6	6.1-7	6.1-14	6.1-15
6.2-4	6.2-9	6.2-11	6.2-14	6.2-29	6.2-31	6.2-32	6.3-28
6.3-31	8.1-22	8.1-56	8.2-4	8.2-6	8.2-12	8.2-14	

insects
3.1-1	3.1-2	3.1-3	4.1-2	4.1-5	4.1-8	4.5-1	4.5-4
5.9-1	5.9-2	5.9-8					

iodine mfr.
  2.4-2

ion xch. resin mfr.
  8.1-53   8.1-54

iron pellet mfr.
  2.3-5

lacquer mfr.
2.2-2	2.4-3	2.4-4	3.5-1	3.5-5	4.1-1	4.1-7	4.2-1
4.8-24	5.1-1	5.1-2	5.1-4	5.2-2	5.5-8	5.12-1	

landfill
3.1-57	3.5-1	3.5-2	3.6-2	3.6-3	4.5-11	5.15-1	8.1-3
8.1-4	8.1-20	8.1-27	8.1-29	8.1-30	8.2-1		

lead mfr.
2.4-16	2.5-11	2.5-29	2.5-30	2.5-31	2.5-33	2.5-34	2.5-36
2.5-37	2.5-38	2.5-40	2.5-41				

lead smelting
  2.3-4

lightning
  2.1-9

lime mfr.
2.3-18	2.5-10	2.5-11	2.5-16	2.5-24	2.5-41	2.5-42	2.5-43

litho. coater
4.1-1	4.1-2	4.1-5	4.1-7	4.2-1

magnesium mfr.
  2.5-10

paint
    2.5-2

pesticide

6.2-39	6.2-40	6.2-43	8.3-1	8.3-2	8.3-3	8.3-4	8.3-5
8.3-6	8.3-7	8.3-8	8.3-9	8.3-10	8.3-11	8.3-12	8.3-13
8.3-17	8.3-19	8.3-20	8.3-21	8.3-22	8.3-23	8.3-24	8.3-25
8.3-26	8.3-27	8.3-28	8.3-29	8.3-30	8.3-31	9.2-1	9.2-2
9.2-3	9.2-5	9.2-6	9.2-7	9.2-9	9.2-10	9.2-12	

petrochem. mfr.
    2.5-45   8.1-28

petrol. vapor
    2.2-4

petroleum mfr.

2.2-2	2.3-1	2.3-7	3.1-1	3.1-2	3.1-3	3.1-4	3.1-5
3.1-6	3.1-10	3.1-11	3.1-13	3.1-14	3.1-29	3.1-44	3.1-65
3.2-1	3.2-2	3.2-3	3.2-5	3.2-7	3.2-14	3.2-15	3.2-16
3.2-17	3.2-18	3.4-7	3.4-8	3.4-19	3.4-34	3.4-35	3.5-1
3.5-2	3.7-18	3.7-41	3.7-64	3.7-86	3.7-96	3.7-101	3.7-106
3.7-130	3.7-130	3.7-136	4.1-1	4.1-2	4.1-5	4.1-7	5.5-8
5.5-9	5.9-9	5.16-2	7.1-1	7.1-2	7.2-1	7.2-16	7.2-17
7.3-2							

petroleum stor.

4.5-1	5.5-9	5.9-1	5.9-2	5.9-6	5.9-9	5.9-10	6.1-1
8.1-5							

phthal anhyd. mfr.

3.5-3	3.5-59	3.6-1	4.7-30	4.8-24	4.8-60	5.4-1	5.17-30

phthalic acid mfr.
    4.1-1   4.1-2   4.5-1

plant volatile

2.2-7	3.1-11	3.1-13	3.1-14	3.1-29	3.3-4	3.3-10	3.3-11
3.3-13	3.3-14	3.3-15	3.3-16	3.3-17	3.3-18	3.3-19	3.4-8
3.4-19	3.4-44	3.5-1	3.5-2	3.5-3	3.5-6	3.5-7	3.5-8
3.5-14	3.5-15	3.5-16	3.5-21	3.5-30	3.5-33	3.5-35	3.5-37
3.5-39	4.1-1	4.1-2	4.1-5	4.1-7	4.1-8	4.1-11	4.1-12
4.1-17	4.1-22	4.1-26	4.1-27	4.1-28	4.1-29	4.1-30	4.1-31
4.2-9	4.2-10	4.2-12	4.2-13	4.2-14	4.2-15	4.3-1	4.3-5
4.4-1	4.4-9	4.4-10	4.4-13	4.4-14	4.4-16	4.4-18	4.4-19
4.4-23	4.4-24	4.4-25	4.4-26	4.4-29	4.5-1	4.5-9	4.5-11
4.5-23	4.5-24	4.5-28	4.5-29	4.5-30	4.5-32	4.6-5	4.6-6
4.7-2	4.7-5	4.7-6	4.7-7	4.7-9	4.7-10	4.7-11	4.7-13
4.7-15	4.7-17	4.7-18	4.7-19	4.7-22	4.7-24	4.7-27	4.7-28
4.7-29	4.7-32	4.7-33	4.7-34	4.7-35	4.8-7	4.8-8	4.8-12
4.8-18	4.8-20	4.8-22	4.8-23	4.8-76	4.8-77	4.8-79	5.1-1
5.1-2	5.1-4	5.1-6	5.1-7	5.1-17	5.1-19	5.1-20	5.1-24

5.1-25	5.1-27	5.1-29	5.1-31	5.1-32	5.1-34	5.1-36	5.1-37
5.1-38	5.1-39	5.1-41	5.1-43	5.4-1	5.4-9	5.4-12	5.5-1
5.5-3	5.5-4	5.5-5	5.5-6	5.5-7	5.5-9	5.5-10	5.5-11
5.5-12	5.5-13	5.5-16	5.5-17	5.5-18	5.5-19	5.5-21	5.5-22
5.5-23	5.5-24	5.5-25	5.5-26	5.5-27	5.5-30	5.5-31	5.5-32
5.5-33	5.5-34	5.5-35	5.5-36	5.5-38	5.5-39	5.5-40	5.5-41
5.5-42	5.5-44	5.5-45	5.5-46	5.5-47	5.5-48	5.5-49	5.5-50
5.5-51	5.5-52	5.5-54	5.5-56	5.5-57	5.5-58	5.5-59	5.5-60
5.5-61	5.5-62	5.5-63	5.5-64	5.5-65	5.5-66	5.5-67	5.5-68
5.5-69	5.5-71	5.5-72	5.5-73	5.5-74	5.5-75	5.5-76	5.5-77
5.5-79	5.5-80	5.5-81	5.5-82	5.5-83	5.5-84	5.5-85	5.5-86
5.5-88	5.5-90	5.6-1	5.6-2	5.6-4	5.6-5	5.6-6	5.6-7
5.6-8	5.6-9	5.6-11	5.6-12	5.7-1	5.8-1	5.8-2	5.8-3
5.8-4	5.8-5	5.8-6	5.8-7	5.8-8	5.8-9	5.8-10	5.8-11
5.8-12	5.8-13	5.8-14	5.9-1	5.9-2	5.9-3	5.9-6	5.9-8
5.9-9	5.9-11	5.9-12	5.9-15	5.9-21	5.9-23	5.9-28	5.9-30
5.9-36	5.9-37	5.9-38	5.9-40	5.9-41	5.9-42	5.9-44	5.10-8
5.10-12	5.10-13	5.10-17	5.10-18	5.10-19	5.10-20	5.10-21	5.10-22
5.10-23	5.10-24	5.10-25	5.10-26	5.11-4	5.11-5	5.11-6	5.11-7
5.11-8	5.11-9	5.11-10	5.11-11	5.11-12	5.11-14	5.11-16	5.11-17
5.11-18	5.11-19	5.11-20	5.11-21	5.12-4	5.12-17	5.12-18	5.12-20
5.12-21	5.12-24	5.12-29	5.12-37	5.12-40	5.12-41	5.12-42	5.12-46
5.12-47	5.12-48	5.12-50	5.14-7	5.16-9	5.16-10	5.16-14	5.16-30
5.17-2	5.17-5	5.17-6	5.17-7	5.17-8	5.17-9	5.17-10	5.17-11
5.17-12	5.17-14	5.17-15	5.17-16	5.17-17	5.17-18	5.17-19	5.17-20
5.17-21	5.17-22	5.17-24	5.17-25	5.17-26	5.17-27	5.17-28	6.1-1
6.1-12	6.2-35	6.2-36	6.2-37	6.2-38	6.4-30	6.4-35	7.1-1
7.2-1	7.2-6	7.2-7	7.2-8	7.2-11	7.2-12	7.2-13	7.2-14
8.1-2	8.1-4	8.1-5	9.1-13				

plastics comb.

2.2-2	2.3-1	2.3-7	2.5-7	2.5-8	3.5-59	4.1-1	4.1-2
4.2-1	4.5-1	5.1-1	5.1-2	5.1-4	5.9-1	5.9-2	5.9-3
5.9-6	5.12-1	6.1-1	6.2-31	6.2-32	7.2-16	7.2-17	8.1-10
8.1-12	8.1-18	8.1-28	8.1-44	8.1-45	8.1-51	8.2-1	

plastics mfr.

3.2-18	3.5-21	6.1-6	6.2-21	8.1-27

polymer comb.

2.4-4	3.1-1	3.1-2	3.1-3	3.1-4	3.1-10	3.1-29	3.2-2
3.2-3	3.2-7	3.2-26	3.5-1	3.5-2	3.5-3	3.5-4	3.5-5
3.6-1	8.1-2	8.1-27					

polymer films

4.8-27

polymer mfr

3.5-53	3.4-16	3.4-18	3.5-1	3.5-2	3.5-13	3.5-21	3.5-27
3.5-28	4.8-15	5.5-13	5.6-3	5.13-1	5.13-2	5.16-24	6.1-8
8.1-52							

power trans.
   2.1-9

printing ink mfr.
   5.9-7

printing
   4.1-1   4.1-2   4.5-1   4.5-4   5.9-1   5.9-2   5.9-3   5.9-6

propellant
   2.2-1   2.2-7   2.5-8   3.1-3   3.1-4   3.1-5   5.14-1   8.1-2
   8.1-11   8.1-12   8.1-13   8.1-16   8.1-19   8.1-21   8.1-25   8.1-49

PVC mfr.
   4.8-29   4.8-34   4.8-36   4.8-37   5.5-87

refrigeration
   2.2-2   8.1-11   8.1-12   8.1-13   8.1-14   8.1-15   8.1-16   8.1-23
   8.1-25

refuse comb.
   2.2-2   2.3-7   2.3-23   2.4-1   2.4-4   2.5-2   2.5-9   2.5-10
   2.5-11   2.5-13   2.5-15   2.5-16   2.5-18   2.5-24   2.5-31   3.1-1
   3.2-1   3.2-2   3.2-3   3.2-61   3.4-41   3.5-1   3.6-1   3.6-28
   3.7-4   3.7-18   3.7-41   3.7-64   3.7-74   3.7-86   3.7-96   3.7-101
   3.7-106   3.7-109   3.7-111   3.7-130   3.7-130   3.7-136   4.1-1   4.1-2
   4.4-1   4.5-1   4.8-29   4.8-62   4.8-64   4.8-75   4.8-76   4.8-78
   5.1-1   5.1-2   5.1-41   5.1-43   5.4-1   5.4-13   5.9-1   5.9-2
   5.12-1   5.12-43   5.12-46   5.16-18   6.1-1   6.1-3   6.1-10   6.4-53
   6.4-54   7.2-16   7.2-17   8.1-2   8.1-4   8.1-5   8.1-18   8.1-28
   8.1-39   8.1-51   8.2-5   8.3-14   8.3-15   8.3-16   8.3-17   8.3-18
   9.2-10

rendering plant
   4.1-5   6.2-7   6.2-18   6.2-19   6.2-25   6.4-36   7.1-1   7.2-1

river water odor
   3.5-21   3.6-1   3.6-18   4.8-7   5.9-26   5.9-35   5.12-49   8.1-55

rock dust
   2.5-9   2.5-11   2.5-13   2.5-13   2.5-15   2.5-16   2.5-18   2.5-21
   2.5-23   2.5-24

rocket
   2.1-7   2.2-2   2.2-10   2.4-3   2.4-4   2.4-10   2.4-17   2.5-5
   2.5-6   2.5-7   2.5-8   2.5-11   6.2-1   6.2-5

rubber abrasion
   3.2-19   3.4-36   3.7-96   7.2-16

rubber mfr.
   2.3-1

rubber abrasion
3.4-37

rubber mfr.
7.2-16

sea salt
2.4-4

sewage tmt.

2.2-2	2.3-1	2.5-2	3.5-1	4.1-2	4.1-5	5.1-2	5.9-3
5.9-6	5.9-8	6.2-7	6.2-8	6.2-13	6.2-17	6.2-18	6.2-20
6.2-22	6.2-28	7.1-1	7.1-2	7.1-3	7.1-5	7.1-8	7.1-9
7.2-1	7.2-6	8.1-4	9.1-15	9.1-16			

sewage treatment
2.3-1   8.1-4

skunk odor
7.1-6

soil
2.1-6

solvent

3.1-44	3.1-57	3.2-2	3.2-22	3.3-1	3.4-19	3.5-1	3.5-2
3.5-4	3.5-8	3.5-13	3.5-21	3.5-22	3.5-33	3.5-35	3.5-41
4.2-5	4.5-1	4.5-4	4.5-7	4.5-10	4.5-11	4.5-13	4.5-20
4.5-26	4.5-27	4.7-3	4.7-12	5.5-8	5.5-12	5.5-13	5.5-14
5.5-15	5.5-29	5.5-43	5.8-1	5.9-1	5.9-2	5.9-3	5.9-6
5.9-8	5.9-9	5.10-2	5.10-3	5.11-3	5.12-1	5.14-3	5.14-4
5.14-5	5.16-6	5.16-12	6.2-4	6.2-10	6.2-13	6.3-6	6.3-28
6.4-3	8.1-2	8.1-3	8.1-4	8.1-5	8.1-18	8.1-20	8.1-23
8.1-24	8.1-27	8.1-28	8.1-29	8.1-30	8.1-31	8.1-35	8.1-36
8.2-1							

starch mfr.

2.3-1	2.3-7	5.1-2	5.1-4	5.1-6	7.1-1	7.2-1	7.2-16
7.2-17							

steel mfr.

2.2-3	2.3-11	2.4-3	2.5-10	2.5-11	2.5-13	2.5-13	2.5-16
2.5-20	2.5-21	2.5-23	2.5-24	2.5-27	2.5-28	2.5-29	2.5-38
6.1-1							

steel molds
3.7-96

sulfur mfr.
2.3-6

syn. rubber mfr.

3.1-10	3.1-49	3.2-14	3.2-16	3.2-18	3.2-36	4.2-1	6.1-2
6.1-6							

synthetic fibre mfr.
   2.3-1   7.2-16   7.2-17

tap water odor
   5.11-5   5.12-1   5.12-3   5.12-55

titanium mfr.
   2.3-6   2.4-1   2.4-4   2.4-15   2.4-22   2.5-1   2.5-10   2.5-11
   2.5-13   2.5-16   2.5-18   2.5-19   2.5-20   2.5-22   2.5-24

tobacco smoke
   2.2-2   2.2-6   2.3-1   2.5-7   2.5-8   3.1-1   3.1-2   3.1-3
   3.1-4   3.1-5   3.1-46   3.1-53   3.1-57   3.1-59   3.1-65   3.1-69
   3.1-70   3.1-71   3.1-73   3.1-74   3.1-77   3.1-79   3.1-99   3.1-100
   3.1-101   3.1-102   3.1-103   3.1-104   3.1-105   3.1-106   3.1-107   3.1-108
   3.1-109   3.1-110   3.1-111   3.1-112   3.2-1   3.2-2   3.2-3   3.2-5
   3.2-16   3.2-18   3.2-19   3.2-36   3.2-51   3.2-81   3.2-82   3.2-84
   3.2-86   3.2-90   3.3-1   3.3-2   3.3-5   3.3-20   3.3-21   3.4-11
   3.4-21   3.4-24   3.4-25   3.5-1   3.5-2   3.5-3   3.5-7   3.5-8
   3.5-13   3.5-21   3.5-22   3.5-29   3.5-33   3.5-44   3.5-45   3.5-49
   3.5-53   3.5-54   3.5-56   3.5-57   3.5-58   3.5-59   3.5-60   3.5-62
   3.6-1   3.6-2   3.6-3   3.6-4   3.6-7   3.6-10   3.6-11   3.6-12
   3.6-13   3.6-19   3.6-23   3.6-24   3.6-26   3.6-28   3.6-29   3.7-1
   3.7-3   3.7-4   3.7-5   3.7-10   3.7-12   3.7-13   3.7-14   3.7-16
   3.7-17   3.7-18   3.7-23   3.7-24   3.7-25   3.7-26   3.7-27   3.7-30
   3.7-31   3.7-32   3.7-32   3.7-41   3.7-44   3.7-45   3.7-47   3.7-49
   3.7-50   3.7-51   3.7-52   3.7-53   3.7-54   3.7-55   3.7-56   3.7-58
   3.7-59   3.7-60   3.7-62   3.7-64   3.7-67   3.7-68   3.7-69   3.7-70
   3.7-71   3.7-72   3.7-73   3.7-75   3.7-79   3.7-80   3.7-81   3.7-85
   3.7-86   3.7-89   3.7-90   3.7-91   3.7-96   3.7-99   3.7-100   3.7-101
   3.7-106   3.7-107   3.7-110   3.7-111   3.7-114   3.7-115   3.7-116   3.7-117
   3.7-118   3.7-119   3.7-123   3.7-124   3.7-125   3.7-127   3.7-130   3.7-130
   3.7-136   4.1-1   4.1-2   4.1-4   4.1-5   4.1-6   4.1-7   4.1-8
   4.1-20   4.2-1   4.2-3   4.3-1   4.3-4   4.4-1   4.5-1   4.5-3
   4.5-4   4.5-5   4.5-9   4.5-15   4.5-16   4.5-17   4.5-18   4.5-26
   4.5-33   4.8-1   4.8-2   4.8-3   4.8-4   4.8-5   4.8-6   4.8-10
   4.8-11   4.8-25   4.8-27   4.8-58   4.8-59   4.8-65   4.8-66   4.8-80
   5.1-1   5.1-2   5.1-3   5.1-4   5.1-5   5.1-6   5.1-8   5.1-10
   5.1-12   5.1-13   5.1-14   5.1-19   5.1-21   5.1-24   5.1-27   5.1-29
   5.1-37   5.1-38   5.1-39   5.1-40   5.1-41   5.1-42   5.1-43   5.1-44
   5.1-45   5.1-46   5.1-47   5.1-48   5.1-49   5.1-50   5.1-51   5.1-52
   5.1-53   5.1-54   5.1-55   5.1-56   5.1-57   5.2-7   5.2-9   5.2-10
   5.4-1   5.4-4   5.5-8   5.5-9   5.5-52   5.5-63   5.5-70   5.5-78
   5.5-81   5.9-1   5.9-2   5.9-5   5.9-7   5.9-14   5.10-1   5.12-1
   5.12-2   5.12-3   5.12-4   5.12-6   5.12-10   5.12-11   5.12-24   5.12-25
   5.12-26   5.12-27   5.12-28   5.12-31   5.12-32   5.12-37   5.12-48   5.12-53
   5.12-54   5.12-56   5.12-57   5.12-58   5.12-59   5.14-5   5.14-6   5.16-11
   5.16-15   5.16-16   5.16-20   5.16-24   5.16-25   5.16-27   5.16-29   5.17-7
   5.17-32   6.1-1   6.1-3   6.2-2   6.2-3   6.2-6   6.2-7   6.2-8
   6.2-10   6.2-26   6.2-27   6.2-31   6.2-32   6.3-1   6.3-30   6.4-1
   6.4-9   6.4-10   6.4-11   6.4-12   6.4-14   6.4-16   6.4-17   6.4-19

volcano

2.1-1	2.1-3	2.1-6	2.1-7	2.1-8	2.1-10	2.2-1	2.2-2
2.2-6	2.3-1	2.3-6	2.3-7	2.3-11	2.4-3	2.4-4	2.4-5
2.4-6	2.5-2	2.5-7	2.5-8	2.5-46	3.1-1	3.1-2	3.1-3
3.1-5	3.1-10	3.2-2	3.2-3	3.2-7	3.2-9	3.2-16	3.4-19
3.4-20	3.5-1	3.5-2	4.1-2	4.1-5	4.2-2	4.2-3	4.5-1
4.5-4	5.9-1	5.9-2	5.9-3	5.9-6	7.2-16	7.2-17	8.1-2
8.1-10	8.1-11	8.1-12	8.1-23	8.1-26	8.1-29	8.1-32	8.1-38

vulcanization

3.4-39	3.4-42	3.4-43	3.5-1	3.5-2	3.5-4	3.5-5	3.5-13
3.5-21	3.6-2	3.6-9	3.6-14	6.2-34	7.3-12		

water treatment

2.3-1	3.1-1	3.1-2	3.1-3	3.1-4	3.1-10	8.1-4

water works emiss.

4.8-29	4.8-32	4.8-33

whiskey mfr.

5.5-9	5.5-17	5.9-2	5.9-3	5.9-9	5.9-12

wood comb.

4.4-24	4.4-26	4.4-27	4.4-28

wood pulping

2.3-1	2.3-2	2.3-16	2.3-17	2.3-18	2.4-1	2.4-12	2.5-7
2.5-11	2.5-13	2.5-16	2.5-24	2.5-41	2.5-42	2.5-43	3.1-1
3.1-2	3.1-3	3.2-2	3.2-3	3.3-1	3.3-2	3.3-3	3.3-5
3.3-6	3.3-7	3.3-8	3.3-9	3.3-10	3.3-12	3.3-13	3.5-2
3.5-35	4.4-24	4.5-1	4.5-4	4.5-6	4.5-10	4.5-11	4.5-18
4.5-23	4.7-8	4.8-14	5.9-1	5.9-2	5.9-3	5.9-6	5.9-8
5.9-9	5.9-15	5.9-20	5.10-22	5.11-5	5.11-9	5.11-12	5.12-1
5.12-2	5.12-3	5.12-4	5.12-33	5.16-16	5.17-7	7.1-1	7.1-2
7.1-3	7.2-1	7.2-2	7.2-3	7.2-7	7.3-2		

zinc mfr.

2.3-3	2.4-1	2.4-9	2.4-27	2.5-9	2.5-10	2.5-11	2.5-13
2.5-16	2.5-27	2.5-29	2.5-35				

APPENDIX B
CHEMICAL PRODUCTS AND PRECURSORS

The compounds in this appendix are those for which gas phase atmospheric chemical reactions leading to their formation have been identified. The number and name of each product are given below, together with the species numbers of those compounds that are its precursors. The specific reaction linking each precursor and product is given in the appropriate precursor table.

| 2.1-7 | Hydrogen | | | |
| | 3.1-1 | 3.2-2 | 3.4-5 | 4.1-1 |

| 2.1-9 | Ozone |
| | 2.2-6 |

| 2.1-12 | Hydroxyl radical | | | | | |
| | 2.1-11 | 2.2-9 | 2.2-10 | 3.2-1 | 3.2-4 | 3.2-5 | 4.1-1 |

| 2.2-5 | Nitric oxide | | | | | | |
| | 2.2-2 | 2.2-6 | 2.2-7 | 2.2-9 | 6.1-1 | 6.1-3 | 6.2-6 |

| 2.2-6 | Nitrogen dioxide | | |
| | 2.1-9 | 2.2-5 | 2.2-10 |

| 2.2-8 | Dinitrogen pentoxide |
| | 2.2-6 |

| 2.2-9 | Nitrous acid |
| | 2.2-5 |

| 2.2-10 | Nitric acid | | |
| | 2.2-6 | 2.2-8 | 4.1-2 |

| 2.2-12 | Ammonium nitrate | |
| | 2.2-2 | 2.2-10 |

| 2.3-7 | Sulfur dioxide | | | | | |
| | 2.3-1 | 7.1-1 | 7.2-1 | 7.2-16 | 7.2-17 | 7.3-2 |

| 2.3-8 | Sulfur trioxide |
| | 2.3-7 |

| 2.3-11 | Sulfuric acid | |
| | 2.3-7 | 2.3-8 |

INDEX